HARMFUL CYANOBACTERIA

AQUATIC ECOLOGY SERIES

Volume 3

The titles published in this series are listed at the end of this volume.

HARMFUL CYANOBACTERIA

Edited by

JEF HUISMAN
Institute for Biodiversity and Ecosystem Dynamics, University of Amsterdam, The Netherlands

and

HANS C.P. MATTHIJS
Institute for Biodiversity and Ecosystem Dynamics, University of Amsterdam, The Netherlands

and

PETRA M. VISSER
Institute for Biodiversity and Ecosystem Dynamics, University of Amsterdam, The Netherlands

Springer

A C.I.P. Catalogue record for this book is available from the Library of Congress.

ISBN 1-4020-3009-6 (HB)
ISBN 1-4020-3022-3 (e-book)

Published by Springer,
P.O. Box 17, 3300 AA Dordrecht, The Netherlands.

Sold and distributed in North, Central and South America
by Springer,
101 Philip Drive, Norwell, MA 02061, U.S.A.

In all other countries, sold and distributed
by Springer,
P.O. Box 322, 3300 AH Dordrecht, The Netherlands.

Printed on acid-free paper

Cover photo:
Dense bloom of the cyanobacterium /Microcystis/, Wann See, Germany
(photo taken by Gertrud Schlag).

springeronline.com

Printed in the Netherlands.

To Professor Luuc R. Mur

For his leading contributions
to European research on harmful cyanobacteria
For his guidance of a new generation of aquatic scientists

AQUATIC ECOLOGY SERIES

Aquatic ecology is an extraordinarily broad and diverse discipline. Aquatic ecology is the study of the functional relationships and productivity of organisms and communities of waters as regulated by their physical, chemical, and biotic environment. The marine environment extends broadly from the complex land-water coastal environments to the open ocean. Limnology encompasses all inland aquatic environments, including streams, rivers, lakes, reservoirs, and wetlands. Research has accelerated in certain areas and been less active in others. Reassessments and syntheses are stimulating to the discipline as a whole, as well as enormously useful to students and researchers in ecological sciences.

A series of succinct monographs and specialized evaluations in aquatic ecology has been developed. Subjects covered are topical (e.g., lake hydrodynamics, microbial loop in aquatic ecosystems) rather than broad and superficial.

The treatments must be comprehensive and state-of-the-art, whether the topic is at the biochemical, mathematical, population, community, or ecosystem level. The objectives are to advance the topics by the development of arguments, with documented support, that generate new insights, concepts, theories to stimulate thought, ideas, directions, controversies. The books are intended for mature as well as emerging scientists to stimulate intellectual leadership in the topics treated.

Receipt of manuscripts approximately 18 months after an agreement is desired, for publication within 10 months thereafter. Electronic submission is essential with hardcopy. Format and manuscript guidance will be provided.

For further information and book proposal details please contact:

Prof. Robert G. Wetzel, Series Editor
Department of Environmental Sciences
and Engineering
The University of North Carolina
Chapel Hill, North Carolina 27599-7431
USA
Email: rwetzel@unc.edu
Phone: 919 + 843-4916

Dr. Anna Besse, Publishing Editor
Aquatic and Biogeosciences
Springer
Van Godewijckstraat 30
P. O. Box 17
3300 AA Dordrecht
The Netherlands
Email: anna.besse@springer-sbm.com
Phone: 31 (0) 651 33 86 01

CONTENTS

PREFACE

Innocently, the white dog looks up: '*What 's all that fuss about? Why am I called back ashore, when I 'm having so much fun in this green soup?*' Indeed, it is a wise decision of the owner to call this dog back ashore without delay. The green soup the dog is standing in is a dense surface bloom of *Microcystis*. The cyanobacterium *Microcystis* is a cosmopolitan species of freshwater lakes. It can produce microcystins, a toxic substance that has caused numerous illnesses and deaths of dogs, cattle, birds, and even humans.

Are all cyanobacteria harmful? No, not at all. Cyanobacteria are phototrophic microorganisms that convert sunlight into organic biomass by means of photosynthesis. As such, cyanobacteria are the aquatic equivalents of grasses and trees on land. In fact, cyanobacteria were the first organisms in the evolutionary history of our planet to produce oxygen. They are the evolutionary ancestors of all modern plants. Together with the eukaryotic phytoplankton, cyanobacteria form the basis of the aquatic food web.

However, a few cyanobacterial species are capable of producing powerful toxins. Several of these harmful species are quite widespread. They occur throughout the world, especially in fresh and brackish waters, and may occasionally reach high population densities, as the cover of this book illustrates. Waters dominated by such harmful cyanobacteria are often closed for recreation, and cannot be used for the intake of drinking water or agricultural purposes. During the past few years, much new knowledge on harmful cyanobacteria has been acquired, owing to scientific advances in molecular biology, genomics, and the computational sciences and also owing to an increased recognition of the problems caused by harmful cyanobacteria. In fact, it is fair to say that we now know a lot more about harmful cyanobacteria than only five years ago. The aim of this book, therefore, is to provide an up-to-date review that makes this recently acquired knowledge accessible to a broad readership.

The first chapter (chapter 1) in this book introduces the subject, and explains the problems associated with harmful cyanobacteria. Two subsequent chapters (chapters 2 and 3) focus on the molecular biology and physiology of toxin production, with emphasis on the genetic regulation and variability in toxin production. Next, modern approaches are discussed that allow rapid assessment of the extent to which nutrients, such as nitrogen and phosphorus, limit the growth of cyanobacteria (chapters 4 and 5). The subsequent chapter (chapter 6) reviews our current knowledge of the physiology and ecology of *Microcystis*, one of the most widespread and extensively studied harmful species. This is followed by a state-of-the-art review of the population dynamics of harmful cyanobacteria, with particular attention to environmental factors that favour cyanobacteria over other phytoplankton species (chapter 7). Chapter 8 introduces novel techniques in remote sensing, which allow monitoring of the development of cyanobacterial blooms from aircraft and satellites. The final chapter (chapter 9) discusses how improvements in water management, based on the new Water Safety Plans introduced by the World Health Organization, may reduce human exposure to harmful cyanobacteria.

The idea for this book came up during a farewell symposium, organized at the University of Amsterdam in September 2003, on the occasion of the retirement of Professor Luuc Mur. For more than 40 years, Luuc Mur has been active in phytoplankton biology, first as lecturer and later as Professor in Aquatic Environmental Biology at the University of Amsterdam. He has educated numerous undergraduate and graduate students on water management, the marvellous beauty of the phytoplankton, and the potential trouble of harmful cyanobacteria. Many of these former students are still very active in aquatic research and water management today. During the 1970s and 1980s, Luuc Mur was among the first scientists to recognize that advances in microbiology could be made useful in aquatic ecology. In particular, phytoplankton species such as the cyanobacteria and green algae could be studied in the laboratory using continuous-culture techniques. This culture technique enabled detailed investigation of the growth kinetics of these microorganisms under controlled laboratory conditions. In the 1980s and 1990s, with the increasing eutrophication of many European lakes, the problems caused by harmful cyanobacteria became widely apparent. With funding provided by the fourth framework program of the European Committee, Luuc Mur initiated two research projects (CYANOTOX and TOPIC) to study the toxicology, molecular biology, physiology, and ecology of harmful cyanobacteria. These two research projects brought together a diverse group of European scientists that have stood at the basis of much of what is currently known about harmful cyanobacteria. It is our pleasure - on behalf of all colleagues, students and friends - to thank Luuc Mur for his leading contributions and guidance of a new generation of aquatic scientists.

This book would not have been possible without the financial support provided by the Earth and Life Sciences Foundation (ALW), the Netherlands Organization for Scientific Research (NWO), and the University of Amsterdam.

Special thanks go to all chapter authors, whose work provides the backbone of this book, and to all reviewers that have been of great help in sharpening the chapters. We thank Bibi Krot and Pascale Thiery for helping us in organizing the symposium, and for their invaluable assistance during the development of this book. We thank Gertrud Schlag for providing the beautiful photograph that illustrates the cover, and Ellen Spanjaard for the cover design. Last but not least, we are most grateful to Professor Robert Wetzel and the staff at Springer, especially Anna Besse-Lototskaya and Judith Terpos, for their stimulating guidance.

Jef Huisman, Hans C.P. Matthijs & Petra M. Visser (editors)
Amsterdam, August 2004.

CONTRIBUTORS

THOMAS BÖRNER is Professor of Genetics at the Humboldt University in Berlin. His research focuses on plant molecular biology, and in particular on the molecular biology of harmful cyanobacteria. Address: Institut für Biologie, Humboldt Universität, Chausseestrasse 117, D-10115 Berlin, Germany.

INGRID CHORUS is research scientist at the Federal Environmental Agency Germany in Berlin. Her research spans many aspects of harmful cyanobacteria, including detection methods, toxin production and water management strategies. Address: Institute for Water, Soil and Air Hygiene, Federal Environmental Agency Germany, Corrensplatz 1, D-14195 Berlin, Germany.

GEOFFREY A. CODD is Professor of Microbiology at the University of Dundee. His research focuses on the analysis, biology, and toxicology of cyanobacterial toxins, with applications in water-quality management and health. Address: Division of Environmental and Applied Biology, School of Life Sciences, University of Dundee, Dundee DD1 4HN, United Kingdom.

MARCO DIGNUM is postdoctoral researcher at the Netherlands Institute of Ecology (NIOO-KNAW). He works on the mechanisms and regulation of phosphate uptake by freshwater cyanobacteria. Address: Centre for Limnology, Netherlands Institute of Ecology, Rijksstraatweg 6, 3631 AC Nieuwersluis, Netherlands.

ELKE DITTMANN is Junior Professor of Molecular Ecology at the Humboldt University in Berlin. Her research addresses the biochemistry of peptide and polyketide biosynthesis in cyanobacteria, and the functional role of cyanobacterial secondary metabolites. Address: Institut für Biologie, Humboldt Universität, Chausseestrasse 117, D-10115 Berlin, Germany.

HERMAN J. GONS is senior scientist at the Netherlands Institute of Ecology (NIOO-KNAW). His research interests include the role of viruses in plankton communities, and the application of optical teledetection to study spatial distributions of cyanobacteria. Address: Centre for Limnology, Netherlands Institute of Ecology, Rijksstraatweg 6, 3631 AC Nieuwersluis, Netherlands.

HANS HAKVOORT is senior advisor at the Dutch Directorate-General for Public Works and Water Management. His work specializes on the application of geographical information systems for monitoring of water quality. Address: Geo-information and ICT Department, PO Box 5023, 2600 GA Delft, Netherlands.

JEF HUISMAN is Professor of Aquatic Microbiology at the University of Amsterdam. His research focuses on the population dynamics of phytoplankton, using mathematical models, lab experiments, and field research. Address: Aquatic Microbiology, Institute for Biodiversity and Ecosystem Dynamics, University of Amsterdam, Nieuwe Achtergracht 127, 1018 WS Amsterdam, Netherlands.

FLORENCE D. HULOT is assistant professor at the Ecole Normale Supérieure in Paris. She works on species interactions and biodiversity in aquatic ecosystems, using mathematical models and experimental research. Address: Laboratoire d'Ecologie, Ecole Normale Supérieure, 46 rue d'Ulm, 75230 Paris cedex 05, France.

BAS W. IBELINGS is postdoctoral researcher at the Netherlands Institute of Ecology (NIOO-KNAW). His research interests include the ecophysiology of freshwater phytoplankton, and the evolutionary dynamics of aquatic communities. Address: Centre for Limnology, Netherlands Institute of Ecology, Rijksstraatweg 6, 3631 AC Nieuwersluis, Netherlands.

W. EDWIN A. KARDINAAL is PhD student at the University of Amsterdam. His research addresses the dynamics and toxin production of freshwater cyanobacteria. Address: Aquatic Microbiology, Institute for Biodiversity and Ecosystem Dynamics, University of Amsterdam, Nieuwe Achtergracht 127, 1018 WS Amsterdam, Netherlands.

HENDRIKUS J. LAANBROEK is director of the Centre for Limnology of the Netherlands Institute of Ecology (NIOO-KNAW), and Professor in Soil Biology at the University of Nijmegen. His research focuses on the role of microorganisms in biogeochemical cycles. Address: Centre for Limnology, Netherlands Institute of Ecology, Rijksstraatweg 6, 3631 AC Nieuwersluis, Netherlands.

JAIME LINDSAY is PhD student at the University of Dundee. Her research includes investigations into the characterisation and significance of cyanobacterial lipopolysaccharides and endotoxins. Address: Division of Environmental and Applied Biology, School of Life Sciences, University of Dundee, Dundee DD1 4HN, United Kingdom.

HANS C.P. MATTHIJS is associate professor at the University of Amsterdam. He investigates the molecular biology and physiology of cyanobacteria, with special emphasis on nutrient acquisition and photosynthesis. Address: Aquatic Microbiology, Institute for Biodiversity and Ecosystem Dynamics, University of Amsterdam, Nieuwe Achtergracht 127, 1018 WS Amsterdam, Netherlands.

JAMES S. METCALF is postdoctoral researcher at the University of Dundee. His current research interests include the chromatographic and molecular biological analysis of cyanobacterial toxins in a wide range of biological and clinical materials. Address: Division of Environmental and Applied Biology, School of Life Sciences, University of Dundee, Dundee DD1 4HN, United Kingdom.

LOUISE F. MORRISON is PhD student at the University of Dundee. Her research includes investigations into the occurrence and environmental fate of cyanobacterial toxins. Address: Division of Environmental and Applied Biology, School of Life Sciences, University of Dundee, Dundee DD1 4HN, United Kingdom.

LUUC R. MUR is emeritus professor at the University of Amsterdam. His research focuses on phytoplankton biology and water management. He has initiated several European research programmes on harmful cyanobacteria. Present address: Aquatic Microbiology, Institute for Biodiversity and Ecosystem Dynamics, University of Amsterdam, Nieuwe Achtergracht 127, 1018 WS Amsterdam, Netherlands.

ROEL PEL is senior scientist at the Netherlands Institute of Ecology (NIOO-KNAW). His research focuses on the applications of stable isotopes in microbial ecology. Address: Centre for Limnology, Netherlands Institute of Ecology, Rijksstraatweg 6, 3631 AC Nieuwersluis, Netherlands.

STEEF W.M. PETERS is senior scientist at the Free University in Amsterdam. He is an expert on remote sensing and geographical information systems, with a specific interest in aquatic ecosystems. Address: Institute for Environmental Studies, Free University, De Boelelaan 1087, 1081 HV Amsterdam, Netherlands.

ANTON F. POST is senior scientist at the Interuniversity Institute for Marine Sciences in Eilat, Israel, and professor at the Hebrew University. He investigates the molecular ecology and genomics of marine cyanobacteria, with a specific focus on nutrient stress responses. Address: Interuniversity Institute for Marine Sciences, Coral Beach, PO Box 469, Eilat 88000, Israel.

STEFAN G.H. SIMIS is PhD student at the Netherlands Institute of Ecology (NIOO-KNAW). The focus of his work is on the remote sensing of freshwater cyanobacteria, and the impact of viruses on cyanobacterial blooms. Address: Centre for Limnology, Netherlands Institute of Ecology, Rijksstraatweg 6, 3631 AC Nieuwersluis, Netherlands.

PETRA M. VISSER is assistant professor at the University of Amsterdam. She investigates the ecophysiology and population dynamics of freshwater cyanobacteria. Address: Aquatic Microbiology, Institute for Biodiversity and Ecosystem Dynamics, University of Amsterdam, Nieuwe Achtergracht 127, 1018 WS Amsterdam, Netherlands.

ANTHONY E. WALSBY is Melville Wills Professor of Botany at the University of Bristol. His research focuses on the genetics, physiology and ecology of planktonic cyanobacteria, using a combination of molecular approaches, field research, and mathematical modeling. Address: School of Biological Sciences, University of Bristol, Bristol BS8 1UG, United Kingdom.

FIONA YOUNG is PhD student at the University of Dundee. Her research interests include the localisation of cyanobacterial toxins and their potential role in cyanobacteria. Address: Division of Environmental and Applied Biology, School of Life Sciences, University of Dundee, Dundee DD1 4HN, United Kingdom.

CHAPTER 1

HARMFUL CYANOBACTERIA

From mass mortalities to management measures

Geoffrey A. CODD, Jaime LINDSAY, Fiona M. YOUNG, Louise F. MORRISON & James S. METCALF

1 Introduction

When the first triennial international conferences on photosynthetic prokaryotes, including cyanobacteria, began in the 1970's, their packed and novel proceedings included no contributions on the toxicology or health hazards of cyanobacteria. In 2004, the "Sixth International Conference on Toxic Cyanobacteria" in Bergen (another triennial series) is expected to attract several hundred participants, all working with, or concerned in other ways, with cyanobacterial toxins, from their molecular biology, to risk management. The drivers accounting for the rapid increase in knowledge about cyanobacterial toxins in the intervening decades have included: fundamental research; the increasing availability of enabling technologies e.g. analytical techniques and molecular biology tools; a "need to know" from the health and environmental agencies, the water industry and water-users; and several human and animal poisoning incidents associated with cyanobacterial toxins to which modern methods have been applied.

Cyanobacteria are equipped to flourish in aquatic environments where they can produce blooms, scums and mats (Mur, 1983; Mur et al., 1999). They also produce a diverse range of toxins (Codd et al., 1989, 1999; Namikoshi and Rinehart, 1996; Carmichael 1997, 2001; Duy et al., 2000; Sivonen and Jones, 1999; Sivonen 2000). The cyanobacterial toxins include human and animal health hazards which can present risks of illness and mortality at environmentally-encountered concentrations (Falconer, 1996, 1998; Kuiper-Goodman et al., 1999; Codd et al., 1999).

This review includes early and recent examples of human and animal illnesses and deaths associated with cyanobacterial mass populations, in some cases attributable to their toxins. We briefly consider cyanobacterial toxins among the plethora of other cyanobacterial bioactive compounds and whether the main cyanobacterial hazards to human health are now identified. Steps to mitigate the

J. Huisman, H.C.P. Matthijs and P.M. Visser (eds.), Harmful Cyanobacteria, 1–23.

harmful effects of cyanobacterial populations and toxins on animal and human health, on the basis of the then-available knowledge, have been developed several times over the past 130 years (Codd et al., 1994). Management measures developed and implemented from 1989 (NRA, 1990) to the present are briefly considered. Finally, some remaining needs for the risk management of harmful cyanobacteria and their toxins are identified.

2 Mortalities and illnesses associated with cyanobacterial toxins

2.1 HISTORICAL EXAMPLES

Occasional descriptions by travellers, historians and at least one playwright (W. Shakespeare: The Merchant of Venice) indicate that an awareness of plankton scums and blooms has existed in Europe for at least two millennia (see Fogg et al., 1973; Codd and Beattie, 1991). The description of Llangorse Lake, Wales, by Geraldus Cambrensis in 1188 indicates the presence of buoyant cyanobacteria: "The lake has many miraculous properties --- it sometimes turns bright green, and in our days it has been known to become scarlet, not all over, but as if blood were flowing along certain currents and eddies" (see Codd and Beattie, 1991). There are also indications of traditional knowledge of the toxicity of cyanobacterial blooms and scums among aboriginal peoples in North America, Africa and Australia (e.g. Hayman, 1992).

2.1.1 Animals

Early publications from Denmark (Hald, 1833), Australia (Francis, 1878) and Pomerania, then West Prussia now Poland (Benecke, 1884), report farm livestock deaths, and bird-kills after drinking from blooms or scums, with fish-kills also at some sites (Table 1). Contemporary investigations and a developing awareness of the health hazards of cyanobacterial blooms occurred in the USA (e.g. Arthur, 1883; Stalker, 1887). The Danish poisonings were clearly not single events since the local people referred to the lakes as being "sick" when they bore blooms and scums (Hald, 1833; Moestrup, 1996). The keynote observations of Francis (1878) included survival times of sheep, cattle, horses, pigs and dogs after drinking from *Nodularia spumigena* scum, *post mortem* examinations after sheep poisonings consistent with cyanobacterial toxicosis, and the reproduction of the effects after quantitative oral dosing of a sheep with the scum. Although bloom and scum development at Lake Barlewice usually occurred annually, Benecke (1884) noted that harmful effects were not observed in all cases, leading him to suggest the occurrence of toxic and non-toxic forms. These Nineteenth Century investigations were made under official auspices and /or were communicated to relevant professional organizations. Hald (1833) reported by order of the Danish king to the "Royal Farm Household Society"; Francis (1878) carried out his investigations at the request of the Chief Inspector of Sheep to the (then) Commissioner of Crown Lands in South Australia (Codd et al., 1994); and Benecke (1878) reported to the Society of Fisheries of the Provinces of East and West Prussia and asked for the help of its members in sample collection and in the reporting of blooms and associated "disease".

Numerous later examples of animal deaths related to cyanobacterial blooms over the past century are listed by Schwimmer and Schwimmer (1968) and Yoo et al. (1995). Domestic animals affected have included cattle, pigs, horses, water buffalo, sheep, dogs, cats and poultry (chickens, ducks, turkeys, geese). Wild animal mortalities have included deer, muskrats, frogs, salamanders, skunk, mink, squirrels, bats, rhinoceros, zebra, giraffe, a wide range of wild birds (ducks, swans, coots, gulls, hawks, pheasants, songbirds) and honeybees. Poisonings have most typically occurred during drinking, although in the case of wild animal deaths at higher trophic levels (e.g. muskrats), it is possible that exposure to cyanobacterial toxins may also have occurred via the food chain (Eriksson et al., 1986). Cyanobacterial toxins have been identified, alongside characteristic behavioural effects or pathology, in cases of mass mortalities of wild and farmed fish (e.g. Rodger et al., 1994). However, the toxins are likely to be among a range of toxicants and physiological stresses that are presented to the fish by cyanobacterial blooms, especially during bloom senescence. These include oxygen deficit, high pH and ammonia release due to bloom proteolysis.

Table 1. Examples of early reports of animal deaths associated with the ingestion of cyanobacterial blooms and scums.

Location	***Cyanobacteria reported***	***Animal deaths***	***Reference***
Jutland, Denmark, Lakes Sunds sø, Gjødstrup sø, Flynder sø, Stubbergårds sø	Scums, no identification	Cattle, fish	Hald (1833)[a]
South Australia, Lake Alexandrina	Scums, *Nodularia spumigena*	Sheep, cattle, horses, pigs, dogs	Francis (1878)
Pomerania, Poland, Lake Barlewice	Scums, *Microcystis aeruginosa*, *Anabaena flos-aquae*, "*Limnochlide flos-aquae*" (*Aphanizomenon flos-aquae*)	Foals, other farm animals, ducks, chickens, pigeons, fish	Benecke (1884)

[a] See Moestrup (1996)

2.1.2 Humans

It is not clear whether the traditional practice of bank-side filtration by New South Wales Aborigines to produce drinking water from billabongs containing cyanobacterial scum (Hayman, 1992) was a response to adverse human or animal health effects, or solely for aesthetic reasons. A further episode of 200 sheep deaths in 1880 at Lake Alexandrina was accompanied by a human illness (Codd et al., 1994) and Benecke (1884) reported human skin irritations after local people were in contact with scums at Lake Barlewice. Early reports of allergenic reactions after swimming in cyanobacterial bloom (freshwater lake, Wisconsin) followed up by positive intradermal skin tests, and of skin irritation among swimmers due to marine *Lyngbya majuscula* (Hawaii), were summarised by Schwimmer and Schwimmer (1968).

2.2 RECENT EXAMPLES

Understanding of the true incidence of mortalities and illness continues to be restricted by inadequate recognition and case definition, analytical epidemiology and notification (Skulberg et al., 1994; Hunter, 1994). These limitations have been reduced to some extent over recent years, where cases have been investigated and a supporting research base in cyanobacterial toxicology has developed. However, the deficiencies remain in many parts of the world where exposure to toxigenic cyanobacterial mass populations occurs.

2.2.1 Animals

An example of the inadequate recognition of cyanobacterial poisoning hazards has been presented several times over recent years by mat-forming cyanobacteria. Dog deaths in Scotland, Ireland and New Zealand have occurred after eating fragments of anatoxin-a-producing cyanobacterial mats and drinking the water nearby (Edwards et al., 1992; James et al., 1997; Hamill, 2001). The neurotoxic material, including benthic *Oscillatoria* and *Phormidium*, can be overlooked during waterbody inspection since it is brown-black, rather than blue-green. This is due to the presence of cyanobacterial phycoerythrin, in addition to phycocyanin. Shoreline accumulations of the mats are, however, prone to scavenging by dogs, which appear to be attracted to the odorous material (Codd et al., 1992). Further benthic cyanobacterial neurotoxicoses involving three dogs, of which two died, occurred at a Dunfermline (Scotland) freshwater lake in May 2003 (Metcalf, Morrison and Codd, unpublished). Anatoxin-a- and microcystin-containing cyanobacterial mats have accounted for numerous cattle deaths at small lakes in alpine summer pastures in Switzerland (Mez et al., 1997).

Cattle died after drinking water containing a *Cylindrospermopsis raciborskii* bloom at a farm pond in Queensland, Australia in 1997. The subsequent findings of Saker et al. (1999), including liver histopathology and cylindrospermopsin in the bloom and in a derived laboratory isolate, appear to be the first report of animal poisonings attributable to *C. raciborskii* and cylindrospermopsin.

The role of cyanobacterial toxins in recent mass mortalities among three of the world's six species of flamingo (Table 2) may appear anomalous upon first consideration. Flamingos are planktivorous birds which can subsist partly or entirely on cyanobacteria. Indeed, there are no present indications that planktonic *Arthrospira* and *Spirulina* spp., major cyanobacterial components of the flamingo diet, produce potent toxins such as the microcystins or anatoxin-a, found in these mass mortalities. Rather, the circumstances leading to the flamingo deaths each require special explanation. The deaths of the Chilean Flamingos, attributed to microcystins, occurred in captivity after exposure to *Microcystis* spp. in their pond at a visitor exhibit (Chittick et al., 2002). Greater Flamingo chick deaths, also attributed to microcystins, occurred at a national wetlands park lagoon after the sudden development of a bloom dominated by *Microcystis aeruginosa* and *Anabaena flos-aquae*. Artificial measures to maintain water levels and eutrophication problems in the park are reported (Alonso-Andicoberry et al., 2002).

*Table 2. Mass mortalities of flamingo species (*Phoenicopterus*) after exposure or ingestion of cyanobacterial biomass.*

Location	***Flamingo species***	***Numbers dying***	***Candidate toxigenic cyano-bacteria***[b]	***Presence of cyanobacterial toxins***[c] ***in:***		***Reference***
				Blooms, mats, scum	***Bird tissues, excreta***	
Zoo pond, Orlando, Florida, USA	Chilean (*P. chilensis*)	10 (adults)	*M*	MC	MC	Chittick et al. (2002)
Wetland lagoon, Doñana National Park, Spain	Greater (*P. ruber*)	579 (out of 943 chicks)	*M, Ana, Osc*	MC	MC	Alonso-Andicoberry et al. (2002)
Lakes Nakuru and Bogoria, Kenya	Lesser (*P. minor*)	Thousands[a]	*Ana, An, Osc, Ph,* (others ?)	MC, ANA	MC, ANA	Ballot et al. (2002) Krienitz et al. (2003)

[a] Tens of thousands, 1991-1999 (see Krienitz et al., 2003); [b] *M, Microcystis; Ana, Anabaena; Osc, Oscillatoria; An, Anabaenopsis; Ph, Phormidium;* [c] MC, microcystins; ANA, anatoxins.

Table 3. Examples of human exposures to cyanobacterial blooms and toxins, with associated health outcomes.

Year	*Location (source[a])*	*Cyano-bacteria[b]*	*Toxins[c]*	*Health outcomes[d]*	*References*
Drinking water					
1. 1975	USA (DWR)	*S, L, Ph*	?	GI	Lippy and Erb (1976), Keleti (1981)
2. 1979	Australia (DWR)	*C*	CYN	GI, LD, KD, ID	Byth (1980), Griffiths et al. (1998)
3. 1981	Australia (DWR)	*M*	MC	LD	Falconer et al. (1983)
4. 1972 – ca. 1990	China (SWR)	*M*	MC	PLC, D	Yu (1989, 1995)
5. 1988	Brazil (DWR)	*M, Ana*	?	GI, D	Teixeira et al. (1993)
6. 1994	Sweden (DWR, CR)	*Pl*	MC	GI, F, AP, MP	Annadotter et al. (2001)
Recreational/occupational water contact					
7. 1989	UK (FR, SW, KY)	*M*	MC	GI, ST, BM, V, AP, F, PC	Turner et al. (1990)
8. 1995	Australia (FR, SW, BA)	*M, Ana, Aph, Nod*	Hepato-toxins	GI, FLS, BM, F, EE	Pilotto et al. (1997)
9. 1996	UK (FR, BO)	*Pl*	MC	R, F	G. A. Codd & K. A. Beattie, Environment Agency report
10. 1996-1998	Australia (MR, SW, BA, FI)	*L*	?	CD, EE, RI	Dennison et al. (1999)

Table 3. Continued

Year	*Location (source[a])*	*Cyano-bacteria[b]*	*Toxins[c]*	*Health outcomes[d]*	*References*
Haemo-dialysis					
11. 1974	USA (DC)	present	LPS	F, MY, C, V	Hindman et al. (1975)
12. 1996	Brazil (DC)	present	MC, CYN	VID, TIN, N, V, LD, D	Jochimsen et al. (1998), Pouria et al. (1998), Carmichael et al. (2001), Azevedo et al. (2002)
13. 2001	Brazil (DC)	*Ana, Mic*	MC	Same as in 12 (no D reported)	Carmichael et al. (2002)

[a] DWR, treated water from drinking water reservoir; SWR, raw water from surface sources; CR, contaminated with raw river water; FR, freshwater; SW, swimming; KY, kayaking; BA, bathing; BO, boating; MR, marine; FI, fishing; DC, dialysis clinic water.
[b] *S, Schizothrix*; *L, Lyngbya*; *Ph, Phormidium*; *C, Cylindrospermopsis*; *M, Microcystis*; *Ana, Anabaena*; *Pl, Planktothrix*; *Aph, Aphanizomenon*; *Nod, Nodularia.*
[c] CYN, cylindrospermopsin; MC, microcystin(s); LPS, lipopolysaccharide endotoxin(s).
[d] D, deaths; GI, gastroenteritis; LD, liver damage; KD, kidney damage; ID, intestinal damage; PLC, primary liver cancer; F, fevers; AP, abdominal pain; MP, muscular pain; ST, sore throat; BM, blistered mouth; V, vomiting; PC, pulmonary consolidation; FLS, flu-like symptoms; EE, eye and/or ear irritation; R, rashes; CD, contact dermatitis; RI, respiratory irritation; MY, myalgia: C, chills; VID, visual disturbance: TIN, tinitus: N, nausea.

The circumstances leading to mass deaths of Lesser Flamingos at the Kenyan lakes are complex since feeding, drinking and breeding all occur at different lakes. An early indication of neurotoxicosis, consistent with anatoxin-a poisoning, was provided by the appearance of opisthotonus (convulsions of extremities and arched-back neck) in several of the dead birds. Anatoxin-a and four microcystins (MC-LR, -RR, -YR, -LF) were quantified in cyanobacterial samples from Lakes Nakuru and Bogoria and in the livers, stomach and gut contents and excreta of dead birds (Ballot et al., 2002; Krienitz et al., 2003). Candidate sources of the cyanobacterial toxins at these lakes (Table 2) include planktonic species in Lake Nakuru (*Anabaenopsis, Anabaena*). These occur as lesser components in the blooms dominated by (presumed benign) *Arthrospira fusiformis*. Exposure to microcystins and anatoxin-a also occurs at freshwater drinking sites on the L. Bogoria shoreline. At these freshwater hot springs, the flamingos ingest fragments of cyanobacterial mats containing the hepatotoxins and neurotoxin. The potentially toxigenic hot spring mat-forming cyanobacteria include *Phormidium* and *Oscillatoria* spp..

2.2.2 Humans

A small number of publications on human illnesses and deaths associated with exposure to cyanobacterial cells and toxins is available covering the past 25 years (for reviews, see Carmichael, 1997, 2001; Falconer, 1998; Codd et al., 1999; Kuiper-Goodman et al., 1999; Duy et al., 2000). These and additional episodes, including skin and respiratory irritations among swimmers and fishermen on the Queensland coast from 1996-1998, and a further outbreak of illness among patients at a haemodialysis clinic in Rio de Janeiro in 2001, are summarised in Table 3. The serious health incidents, most notably the deaths of 52 haemodialysis patients up to Dec. 1996, out of over 100 affected at a clinic in one episode (Table 3, item 12), have increased the attention and co-operation of clinicians, epidemiologists, public health professionals, water scientists and cyanobacteriologists, to begin to identify needs, formulate and implement policies to reduce health risks presented by cyanobacterial toxins. Indications of the extent of health problems due to recreational exposure are still limited: partly because the reporting of small episodes of relatively mild illness (skin irritation, gastrointestinal upsets) associated with recreational contact with cyanobacterial blooms, if recognised at all, may only be to local and national health and environmental authorities, rather than to primary journals.

3 Cyanobacterial toxins and other cyanobacterial bioactive metabolites

3.1 DEFINITIONS AND TERMINOLOGY

> "When *I* use a word", Humpty Dumpty said, "it means just what I choose it to mean – neither more nor less."
>
> Lewis Carroll, *Through the Looking Glass.*

How do we define a cyanobacterial toxin? There is purpose rather than pedantry in this question, since the term is now used in practical contexts e.g. in water treatment and health risk management, and the understanding of the natural function(s) of cyanobacterial toxins is at an early stage. A classical definition of toxins, e.g. that they are "Microbial products or components that can injure another cell or organism at low concentrations" (Prescott et al., 2002), when extended to include plant and animal products, only partly applies to those of cyanobacteria. Beyond doubt, cyanobacteria produce potent compounds such as microcystin-LR, saxitoxins and anatoxin-a, which cause injury in vertebrates at low doses (micrograms per kg body weight: see Carmichael, 1997; Falconer, 1998; Codd et al., 1999; Kuiper-Goodman et al., 1999) and these compounds are rightly defined as toxins. Indeed, fish embryo development and survival can be impaired during exposure to microcystins at concentrations (0.5 to 13.0 micrograms per litre) found in lakes and reservoirs (Oberemm et al., 1999; Palikova et al., 2003).

However, the growth habits of toxigenic cyanobacteria, including the ability of gas-vacuolate planktonic genera to form scums (e.g. *Microcystis, Anabaena, Nodularia, Planktothrix, Aphanizomenon, Cylindrospermopsis, Trichodesmium*), and of benthic genera to form mats (e.g. *Lyngbya, Phormidium, Oscillatoria, Schizothrix*) can result in high local concentrations of cyanobacterial biomass in grams dry wt per litre, and toxins such as microcystins and anatoxins in mg per litre. In these scenarios, adult or juvenile vertebrates and invertebrates may come into contact with, or ingest, above-threshold concentrations or doses of relatively low toxicity material, leading to functional or, developmental impairment, injury, or mortality. Thus cyanobacterial toxins may include products or components that are only injurious at higher concentrations. This widens our concepts of the diversity, production and significance of cyanobacterial toxins.

3.2 CYANOBACTERIAL TOXINS: THE PRIORITY HAZARDS TO HUMAN AND ANIMAL HEALTH

Human and animal case histories, experimental toxicity assessments with susceptible animal species using purified cyanobacterial metabolites, and risk assessments, have identified microcystins, nodularins, anatoxins, saxitoxins, and cylindrospermopsins as priority hazards to human and animal health (Table 3; Codd and Beattie, 1991; Falconer, 1996, 1998; Hunter, 1994; Yoo et al., 1995; Carmichael, 1997, 2001; Codd et al., 1999; Sivonen and Jones, 1999; Kuiper-Goodman et al., 1999; Duy et al., 2000). These sources also provide information on molecular structures, toxicities, and modes of action, from the whole animal to the molecular interaction level, and are summarised in Table 4. Research continues on all of these toxin groups. Recent findings have included: (i) chromosome loss and DNA strand breakage by cylindrospermopsin (Humpage et al., 2000; Shen et al., 2002); (ii) inhibition of nuclear protein phosphatase by microcystin with changes in phosphorylation status of the p53 tumour-suppressor protein (Guzman et al., 2003); (iii) conjugation of nodularin and microcystins to glutathione by glutathione *S*-transferases (GST), as an initial step in cyanobacterial hepatotoxin detoxication in the brine shrimp *Artemia salina* (Beattie et al., 2003). In Chapter 2 of this volume, Börner and Dittmann discuss exciting advances in the molecular genetics of cyanobacterial toxins.

Only one cyanobacterial organophosphorus toxin, the neurotoxin anatoxin-a(s) (Matsunaga et al., 1989) remains known after recent years of intensive cyanobacterial toxin characterization. However, known numbers of structural variants of other toxin classes, especially microcystins (71 variants to date; Fig. 1) have increased. The growth of the latter has been due to a large extent to the intensity and effectiveness of the analytical chemistry which has been applied (Namikoshi and Rinehart, 1996; Harada, 1996).

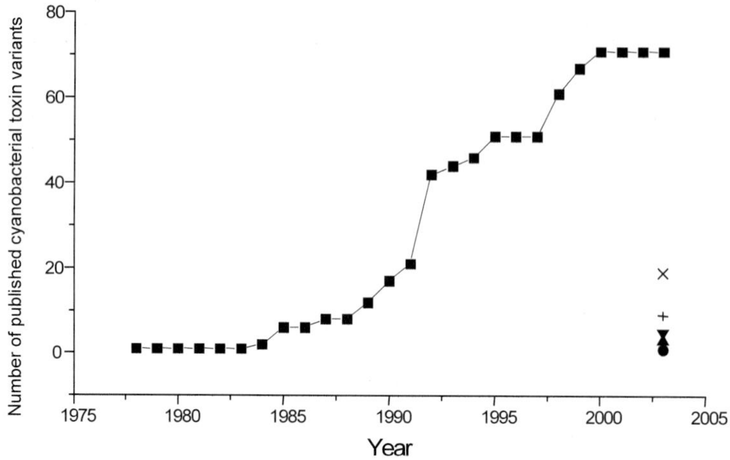

Figure 1. Number of published microcystin variants to date, alongside numbers of variant(s) currently known for other classes of cyanobacterial toxins. ■, *microcystin;* ×, *saxitoxin;* ●, *anatoxin-a(s);* ▼, *anatoxin-a;* ▲, *cylindrospermopsin;* +, *nodularin.*

Recent additions to the list of toxigenic genera indicate that the true extent of toxin production among the cyanobacteria, particularly bloom-formers, is not fully apparent (Table 4). These include: cylindrospermopsin production by *Anabaena* sp. (Schembri et al., 2001), and *Raphidiopsis curvata* (Li et al., 2001); production of saxitoxins by *Cylindrospermopsis raciborskii* (Lagos et al., 1999; Molica et al., 2002); and the production of anatoxin-a and homoanatoxin-a by *Raphidiopsis mediterranea* (Namikoshi et al., 2003).

3.3 OTHER BIOACTIVE CYANOBACTERIAL METABOLITES: TOXINS TO TARGET ORGANISMS AND NATURAL FUNCTIONS

Belatedly compared to those of e.g. actinomycetes and fungi, cyanobacterial secondary metabolites are now receiving increasing attention. They include diverse volatile metabolites causing taste and odour problems in water supplies (Juttner, 1995; Watson, 2003). Bioassay-guided screening programmes of cyanobacteria from fresh- and marine waters and terrestrial sources, with a low rate of rediscovery, are revealing novel metabolites with antimicrobial and cytotoxic activities of therapeutic potential (Moore, 1996; Namikoshi and Rinehart, 1996; Radau, 2000). The role of cyanobacterial secondary metabolites in aquatic chemical ecology, including allelopathy, is also a growing area of interest (Schagerl et al., 2002; Gross, 2003; Cembella, 2003; Legrand et al., 2003). The foregoing reviews include reports on products that inhibit or stimulate the activities of other photoautotrophs including higher plants, algae and cyanobacteria (allelochemicals), and of bacteria, fungi,

Table 4. Principle groups of cyanobacterial toxins and their sources[a]

Toxin	*Number of structural variants*	*Structure and activity*	*Toxigenic genera*[b]
Hepatotoxins			
Microcystins	71	Cyclic heptapeptides; hepatoxic, protein phosphatase- inhibition, membrane integrity and conductance disruption, tumour promoters	*Microcystis, Anabaena, Nostoc, Anabaenopsis, Planktothrix, Oscillatoria, Hapalosiphon*
Nodularins	9	Cyclic pentapeptides; hepatotoxins, protein phosphatase- inhibition, membrane integrity and conductance disruption, tumour promoters, carcinogenic	*Nodularia, Theonella* (sponge-containing cyanobacterial symbionts)
Cylindro-spermopsins	3	Guanidine alkaloids; necrotic injury to liver (also to kidneys, spleen, lungs, intestine), protein synthesis-inhibitor, genotoxic	*Cylindrospermopsis, Aphanizomenon, Umezakia, Anabaena*[c], *Raphidiopsis*[d]
Neurotoxins			
Anatoxin-a (including homoanatoxin-a)	5	Alkaloids; postsynaptic, depolarising neuromuscular blockers	*Anabaena, Oscillatoria, Phormidium, Aphanizomenon, Rhaphidiopsis*[e]
Anatoxin-a (s)	1	Guanidine methyl phosphate ester; inhibits acetylcholinesterase	*Anabaena*
Saxitoxins	20	Carbamate alkaloids, sodium channel-blockers	*Aphanizomenon, Anabaena, Lyngbya, Cylindrospermopsis*[f], *Planktothrix*

Table 4 continued

Toxin	*Number of structural variants*	*Structure and activity*	*Toxigenic genera*[b]
Dermatotoxins and cytotoxins			
Lyngbyatoxin-a	1	Alkaloids; inflammatory agents, protein kinase C activators	*Lyngbya, Schizothrix, Oscillatoria*
Aplysiatoxins	2	Alkaloids; inflammatory agents, protein kinase C activators	*Lyngbya, Schizothrix, Oscillatoria*
Endotoxins			
Lipopoly-saccharide	Many	Lipopolysaccharides; inflammatory agents, gastrointestinal irritants	All?

[a], updated from Carmichael (1997); Falconer (1998); Codd et al. (1999); Sivonen and Jones (1999). [b] not all species within a genus, or strains within species, produce the particular toxin, except for lipopolysaccharide which is a marker from Gram negative prokaryotes including cyanobacteria. [c] Schembri et al. (2001). [d] Li et al. (2001). [e] Namikoshi et al. (2003) [f] Lagos et al. (1999), Molica et al. (2002)

protozoa and invertebrates. Examples of cyanobacterial products that inhibit the growth of other cyanobacteria and microalgae and/or aquatic macrophytes include cyanobacterins, fischerellins, and nostocyclamides. A further group of metabolites which adversely affect phytoplankton-grazers includes di(hydroxymethyl) dihydroxypyrrolidone, from *Cylindrospermum,* which inhibits crustacean digestive glucosidases (Juttner and Wessel, 2003), and microviridin J from *Microcystis,* which inhibits moulting in *Daphnia galeata* and *Daphnia pulicaria* (Kaebernick et al., 2001). Microviridin J is a 13 amino acid-acetylated tricyclic peptide (Rohrlack et al., 2003), which inhibits protein-chitin hardening. From the standpoint of the injured or impaired biota, these secondary metabolites are cyanobacterial toxins.

Allelopathic actions of the toxins of high human health significance have also been reported recently. These include the growth-inhibition of *Chlamydomonas reinhardtii* (Kearns and Hunter, 2001) and of *Ceratophyllum demersum* by microcystin-LR (Pflugmacher, 2002). Inhibition of photosynthetic oxygen evolution by the aquatic macrophytes *C. demersum, Myriophyllum spicatum, Elodea canadensis, Phragmites australis* (Pflugmacher, 2002) by microcystin-LR is consistent with the inhibition of whole-leaf photosynthetic CO_2 photoassimilation after topical application of the toxin to French bean plants (Abe et al., 1996).

As with potent microalgal toxins (Cembella, 2003), the natural functions of the cyanobacterial toxins of high significance to human health are unclear. However, it is possible that they may serve as defence or deterrent agents against grazing, or as intra- and extracellular differentiation or signalling compounds, among others (Codd, 1995). Consideration of e.g. the microcystins, anatoxins and saxitoxins in a continuum of bioactive cyanobacterial metabolites seems appropriate and useful in helping to understand their natural function(s).

4 Management measures and needs

4.1 MANAGEMENT MEASURES: FORMULATION AND IMPLEMENTATION

Requirements for multidisciplinary input to investigate the occurrence, properties, production, fates and significance of cyanobacterial toxins, and for policies on permissible concentrations of the toxins in waters used for human access and drinking have been advocated in the modern era for some time (Codd et al., 1989; Falconer, 1989). Several task groups have been convened over the past 14 years to address the risk management requirements. These have included local, state and national groups with examples from the UK, Australia, Norway, the USA, South Africa, the Netherlands, Germany, France and Poland.

NRA (1990), NSWBGATF (1992), M-DBMC (1994) and Yoo et al. (1995) give examples of the work of such task groups including: **(a)** situation assessment: historical, current, potential; **(b)** the prioritization of needs for action: emergency, remedial, preventative; **(c)** identification of potential control options: short- and long-term, ideally with incorporation into a decision-making system; **(d)** environmental and economic appraisal, in some cases, of the technically-feasible control options; **(e)** selection of control options and incorporation into action plans.

These stages have been included in international contributions to cyanobacterial toxin risk management by working groups of the World Health Organization (WHO, 1998, 2003; see also Chorus and Bartram, 1999). From case histories (human and animal), available toxicity data from laboratory trials, and exposure assessments, a valuable start has been possible to the identification of Guideline Values for microcystin-LR in drinking water, and of Guideline Levels for the use of recreational waters which may contain toxigenic cyanobacterial populations.

Cyanobacterial toxin risk management action plans should be adaptable to local circumstances and be responsive to advances in research, risk assessment and to the monitoring of their effectiveness. Information on the results of subsequent phases of risk management, namely: **(f)** implementation of action plans; **(g)** assessment of the effectiveness of the implemented plans; and **(h)** plan modification if necessary, with implementation of revised action plans, may be the subject of future symposia on water quality and health protection. This is in line with the generic “Water Safety Plan” approach promoted in the latest revision of WHO Guidelines for Drinking-water quality (see Chorus, Chapter 9 in this volume).

4.2 REQUIREMENTS

4.2.1 Cyanobacterial toxin detection and analysis

Considerable advances have been made in the detection and analysis of cyanobacterial toxins (Meriluoto, 1997; Codd et al., 2001; Nicholson and Burch, 2001; Metcalf and Codd, 2003). However, numerous needs remain. These include an international requirement for purified, and certified, quantitative analytical standards, which should be distinguished from reagents supplied for research purposes. Certified reference materials e.g. cyanobacterial biomass and other matrices containing defined toxins at known concentrations, are also needed to facilitate interlaboratory calibration exercises (e.g. Fastner et al., 2002) and for the further development of analytical competence and accountability. Indeed for some classes of cyanobacterial toxins, e.g. cylindrospermopsins or anatoxin-a(s), no materials are commercially available at all.

Although progress has been made with microcystins, the extraction efficiencies of these and other toxins from cyanobacterial biomass and the tissues of exposed organisms also need to be optimised and/or defined. Optimised estimations of cyanobacterial toxin concentrations in pathological and clinical materials are highly useful in estimating tolerable body burdens for risk assessment.

Early warning systems for the detection of toxigenic cyanobacterial populations are required by water managers and engineers in drinking water utilities, recreational facilities etc., to give warning before the populations develop to unacceptable health risk proportions. Polymerase chain reaction (PCR) and DNA microarray tests are being developed for the detection of marker genes, e.g. for the peptide and polyketide synthases for microcystin, nodularin and cylindrospermopsin biosynthesis. They can provide an indication, at the level of the single colony or filament, of the potential for toxin production. Early detection of actual toxin production by single colonies and filaments is also possible before bloom development by using immunoassays. Quantitative analysis is possible by enzyme-linked immunosorbent (ELISA) assay, with a minimum detection limit (MDL) of about 0.015 ng per colony or filament. By this means (e.g. Fig. 2), we can readily quantify the toxin in single washed colonies of e.g. *Microcystis aeruginosa*. By contrast, the single colonies of *Microcystis flos-aquae* in the same water sample shown contained no, or barely detectable microcystins. Analysis of only a few colonies before bloom formation cannot only determine whether a bloom is likely to be toxic, but may be used to map the extent of toxin production and quantify the contribution of cyanobacterial species, or morphotypes, to the overall pool of microcystins at a particular time. The application of such immunodetection methods in quantifying microcystin contents of colonies and filaments can also be used in the formulation and validation of monitoring thresholds and guideline levels.

4.2.2 Exposure to multiple classes of cyanobacterial toxins

Exposure or dosing trials using single purified toxins, or defined cyanobacterial extracts, with healthy, susceptible subjects (animals, plants, cell-lines) are necessary to obtain data on the toxicity of the toxin of interest. However, in natural and controlled aquatic environments, animals, plants and humans are susceptible to exposure to multiple classes of cyanobacterial toxins, plus other stressors. The latter may include non-cyanobacterial toxins, anthropogenic toxicants e.g. pesticides and metals, and microbial pathogens. Examples of the production of multiple classes of cyanobacterial toxins by single strains in laboratory culture and the co-occurrence of multiple classes in aquatic environments are given in Table 5.

Little is known about the possible additive, synergistic, potentiation or antagonistic effects of exposure to multiple classes of cyanobacterial toxins, or of interactions between the toxins and other stressors. In addition to microcystins and anatoxin-a, the Lesser Flamingo populations at sites of mass mortalities at Kenyan Rift Valley lakes have been exposed to pesticides, heavy metals and avian tuberculosis (see Krienitz et al., 2003, for references). In intra-nasal mouse toxicity trials with microcystin-LR and anatoxin-a, separately and together, co-administration resulted in a synergistic effect (Fitzgeorge et al., 1994).

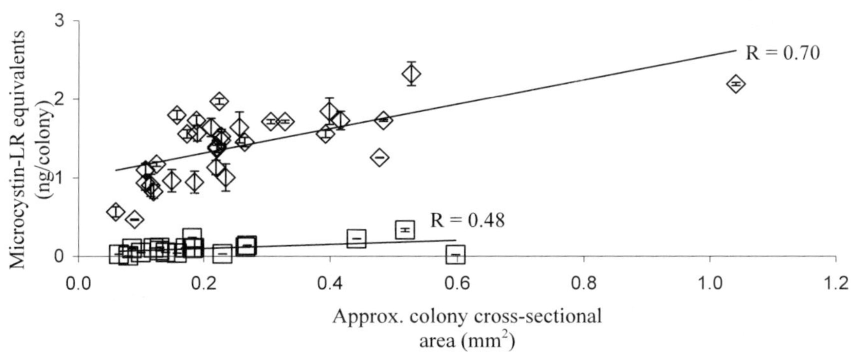

Figure 2. Analysis of 2 species of Microcystis *present in a bloom sample from Hampstead Heath Boating Pond, London (16 Sept 02). Diamond symbols represent single colonies of* M. aeruginosa*, square symbols represent single colonies of* M. flos-aquae*. R values are Spearman's rank order correlation coefficients.*

Table 5. Examples of demonstrated production of multiple classes of cyanobacterial toxins by laboratory strains and co-occurrence in natural populations and water resources.

Source	***Cyanobacterial toxins***[a]	***Reference***
Single strains, lab culture		
Anabaena flos-aquae CYA32 (NRC-44-1)	MCs, ANA	Al-Layl et al. (1988)
Anabaena flos-aquae 525-17	MCs, ANA(s)	Matsunaga et al. (1989)
Microcystis PCC 7820, PCC 7806, UV-017	MCs, LPS	Martin et al. (1989)
Cylindrospermopsis raciborskii CR3	CYN, LPS	Saker and Neilan. (2001), J.Lindsay et al. (in prep.)
Nodularia strains	NOD, LPS	Rapala et al. (2002)
Environmental populations		
Tambo, Switzerland (mats)	MC, ANA	Mez et al. (1997)
Langersee, Gransee, Germany	MC, ANA	Chorus (2001)
Haemodialysis clinic (Caruaru), Water from Tabocas Reservoir, Brazil	MCs, CYN	Carmichael et al. (2001)
Gulf of Gdansk, Poland	NOD, MCs	Mazur et al. (2003)
Lake Bogoria, Kenya (hot spring mats)	MCs, ANA	Krienitz et al. (2003)
Bala Lake, Wales; Lochs Leven and Balgavies, Scotland; Louth Canal, England	MCs, LPS	Best et al. (2002) J. Lindsay et al. (in prep)

a, MC, microcystin; ANA, anatoxin-a; ANA(s), anatoxin-a(S); LPS, lipopolysaccharide endotoxin; CYN, cylindrospermopsin; NOD, nodularin. In the case of LPS, it can be assumed that these endotoxins were present in all of the above examples, although they are only named in the cases where they were specifically reported by the relevant authors.

Several interactions are apparent between cyanobacterial LPS endotoxins and microcystins. Purified LPSs from a range of cyanobacteria inhibit the activities of glutathione *S*-transferases (GSTs) of fertilized zebra fish (*Danio rerio*) eggs and adults, both in vitro and after exposure in vivo of the fertilized eggs (Best et al., 2002). GSTs are involved as an initial step in microcystin detoxication in plants and animals including the brine shrimp *Artemia salina* (Pflugmacher et al., 1998, 1999, 2001; Beattie et al., 2003). It may thus be expected that in vivo exposure of *A. salina* to LPS may increase the susceptibility of the animals to a lethal dose of microcystin. However, when *A. salina* and *Daphnia galeata* are exposed to a sublethal dose of purified *Microcystis* LPS (< 0.1% of LPS LC_{50}), either 24-hr before, or simultaneously with exposure to lethal concentrations of microcystin-LR, a marked decrease in toxicity occurs (2.5- to 167-fold increases in LC_{50}; Table 6). These findings indicate that LPS may participate in multiple aspects of microcystin toxicity, for example in detoxification systems, in addition to the inhibition of GSTs. Further research is needed to understand, and indeed to recognise, the complex effects of sequential- and co-exposure to different cyanobacterial toxins, the responses and net outcomes.

Table 6. Exposure of Artemia salina *and* Daphnia galeata *to purified cyanobacterial LPS and purified microcystin-LR.*

Exposure conditions	***LC_{50} (µg microcystin-LR mL^{-1})[a]***	
	A. salina[e]	***D. galeata[f]***
Microcystin-LR	2.0	0.003
Pre-exposure to LPS[b], then microcystin-LR	8.0	0.50
Simultaneous addition of LPS[c] with microcystin-LR	4.9	0.30
Post-addition of LPS[d], after microcystin-LR	2.1	0.08
	LC_{50} (µg LPS mL^{-1})	
	A. salina[e]	***D. galeata[f]***
LPS	100.0	11.9

a, LC_{50} is the 'lethal concentration' at which 50% of the population dies. LC_{50}'s were determined over 72-hr; values shown are the means of 6 determinations; **b,** 24 hr at 2 ng mL^{-1} before microcystin-LR; **c,** at 2 ng mL^{-1}; **d,** LPS added to 2 ng mL^{-1}, 24 hr after microcystin-LR; **e,** nauplii used (fourth instar); **f,** used 7 days after hatching.

5 Concluding remarks

Cyanobacterial toxins present continuing challenges in fundamental research, in the health protection of water-users and in the conservation of water resources and habitats. Much now depends on the translation of basic knowledge in cyanobacterial ecophysiology and toxicology into practical measures for the protection of water resources and health. However, for this to be possible, the "enabling technology" needs to be supported by "enabling science". This needs to include a sound understanding of the growth dynamics and ecophysiology of toxigenic (and non-toxigenic) cyanobacterial cells and populations, from the rigorous control possible in chemostat and turbidostat cultures, to their growth dynamics in competitive and dominant modes in natural environments. By his research, teaching and inspiration over 40 years, Luuc Mur has been responsible to a very large extent for engendering the high levels of understanding and expertise available today in these important fields. It is our pleasure here to recognise this.

6 Acknowledgements

We thank the following for financial support: the UK Natural Environment Council and the European Commission (QLRT-2001-02634, PEPCY; and EVK1-2002-00107, TOXIC).

7 References

Abe, T., Lawson, T., Weyers, J.D.B. and Codd, G.A. (1996) Microcystin-LR inhibits photosynthesis of *Phaseolus vulgaris* primary leaves: implications for current spray irrigation practice, *New Phytologist* **133**, 651-658.

Al-Layl, J.K., Poon, G.K. and Codd, G.A. (1988) Isolation and purification of peptide and alkaloid toxins from *Anabaena flos-aquae* using high performance thin-layer chromatography, *Journal of Microbiological Methods* **7**, 251-258.

Alonso-Andicoberry, C., Garcia-Villada, L., Lopez-Rodas, V. and Costas, E. (2002) Catastrophic mortality of flamingos in a Spanish national park caused by cyanobacteria, *Veterinary Record* **151**, 706-707.

Annadotter, H., Cronberg, G., Lawton, L.A., Hasson, H-B., Göthe, U. and Skulberg, O.M. (2001) An extensive outbreak of gastroenteritis associated with the toxic cyanobacterium *Planktothrix agardhii* (Oscillatoriales, Cyanophyceae) in Scania, south Sweden, in I. Chorus (ed.), *Cyanotoxins – Occurrence, Causes, Consequences*, Springer-Verlag, Berlin, Germany, pp. 200-208.

Arthur, J.C. (1883) Some algae of Minnesota supposed to be poisonous, *Bulletin of the Minnesota Academy of Sciences* **2**, 1-12.

Azevedo, S.M.F.O., Carmichael, W.W., Jochimsen, E.M., Rinehart, K.L., Lau, S., Shaw, G.R. and Eaglesham, G.K. (2002) Human intoxication by microcystins during renal dialysis treatment in Caruaru-Brazil, *Toxicology* **181**, 441-446.

Ballot, A., Pflugmacher, S., Wiegand, C., Kotut, K., Krause, E., Metcalf, J.S., Morrison, L.F., Codd, G.A. and Krienitz, L. (2002) Cyanobacterial toxins, a further contributory cause of mass deaths of flamingos at Kenyan Rift Valley lakes. *Xth International Conference on Harmful Algae*. St. Petersburg, Florida, USA, p. 20.

Beattie, K.A., Ressler, J., Wiegand, C., Krause, E., Codd, G.A., Steinberg, C.E.W. and Pflugmacher, S. (2003) Comparative effects and metabolism of two microcystins and nodularin in the brine shrimp *Artemia salina, Aquatic Toxicology* **62**, 219-226.

Benecke, B. (1884) Wasserblüthe. (Waterblooms). *Berichte des Fischerei-Vereins der Provinzen West- und Ostpreussen, Miscellen* **1**, 11-12.

Best, J.H., Pflugmacher, S., Wiegand, C., Eddy, F.B., Metcalf, J.S. and Codd, G.A. (2002) Effects of enteric bacterial and cyanobacterial lipopolysaccharides, and of microcystin-LR, on glutathione *S*-transferase activities in zebra fish (*Danio rerio*), *Aquatic Toxicology* **60**, 223-231.

Byth, S. (1980) Palm Island mystery disease, *Medical Journal of Australia* **2**, 40-42.

Carmichael, W.W. (1997) The cyanotoxins, in J.A. Callow (ed.), *Advances in Botanical Research*, Vol. 27, Academic Press, London, United Kingdom, pp. 211-256.

Carmichael, W.W. (2001) A mini-review of cyanotoxins; toxins of cyanobacteria (blue-green algae), in W.J. de Koe, R.A. Samson, H.P. van Egmond, J. Gilbert, M. Sabino and W.J. de Koe (eds.), *Mycotoxins and Phycotoxins in Perspective at the Turn of the Millennium*, Ponsen & Looyen, Wageningen, The Netherlands, pp. 495-504.

Carmichael, W.W., Azevedo, S.M.F.O., An, J.S., Molica, R.J.R., Jochimsen, E.M., Lau, S., Rinehart, K.L., Shaw, G.R. and Eaglesham, G.K. (2001) Human fatalities from cyanobacteria: chemical and biological evidence for cyanotoxins, *Environmental Health Perspectives* **109**, 663-668.

Carmichael W.W., Azevedo, S.M.F.O., Servaites, J.C., Yuan, M. and Soares, R.M. (2002) Evaluation of ELISA and HPLC/MS methods for measuring microcystins in human blood serum, in Cyanobacterial toxins, a further contributory cause of mass deaths of flamingos at Kenyan Rift Valley lakes. *Xth International Conference on Harmful Algae*. St. Petersburg, Florida, USA, p. 45.

Cembella, A.D. (2003) Chemical ecology of eukaryotic microalgae in marine ecosystems, *Phycologia* **42,** 420-447.

Chittick, E., Puschner, B., Walsh, M., Gearhart, S., St Leger, J., Skocelas, E. and Branch, S. (2002) Blue-green algae microcystin toxicosis in captive Chilean flamingos, *Proceedings of the American Association of Zoo Veterinarians*, Milwaukee, USA, pp. 115-116.

Chorus, I. (2001) Cyanotoxin occurrence in Germany, in I. Chorus (ed.), *Cyanotoxins – Occurrence, Causes, Consequences*, Springer-Verlag, Berlin, Germany, pp. 5-21

Chorus, I. and Bartram, J. (eds.) *Toxic Cyanobacteria in Water: a Guide to their Public Health Consequences, Monitoring and Management*, Spon, London, United Kingdom.

Codd, G.A. (1995) Cyanobacterial toxins: occurrence, properties and biological significance, *Water Science and Technology* **32**, 149-156.

Codd, G.A. and Beattie, K.A. (1991) Cyanobacteria (blue-green algae) and their toxins: awareness and action in the United Kingdom, *Public Health Laboratory Service Microbiology Digest* **8**, 82-86.

Codd, G.A., Bell, S.G. and Brooks, W.P. (1989) Cyanobacterial toxins in water, *Water Science and Technology* **21,** 1-13.

Codd, G.A., Metcalf, J.S., Ward, C.J., Beattie, K.A., Bell, S.G., Kaya, K. and Poon, G.K. (2001) Analysis of cyanobacterial toxins by physicochemical and biochemical methods, *Journal of AOAC International* **84**, 1626-1635.

Codd, G.A., Edwards, C., Beattie, K.A., Barr, W.M. and Gunn, G.J. (1992) Fatal attraction to cyanobacteria? *Nature* **359**, 110-111.

Codd, G.A., Bell, S.G., Kaya, K., Ward, C.J., Beattie, K.A. and Metcalf, J.S. (1999) Cyanobacterial toxins, exposure routes and human health, *European Journal of Phycology* **34**, 405-415.

Codd, G.A., Steffensen, D.A., Burch, M.D. and Baker, P.D. (1994) Toxic blooms of cyanobacteria in Lake Alexandrina: learning from history, *Australian Journal of Marine and Freshwater Research* **45**, 731-736.

Dennison, W.C., O'Neil, J.M., Duffy, E.J., Oliver, P.E. and Shaw, G.R. (1999) Blooms of the cyanobacterium *Lyngbya majuscula* in coastal waters of Queensland, Australia, in L. Charpy and A.W.D. Larkum (eds.), *Marine Cyanobacteria*, Monaco Musée Océanographique, Monaco, pp. 501-506.

Duy, T.N., Lam, P.K.S., Shaw, G.R. and Connell, D.W. (2000) Toxicology and risk assessment of freshwater cyanobacterial (blue-green algal) toxins in water, *Reviews in Environmental Contamination and Toxicology* **163,** 113-186.

Edwards, C., Beattie, K.A., Scrimgeour, C. and Codd, G.A. (1992) Identification of anatoxin-a in benthic cyanobacteria (blue-green algae) and in associated dog poisonings at Loch Insh, Scotland, *Toxicon* **30,** 1165-1175.

Eriksson, J., Meriluoto, J. and Lindholm, T. (1986) Can cyanobacterial peptide toxins accumulate in aquatic food chains? in F. Megusar and M. Gantar (eds.), *Perspectives in Microbial Ecology. Proceedings of the Fourth International Symposium on Microbial Ecology*, Slovene Society for Microbiology, Ljubljana, pp. 655-658

Falconer, I.R. (1989) Effects on human health of some toxic cyanobacteria (blue-green algae) in reservoirs, lakes and rivers, *Toxicity Assessment* **4**, 175-184.

Falconer, I.R. (1996) Potential impact on human health of cyanobacteria, *Phycologia* **35** (Suppl. 6), 74-79.

Falconer, I.R. (1998) Algal toxins and human health, in J. Hrubec (ed.), *The Handbook of Environmental Chemistry 5 Part C*, Springer-Verlag, Berlin, Germany, pp. 53-82.

Falconer, I.R., Beresford, A.M. and Runnegar, M.T.C. (1983) Evidence of liver damage by toxin from a bloom of the blue-green alga, *Microcystis aeruginosa, Medical Journal of Australia* **1**, 511-514.

Fastner, J., Codd, G.A., Metcalf, J.S., Woitke, P., Wiedner, C. and Utkilen, H. (2002) An international intercomparison exercise for the determination of purified microcystin-LR and microcystins in field material, *Analytical and Bioanalytical Chemistry* **374**, 437-444.

Fitzgeorge, R.B., Clark, S.A. and Keevil, C.W. (1994) Routes of intoxication, in G.A. Codd, T.M. Jefferies, C.W. Keevil and E. Potter (eds.), *Detection Methods for Cyanobacterial Toxins,* The Royal Society of Chemistry, Cambridge, United Kingdom, pp. 69-74.

Fogg, G.E., Stewart, W.D.P., Fay, P. and Walsby, A.E. (1973) *The Blue-Green Algae*. Academic Press, London, United Kingdom.

Francis, G. (1878) Poisonous Australian lake, *Nature* **18**, 11-12.

Griffiths, D.J., Saker, M.L. and Hawkins, P.R. (1998) Cyanobacteria in a small tropical reservoir, *Journal of the Australian Water and Wastewater Association* **25**, 14-19.

Gross, E.M. (2003) Allelopathy of aquatic autotrophs, *Critical Reviews in Plant Sciences* **22**, 313-339.

Guzman, R.E., Solter, P.F. and Runnegar, M.T.C. (2003) Inhibition of nuclear protein phosphatase activity in mouse hepatocytes by the cyanobacterial toxin microcystin-LR, *Toxicon* **41**, 773-781.

Hald, J.C. (1833) *Bidrag til Kundskab om de danske Provindsers nærværende Tilstand I oekonomisk Henseende* (Contributions to our knowledge on the economy of the Danish provinces). Ottende Stykke, Ringkjøbing Amt, Landhusholdningsselskabet, Copenhagen, Denmark.

Hamill, K.D. (2001) Toxicity in benthic cyanobacteria (blue-green algae): first observations in New Zealand, *New Zealand Journal of Marine and Freshwater Research* **35**, 1057-1059.

Harada, K.-I. (1996) Chemistry and detection of microcystins, in M.F. Watanabe, K.-I. Harada, W.W. Carmichael and H. Fujiki (eds.), *Toxic Microcystis*, CRC Press, Boca Raton, USA, pp. 103-148.

Hayman, J. (1992) Beyond the Barcoo: probable human cyanobacterial poisoning in outback Australia, *Medical Journal of Australia* **157**, 794-796.

Hindman, S.H., Favero, M.S., Carson, L.A., Petersen, N.J., Schonberger, L.B. and Solano, J.T. (1975) Pyrogenic reactions during haemodialysis caused by extramural endotoxin, *The Lancet* **2**, 732-734.

Humpage, A.R., Fenech, M., Thomas, P. and Falconer, I.R. (2000) Micronucleus induction and chromosome loss in transformed human white cells indicate clastogenic and aneugenic action of the cyanobacterial toxin, cylindrospermopsin, *Mutation Research* **472**, 155-164.

Hunter, P.R. (1994) An epidemiological critique of reports of human illness associated with cyanobacteria, in G.A. Codd, T.M. Jefferies, C.W. Keevil and E. Potter (eds.), *Detection Methods for Cyanobacterial Toxins,* The Royal Society of Chemistry, Cambridge, United Kingdom, pp. 11-18.

James, K.J., Sherlock, I.R. and Stack, M.A. (1997) Anatoxin-a in Irish freshwater and cyanobacteria, determination using a new fluorimetric liquid chromatographic method, *Toxicon* **35**, 963-971.

Jochimsen, E.M., Carmichael, W.W., An, J.S., Cardo, D.M., Cookson, S.T., Holmes, C.E.M., Antunes, M.B., De Melo Filho, D.A., Lyra, T.M., Barreto, V.S.T., Azevedo, S.M.F.O. and Jarvis, W.R. (1998) Liver failure and death after exposure to microcystins at a haemodialysis center in Brazil, *New England Journal of Medicine* **338**, 873-878.

Juttner, F. (1995) Physiology and biochemistry of odorous compounds from freshwater cyanobacteria and algae, *Water Science and Technology* **25**, 69-78.

Juttner, F. and Wessel, H.P. (2003) Isolation of di(hydroxymethyl) dihydroxypyrrolidine from the cyanobacterial genus *Cylindrospermum* that effectively inhibits digestive glucosidases of aquatic insects and crustacean grazers, *Journal of Phycology* **39**, 26-32.

Kaebernick, M., Rohrlack, T., Christoffersen, K. and Neilan, B.A. (2001) A spontaneous mutant of microcystin biosynthesis: genetic characterization and effect on *Daphnia, Environmental Microbiology* **3**, 669-679.

Kearns, K.D. and Hunter, M.D. (2001) Toxin-producing *Anabaena flos-aquae* induces settling of *Chlamydomonas reinhardtii*, a competing motile alga, *Microbial Ecology* **42**, 80-86.

Krienitz, L., Ballot, A., Kotut, K., Wiegand, C., Pütz, S., Metcalf, J.S., Codd, G.A. and Pflugmacher, S. (2003) Contribution of hot spring cyanobacteria to the mysterious deaths of Lesser Flamingos at Lake Bogoria, Kenya, *FEMS Microbiology Ecology* **43**, 141-148.

Kuiper-Goodman, T., Falconer, I. and Fitzgerald, J. (1999) Human health aspects, in I. Chorus and J. Bartram (eds.), *Toxic Cyanobacteria in Water: a Guide to their Public Health Consequences, Monitoring and Management,* Spon, London, United Kingdom, pp. 113-153.

Lagos, N., Onodera, H., Zagatto, P.A., Andrinolo, D., Azevedo, S.M.F.O. and Oshima, Y. (1999) The first evidence of paralytic shellfish toxins in the freshwater cyanobacterium *Cylindrospermopsis raciborskii,* isolated from Brazil, *Toxicon* **37**, 1359-1373.

Legrand, C., Rengefors, K., Fistarol, G.O. and Graneli, E. (2003) Allelopathy in phytoplankton–biochemical, ecological and evolutionary aspects, *Phycologia* **42**, 406-419.

Lippy, E.C. and Erb, J. (1976) Gastrointestinal illness at Sewickley, PA. *Journal of the American Waterworks Association* **68**, 606-610.

Martin, C., Codd, G.A., Siegelman, H.W. and Weckesser, J. (1989) Lipopolysaccharides and polysaccharides of the cell envelope of toxic *Microcystis aeruginosa* strains, *Archives of Microbiology* **152**, 90-94.

Matsunaga, S., Moore, R.E., Niemczura, W.P. and Carmichael, W.W. (1989) Anatoxin-a(s), a potent anticholinesterase from *Anabaena flos-aquae, Journal of the American Chemical Society* **111**, 8021-8023.

Mazur, H. and Plinski, M. (2003) *Nodularia spumigena* blooms and the occurrence of hepatotoxin in the Gulf of Gdansk, *Oceanologia* **45**, 305-316.

M-DBMC (1994) *The Algal Management Strategy for the Murray-Darling Basin*, Murray-Darling Basin Commission, Canberra, Australia, pp. 33.

Meriluoto, J. (1997) Chromatography of microcystins, *Analytica Chimica Acta* **352**, 277-298.

Metcalf, J.S and Codd, G.A. (2003) Analysis of cyanobacterial toxins by immunological methods, *Chemical Research in Toxicology* **16**, 103-112.

Mez, K., Beattie, K.A., Codd, G.A., Hanselmann, K., Hauser, B., Naegeli, H. and Preisig, H.R. (1997) Identification of a microcystin in benthic cyanobacteria linked to cattle deaths on alpine pastures in Switzerland, *European Journal of Phycology* **32**, 111-117.

Moestrup, Ø. (1996) Toxic blue-green algae (cyanobacteria) in 1833, *Phycologia* **35** (Suppl. 6), 5.

Molica, R., Onodera, H., Garcia, C., Rivas, M., Andrinolo, D., Nascimento, S., Meguro, H., Oshima, Y., Azevedo, S. and Lagos, N. (2002) Toxins in the freshwater cyanobacterium *Cylindrospermopsis raciborskii* (Cyanophyceae) isolated from Tabocas reservoir in Caruaru, Brazil, including demonstration of a new saxitoxin analogue, *Phycologia* **41,** 606-611.

Moore, R.E. (1996) Cyclic peptides and depsipeptides from cyanobacteria: a review, *Journal of Industrial Microbiology* **16**, 134-143.

Mur, L.R. (1983) Some aspects of the ecophysiology of cyanobacteria, *Annals of Microbiology* **134B**, 61-72.

Mur, L.R., Skulberg, O.M. and Utkilen, H. (1999) Cyanobacteria in the environment, in I. Chorus and J. Bartram (eds.), *Toxic Cyanobacteria in Water: a Guide to their Public Health Consequences, Monitoring and Management,* Spon, London, United Kingdom, pp. 14-40.

Namikoshi, M. and Rinehart, K.L. (1996) Bioactive compounds produced by cyanobacteria, *Journal of Industrial Microbiology* **17**, 373-384.

Nicholson, B.C. and Burch, M.D. (2001) *Evaluation of Analytical Methods for Detection and Quantification of Cyanotoxins in Relation to Australian Drinking Water Guidelines*, Australian Cooperative Research Centre for Water Quality and Treatment. Salisbury, South Australia.

NRA (1990) *Toxic Blue-Green Algae.* Water Quality Series 2, National Rivers Authority, London, United Kingdom.

NSWBGATF (1992) *Final Report of the New South Wales Blue-Green Algae Task Force*, Department of Water Resources, Parramatta, Australia.

Oberemm, A., Becker, J., Codd, G.A. and Steinberg, C. (1999) Effects of cyanobacterial toxins and aqueous crude extracts of cyanobacteria on the development of fish and amphibians, *Environmental Toxicology* **14**, 77-88.

Palikova, M., Navratil, S., Marsalek, B. and Blaha, L. (2003) Toxicity of crude extract of cyanobacteria for embryos and larvae of carp (*Cyprinus carpio* L.), *Acta Veterinaria Brno* **72**, 437-443.

Pflugmacher, S., Wiegand, C., Oberemm, A., Beattie, K.A., Krause, E., Codd, G.A. and Steinberg, C.E.W. (1998) Identification of an enzymatically formed glutathione conjugate of the cyanobacterial hepatotoxin microcystin-LR: the first step of detoxication, *Biochimica et Biophysica Acta* **1425**, 527-533.

Pflugmacher, S., Codd, G.A. and Steinberg, C.E.W. (1999) Effects of the cyanobacterial toxin microcystin-LR on detoxication enzymes in aquatic plants, *Environmental Toxicology* **14**, 111-115.

Pflugmacher, S., Wiegand, C., Beattie, K.A., Krause, E., Steinberg, C.E.W. and Codd, G.A. (2001) Uptake, effects and metabolism of cyanobacterial toxins in the emergent reed plant *Phragmites australis* (Cav.) Trin. Ex Steud, *Environmental Toxicology and Chemistry* **20,** 846-852.

Pflugmacher, S. (2002) Possible allelopathic effects of cyanotoxins, with reference to microcystin-LR, in aquatic ecosystems, *Environmental Toxicology* **17**, 407-413.

Pilotto, L.S., Douglas, R.M., Burch, M.D., Cameron, S., Beers, M., Rouch, G.J., Robinson, P., Kirk, M., Cowie, C.T., Hardiman, S. and Moore, S. (1997) Health effects of exposure to cyanobacteria (blue-green algae) during recreational water activities, *Australian and New Zealand Journal of Public Health* **21**, 562-566.

Pouria, S., de Andrade, A., Barbosa, J., Cavalcanti, R.L., Barreto, V.T.S., Ward, C.J., Preiser, W., Poon, G.K., Neild, G.H. and Codd, G.A. (1988). Fatal microcystin intoxication in haemodialysis unit in Caruaru, Brazil, *The Lancet* **352**, 21-26.

Prescott, L.M., Harley, J.P. and Klein, D.A. (2002) *Microbiology*, Fifth Edition, McGraw-Hill, New York, USA.

Radau, G. (2000) Serine proteases inhibiting cyanopeptides, *Pharmazie* **55**, 555-560.

Rodger, H.D., Turnbull, T., Edwards, C. and Codd, G.A. (1994) Cyanobacterial (blue-green algal) bloom associated pathology in brown trout, *Salmo trutta* L., in Loch Leven, Scotland, *Journal of Fish Diseases* **17**, 177-181.

Rohrlack, T., Christoffersen, K., Hansen, P.E., Zhang, W., Czarnecki, O., Henning, M., Fastner, J., Erhard, M., Neilan, B.A. and Kaebernick, M. (2003) Isolation, characterization, and quantitative analysis of microviridin J, a new *Microcystis* metabolite toxic to *Daphnia*, *Journal of Chemical Ecology* **29**, 1757-1770.

Saker, M.L. and Neilan, B.A. (2001) Varied diazotrophies, morphologies and toxicities of genetically similar isolates of *Cylindrospermopsis raciborskii* (Nostocales, Cyanophyceae) from northern Australia, *Applied and Environmental Microbiology* **67**, 1839-1845.

Saker, M.L., Thomas, A.D. and Norton, J.H. (1999) Cattle mortality attributed to the toxic cyanobacterium *Cylindrospermopsis raciborskii* in an outback region of North Queensland, *Environmental Toxicology* **14**, 179-182.

Schagerl, M., Unterrieder, I. and Angeler, D.G. (2002) Allelopathy among Cyanoprokaryota and other algae originating from Lake Neusiedlersee (Austria), *International Reviews in Hydrology* **87**, 365-374.

Schembri, M.A., Neilan, B.A. and Saint, C.P. (2001) Identification of genes implicated in toxin production in the cyanobacterium *Cylindrospermopsis raciborskii, Environmental Toxicology* **16**, 413-421.

Schwimmer, M. and Schwimmer, D. (1968) Medical aspects of phycology, in D.F. Jackson (ed.), *Algae, Man and the Environment*, Syracuse University Press, Syracuse, New York, USA, pp. 279-358.

Shen, X., Lam, P.K.S., Shaw, G.R. and Wickramasingh, W. (2002). Genotoxicity investigation of a cyanobacterial toxin, cylindrospermopsin, *Toxicon* **40,** 1499-1501.

Sivonen, K. and Jones, G. (1999) Cyanobacterial toxins, in I. Chorus and J. Bartram (eds.), *Toxic Cyanobacteria in Water: a Guide to their Public Health Consequences, Monitoring and Management*, Spon, London, United Kingdom, pp. 41-111.

Sivonen, K. (2000) Freshwater cyanobacterial neurotoxins: ecobiology, chemistry and detection, in L.M. Botana (ed.), *Seafood and Freshwater Toxins,* Marcel Dekker, New York, USA, pp. 567-581.

Skulberg, O.M., Codd, G.A. and Carmichael, W.W. (1994) Toxic blue-green algal blooms in Europe, a growing problem, *AMBIO* **13**, 244-247.

Stalker, M. (1887) On the Waterville cattle's disease, in *Department of Agriculture Report 1881/86,* Pioneer Press Co., University of Minnesota. Report # Biennial Report of the Board of Regents No. 4, Suppl. 1, pp. 105-108.

Teixeira, M.G.L.C., Costa, M.C.N., Carvalho, V.L.P., Pereira, M.S. and Hage, E. (1993) Gastroenteritis epidemic in the area of the Itaparica Dam, Bahia, Brazil, *Bulletin of the Pan American Health Organization* **27**, 244-253.

Turner, P.C., Gammie, A.J., Hollinrake, K. and Codd, G.A. (1990) Pneumonia associated with contact with cyanobacteria, *British Medical Journal* **300**, 1440-1441.

Watson, S.B. (2003) Cyanobacterial and eukaryotic algal odour compounds: signals or by-products? A review of their biological activity, *Phycologia* **42**, 332-350.

WHO (1998) *Guidelines for Drinking Water Quality*, second edition, Addendum to Volume 2, Health Criteria and Other Supporting Information. World Health Organisation, Geneva, Switzerland.

WHO (2003) *Guidelines for Safe Recreational Water Environments*, Volume 1 Coastal and Fresh Waters. World Health Organization, Geneva, Switzerland.

Yoo, R.S., Carmichael, W.W., Hoehn, R.C. and Hrudey, S.E. (1995) *Cyanobacterial (Blue-Green Algal) Toxins: A Resource Guide.* American Waterworks Association Research Foundation, Boulder, Colorado, USA.

Yu, S.-Z. (1989) Drinking water and primary liver cancer, in Z.Y. Tang, M.C. Wu and S.S. Xia (eds.), *Primary Liver Cancer,* China Academic Publishers, New York, USA, pp. 30-37.

Yu, S.-Z. (1995) Primary prevention of hepatocellular carcinoma, *Journal of Gastroenterology and Hepatology* **10**, 674-682.

CHAPTER 2

MOLECULAR BIOLOGY OF CYANOBACTERIAL TOXINS

Genetic basis of microcystin production

Thomas BÖRNER & Elke DITTMANN

1 Introduction

Cyanobacteria produce different metabolites including alkaloids, lipopolysaccharides, polyketides and peptides that may act as toxins on other bacteria, lower and higher eukaryotes. Until today, only a few genes involved in the biosynthesis of cyanobacterial toxins have been discovered, all belonging to the group of peptides/polyketides. The way leading to the detection of these genes has been paved by several important discoveries. Microcystins were found to be small circular heptapeptides containing several unusual amino acids (Bishop et al., 1959; Botes et al., 1984; Rinehart et al., 1988). Moore et al. (1991) reported that *Microcystis aeruginosa* PCC 7820 synthesises microcystin-LR from L-methionine, L-phenylalanine, L-glutamic acid, acetate and pyruvate. These precursor studies, together with data from an assay for thio-template activity (Arment and Carmichael, 1996) and with the structure of microcystin, suggested a nonribosomal way of biosynthesis for the peptide backbone and a polyketide pathway for the characteristic Adda ((2*S*, 3*S*, 8*S*, 9*S*)-3-amino-9-methoxy-2,6,8-trimethyl-10-phenyldeca-4,6-dienoic acid) moiety. These two types of biosyntheses were first discovered for the peptide antibiotic gramicidine S and the macrolide antibiotic erythromycin, respectively.

Non-ribosomal peptide synthetases (NRPS) represent a family of large, multifunctional modular enzymes, in which each module is responsible for the activation, modification and condensation of an individual amino acid. These enzymes form large complexes serving as templates for the biosynthesis of peptides as an alternative way to (poly)peptide formation on ribosomes (for review see Marahiel et al., 1997; von Döhren et al., 1997). The large number of bacterial polyketides is synthesised in a similar way through the action of modular polyketide synthases (PKS). The great variety of polyketide structures is achieved by the incorporation of various acyl units that can be further modified by integrated and

J. Huisman, H.C.P. Matthijs and P.M. Visser (eds.), Harmful Cyanobacteria, 25–40.

non-integrated domains and enzymes, respectively (for review see Hopwood, 1997; Khosla, 1997). Even though there was a good chance that microcystins are indeed synthesised on multi-enzyme complexes built up by NRPS and PKS, neither the respective genes nor the enzymes had been described for cyanobacteria. Since NRPS of bacteria and fungi are phylogenetically related and share consensus sequences (Turgay and Marahiel, 1994), an obvious approach to the identification of genes for microcystin biosynthesis was to search for consensus sequences of NRPS-encoding genes in cyanobacteria.

NRPS genes could indeed be identified for the first time in cyanobacteria by investigating microcystin-producing and non-producing *Microcystis* cells (Dittmann et al., 1996; Meissner et al., 1996). In parallel, NRPS genes for the synthesis of anabaenopeptilide had been investigated in *Anabaena* (Rouhiainen et al., 2000). Since toxic as well as non-toxic cells harbour NRPS genes, methods for genetic manipulation of toxin-producing cyanobacteria and for directed mutagenesis of NRPS genes had to be established to facilitate identification of those genes that are needed for microcystin production. The goal was finally attained by knock-out mutagenesis of NRPS genes in *M. aeruginosa* PCC 7806 leading to mutant cells no longer capable of synthesising microcystin (Dittmann et al., 1997). This study was completed by sequencing a large gene cluster coding for the essential components of the microcystin synthetase complex in the *M. aeruginosa* strains PCC 7806 (Tillett et al., 2000) and K-139 (Nishizawa et al., 1999; 2000).

Following the same or a similar approach, it will now be possible to detect the genes for all other cyanobacterial peptide toxins that are synthesised non-ribosomally. The microcystin synthetase (*mcy*) genes have been characterised recently for a *Planktothrix* (Christiansen et al., 2003) and an *Anabaena* strain (Rouhiainen et al., 2004). Furthermore, NRPS and PKS genes could be detected that are involved in the biosynthesis of the related hepatotoxin nodularin (M.C. Moffitt and B.A. Neilan, accession no. AY210783). The search for genes with a role in the biosynthesis of cylindrospermopsin, another hepatotoxin, has also revealed NRPS and PKS genes that occur in toxin-producing strains of *Cylindrospermopsis raciborskii* and *Anabaena bergii*, but not in non-toxic cells (Schembri et al., 2001), though direct evidence for their suggested function is lacking. Here we review studies on the microcystin and nodularin synthetase genes and discuss aspects of their function and evolution.

2 Microcystin synthetase: genes and their function in *Microcystis aeruginosa*

The principle organisation of the gene cluster coding for the individual enzymes involved in microcystin biosynthesis is identical in the two strains of *M. aeruginosa* that have been investigated in more detail (Nishizawa et al., 2000; Tillett et al., 2000), but differs between the genera (see below). The *mcy* gene cluster of *M. aeruginosa* spans about 55kb of the chromosomal DNA and comprises 10 genes embedded in two bidirectionally transcribed operons (Tillett et al., 2000; Kaebernick et al., 2002). Downstream of the promoter region, *mcyA-C* encode three NRPS comprising 5 modules, whereas upstream *mcyD-J* encode polyketide synthases (*mcyD*), hybrid enzymes (*mcyE, G*) and additional tailoring enzymes (*mcyF, I, J*) as well as a component of a putative ABC transporter (*mcyH*) (Fig. 1). Taking into account that the multi-enzyme components of the Mcy complex comprise the minimal set of domains needed for the assembly of seven amino acids and four acetate units plus additional integrated domains with epimerase, methyltransferase and aminotransferase activities, the enzymes encoded by the *mcy* gene cluster are thought to catalyse 45 out of the postulated 48 sequential steps of microcystin biosynthesis (Tillett et al., 2000).

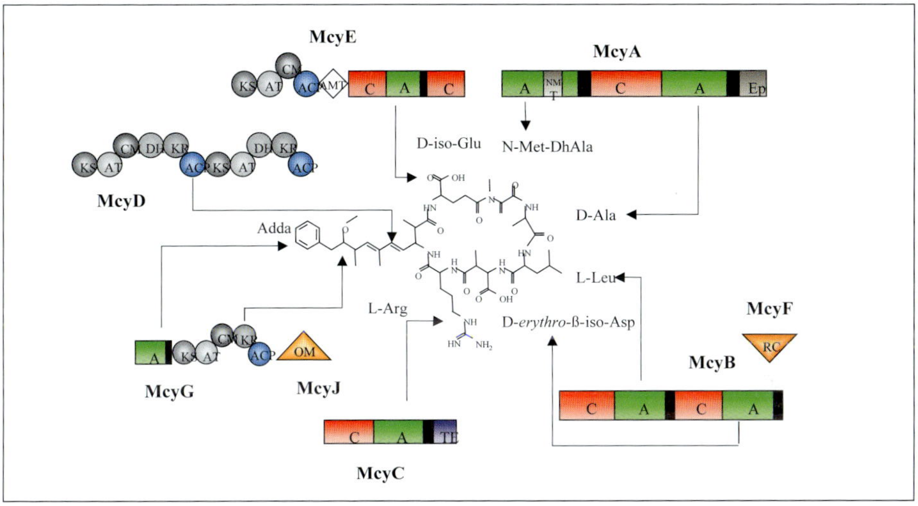

Figure 1. The role of components of the microcystin synthetase (Mcy proteins) deduced from the nucleotide sequence of their genes (Tillett et al., 2000) and functional analyses (Dittmann et al., 1997; Christiansen et al., 2003; Sielaff et al., 2003). PKS domains: AT: acyltransferase; ACP: acyl carrier protein; KS: ß-ketoacyl synthase, KR: ketoacyl reductase; DH: dehydratase; CM: C-methyltransferase; AMT: aminotransferase; NRPS domains: A: aminoacyl adenylation; C: condensation; NMT: N-methyltransferase; Ep: epimerase, TE: thioesterase; McyF: racemase, OM (McyJ): O-methyltransferase. Black bars represent the thiolation motif of NRPS modules. Arrows indicate the assignment of individual proteins to steps of microcystin biosynthesis.

The Adda-D-Glu precursor is synthesised by the activities of at least four enzymes, McyG, J, D and E. McyG is a hybrid enzyme that combines an unusual NRPS adenylation domain with similarity to Acyl-CoA ligases with a typical PKS elongation module. McyG presumably starts microcystin synthesis by activation of phenylacetate followed by transfer to the 4-phosphopantetheine of the first carrier domain. Subsequently, Adda is formed by the four PKS modules of McyG, D and E. McyJ is needed for an O-methylation step. McyE is a hybrid protein composed of PKS and NRPS modules. An integrated domain with similarity to glutamate semialdehyde aminotransferases most likely provides the β-amino group of the

Adda moiety. Finally, the first condensation domain of McyE condenses Adda with the activated glutamate thereby linking the PKS with the NRPS part of microcystin biosynthesis. The NRPS modules of McyA, B and C activate the remaining five amino acids and incorporate them into the growing peptide structure. It has been proposed that the TE domain of McyC (most probably together with a separate enzyme; Christiansen et al., 2003) is responsible for cyclisation and release of microcystin from the synthetase complex. McyF is a racemase thought to provide D-aspartate and most likely also D-methyl-aspartate (Sielaff et al., 2003). The role of McyI remains to be determined (Tillett et al., 2000). Since McyH shows significant similarity to exporters of the ABC transporter family it might play a role in microcystin export (Pearson et al., 2004).

More than 65 structural variants of microcystin have been described. Usually, the individual strains produce more than one microcystin variant (Sivonen and Jones, 1999). This observation raises the question as to whether the biosynthesis of each variant needs its own set of *mcy* genes (then a strain that produces e.g. three variants of microcystin should possess three *mcy* genes clusters) or if one and the same cluster codes for the enzymes needed to synthesise all microcystin variants of a strain. Mutation of single *mcy* genes in different *Microcystis* strains has led in each case to mutant clones that were no longer capable of producing any kind of microcystin. This clearly demonstrates that one *mcy* gene cluster is responsible for the synthesis of all microcystin variants in a given strain (Dittmann et al., 1997; Nishizawa et al., 1999; 2000; Tillett et al., 2000). The substrate specificity of non-ribosomal peptide synthetases is determined by the amino acid sequence of the adenylation domain. Therefore, the ability of strains to produce more than one isoform of microcystin is likely caused by relaxed substrate specificity of the adenylation domains of the involved peptide synthetases (see below 6.). Mikalsen et al. (2003) could identify two groups of sequence variants of an adenylation domain of the *mcyB* gene. Presence of one or the other variant in *Microcystis* strains correlated with the formation of microcystin-LR isoforms and microcystin-RR (in a subgroup in combination with microcystin-LR), respectively. Rouhiainen et al. (2004) observed a higher variability in the specifying amino acids of adenylation domains that are responsible for the incorporation of amino acids at variable positions of microcystins as compared with the other adenylation domains in *mcy* genes (see also 6. below).

3 Distribution of *mcy* genes in cyanobacterial communities

The detection of NRPS genes in cyanobacteria (Meissner et al., 1996) and of *mcy* genes in *Microcystis* (Dittmann et al., 1997) led to investigations into the distribution of these genes among cyanobacteria. Primers deduced from NRPS sequences were used in PCR studies and revealed the presence of NRPS genes in about 75% of 146 axenic strains from the Pasteur Culture Collection representing all traditional botanical orders of cyanobacteria, the *Chroococcales, Pleurocapsales, Oscillatoriales, Nostocales,* and *Stigonematales* (Christiansen et al., 2001). With the help of *mcy*-specific primers, the existence of genes for microcystin biosynthesis was detected in *Microcystis, Planktothrix (Oscillatoria), Anabaena* and *Nostoc,* and of related genes in *Nodularia* (Neilan et al., 1999). At the same time it became evident that some closely related strains (genotypes) did not contain *mcy* genes and that there was a good correlation between the presence of *mcy* genes and the ability to produce microcystin(s) and also between the lack of these genes and the inability to synthesise microcystin(s). Thus, it was recognised that PCR with *mcy*-specific primers provided a way of discriminating between potentially hepatotoxic and non-hepatotoxic strains (genotypes) and a method for detecting potentially harmful cyanobacteria in water bodies (see Ouellette and Wilhelm, 2003 for review). A combination of *mcy*-specific PCR with RFLP analysis readily identifies potential microcystin producers in environmental samples (Fig. 2).

It is anticipated that the knowledge of genes for microcystin production and the biosynthesis of other cyanobacterial toxins will be important for the molecular ecology of harmful cyanobacteria. The sensitivity of PCR will permit the analysis of even single colonies or filaments. Real-time PCR with *mcy*-specific primers will provide data on the quantitative distribution of toxic and non-toxic genotypes in populations (Kurmayer and Kutzenberger, 2003; Vaitomaa et al., 2003). The development of DNA chips has been initiated containing genus-, species-, and also *mcy*-specific sequences. The DNA chips are expected to provide information, by simple hybridisation experiments, on the composition of cyanobacterial

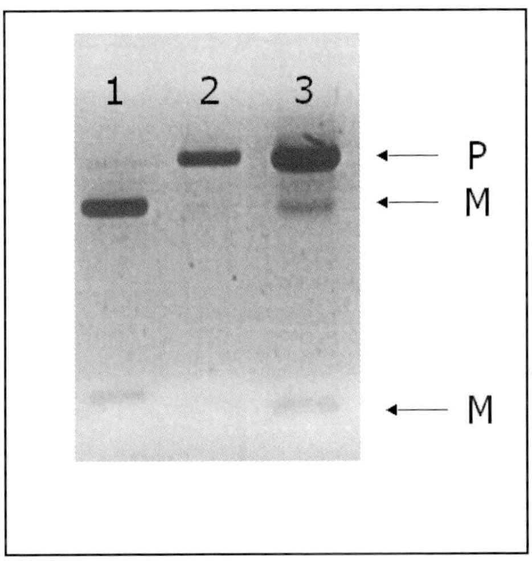

Figure 2. Identification of Microcystis *and* Planktothrix *as microcystin-producers in a water sample from lake Wannsee (Berlin). DNA was isolated from axenic* M. aeruginosa *(lane 1) and* P. agardhii *strains (lane 2) and from the environmental sample (lane 3);* mcy-*specific fragments were amplified by PCR, digested by a restriction endonuclease (*Eco*RV) and electrophoretically separated on an agarose gel. M and P mark the position of* Microcystis- *and* Planktothrix-*specific fragments, respectively. The DNA-based technique revealed the presence of* mcy-*genes from* Microcystis *and* Planktothrix *in the water sample (Hisbergues et al., 2003).*

communities or blooms and their potential toxicity (EU project MidiChip; http://www.ulg.ac.be/ cingprot/midichip/index.htm).

4 Upregulation of *mcy* gene expression by light

Knowledge of the regulation of cyanotoxin biosynthesis may help prevent the appearance of environmental conditions that support toxin production and provide clues on the cellular function of these substances. It has been suggested that several environmental factors influence the biosynthesis of cyanotoxins. For discussion of environmental effects on toxin production see chapter 3 of this volume.

Studies on transcript accumulation of genes encoding the peptide synthetases involved in microcystin biosynthesis in two unrelated strains of *M. aeruginosa* clearly demonstrated a light-dependent increase in the steady state level of those mRNAs (Nishizawa et al., 1999; Kaebernick et al., 2000) (Fig. 3). No data on the effects of other environmental factors on the expression of *mcy* genes have been reported so far. Interestingly, the positive effect of light was only detectable with cells harvested from the early (OD_{750} 0.4 – 0.7) and middle growth phases (OD_{750} 0.84 – 0.87) of batch cultures, but not from the late phase (OD_{750} 1.6 – 2.9)

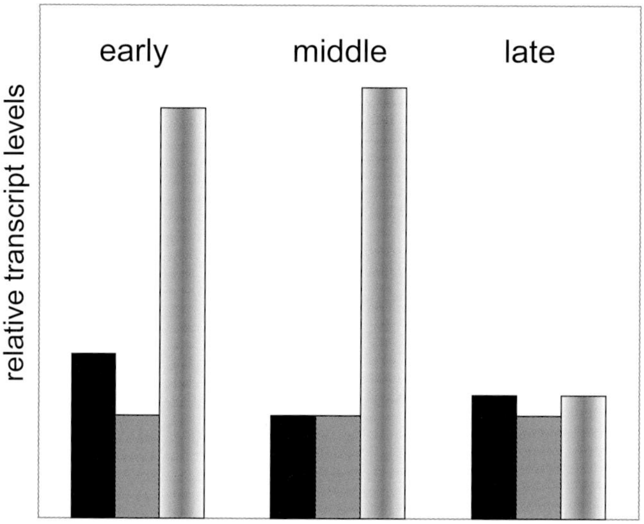

Figure 3. Relative mcyB *transcript levels in* M. aeruginosa *PCC 7806 cells harvested in the early, middle or late growth phase of batch culture. Cultures were illuminated with a photon irradiance of 16 µmol m^{-2} s^{-1} (black column), 31 µmol m^{-2} s^{-1} (grey column), and 68 µmol m^{-2} s^{-1} (shaded column), respectively (Kaebernick et al., 2000).*

(Fig. 3; Kaebernick *et al.* 2000). Shading effects by high cell densities could explain this striking result (Kaebernick *et al.,* 2000). This observation might also indicate an effect of cell density on the transcription of *mcy* genes as known from many genes that are regulated by a quorum sensing mechanism in other bacteria (cf. Dittmann et al., 2001). Studies are under way to discriminate between these alternatives. Since the stimulating effect of light on *mcy* gene transcription could be observed already after a few minutes of illumination of *M. aeruginosa* PCC 7806 cells, light may act directly on transcription and/or transcript stability rather than indirectly, e.g. via a general stimulation of growth (Kaebernick et al., 2000 and unpubl. data). The connection between light, growth and microcystin production is complex and has been the subject of several studies (see chapter 3 of this volume).

The microcystin content could be controlled at the level of transcription, transcript stability, translation, enzyme activities and/or microcystin turnover. Thus, an enhanced *mcy* transcript accumulation after illumination by certain light intensities and qualities (Kaebernick et al., 2000) may not necessarily lead to an enhanced amount of microcystins under the same conditions. However, light has also a marked positive influence on the accumulation of Mcy proteins (shown for McyF in Fig. 4, but also found for other Mcy proteins; M. Hisbergues et al., unpubl. data), which suggests that the enhanced transcript level leads to increased translation. Hence, regulation at the transcriptional level seems to be important for controlling the synthesis of the Mcy proteins and, most probably, also of the toxin.

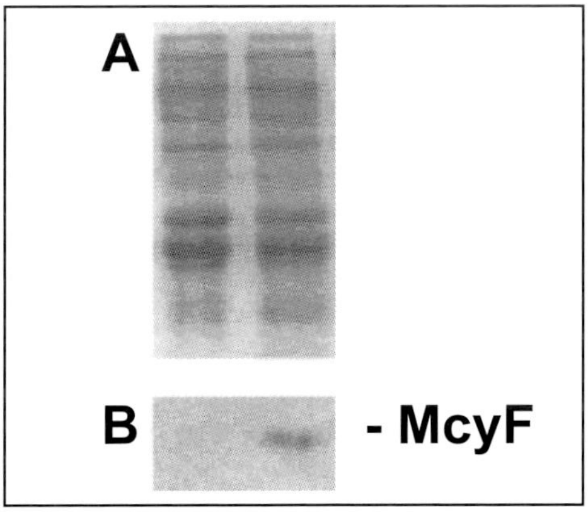

Figure 4. (A) Proteins were isolated from M. aeruginosa *PCC 7806 grown under low light conditions (left lane) or medium light conditions (right lane), electrophoretically separated on SDS polyacrylamide gels and stained by Commassie blue. (B) Western blot of the same gel. McyF was detected by reaction with specific antibodies only in cells grown under medium light conditions (data of M. Hisbergues).*

Studies on the regulation of microcystin production, not to mention the synthesis of other cyanobacterial toxins, are still at their beginning. The presence of a gene encoding a component of a putative ABC transporter in *mcy* gene clusters, and also in the *nda* gene cluster (coding for nodularin biosynthesis), suggests an export of these hepatotoxins from the producing cells. Future analyses of the effects of environmental factors on toxin production will have to consider both intra- and extracellular microcystins.

5 Comparison of *mcy* genes in different genera – evolutionary aspects

Most microcystin-producing cyanobacteria are strains of the genera *Anabaena, Microcystis* and *Planktothrix*, while the only known nodularin producers are strains of the brackish water species *Nodularia spumigena.* Cyanobacteria of these genera possess gas vesicles and show a tendency to form surface water-blooms. According to their positions in phylogenetic trees based on 16S rDNA sequences, the microcystin producing genera are only distantly related (e.g. Lyra et al., 2001; Tillett et al., 2001; Rantala et al., 2004) and differ also from each other in their morphology. *Microcystis* is a unicellular cyanobacterium that forms characteristic colonies in its natural environment (e.g. Otsuka et al., 2000). *Planktothrix* is a filamentous cyanobacterium; *P. rubescens* usually forms distinct layers while *P. agardhii* is more often evenly distributed in the water column, especially in shallow lakes (e.g. Van Liere and Mur, 1980; Kurmayer et al., 2004). Planktic species of *Anabaena* and *Nodularia* are filamentous cyanobacteria that differentiate heterocysts, cells specialised for nitrogen fixation (e.g. Lehtimäki et al., 2000; Lyra et al., 2001; Golden and Yoon, 2003).

The ability to produce microcystin or nodularin shows a patchy distribution among the cyanobacteria. All four genera have members that contain the toxin biosynthesis genes (in most strains correlated with the ability to produce the toxins), and other members that do not. Toxic and non-toxic strains (genotypes) usually coexist in field populations and usually differ with respect to presence *vs*. absence of the genes needed for toxin biosynthesis in their genomes (see section 3 above). Comparisons of the known gene clusters for microcystin and nodularin biosynthesis from the four investigated genera, *Anabaena, Microcystis, Nodularia,* and *Planktothrix*, allow conclusions to be drawn on how this patchy distribution of toxin biosynthesis genes could have evolved among cyanobacteria. In principle, the *mcy* gene clusters might have been distributed by lateral (horizontal) gene transfer between the genera, species and genotypes within a species. If this occurred, genotypes that were non-producers would be expected never to have possessed the genes for the production of peptide toxins. Alternatively, gene clusters might have evolved from common ancestral genes. Had this occurred, many genera, species and strains that harbour no genes for microcystin or nodularin biosynthesis, although phylogenetically related to the producing taxa, must have once had those genes and subsequently lost them. A combination of both alternatives is conceivable. Complete nucleotide sequences of the *mcy* gene clusters are available from *Anabaena* 90, *Microcystis aeruginosa* PCC 7806 and K-139, and *Planktothrix agardhii* CYA 126 in databases; the *nda* gene cluster for nodularin biosynthesis has been sequenced in *Nodularia spumigena* NSOR10 (Fig. 5). Comparison of the structural organisation of the clusters and the sequences of individual genes indicates how microcystin biosynthesis might have evolved.

The multi-enzyme components encoded by the individual gene clusters are similar, but the content and arrangement of the modules and consecutive domains are not identical in the four genera. The general organisation of the clusters and the arrangement of certain genes within the clusters are different, as shown in Figure 5. In *Microcystis*, *Anabaena* and *Nodularia* the genes are transcribed from a central bi-directional promoter region (cf. Kaebernick et al., 2002), whereas in *Planktothrix* all *mcy* genes, except *mcyT,* seem to be transcribed unidirectionally from a promoter located upstream of gene *mcyD*.

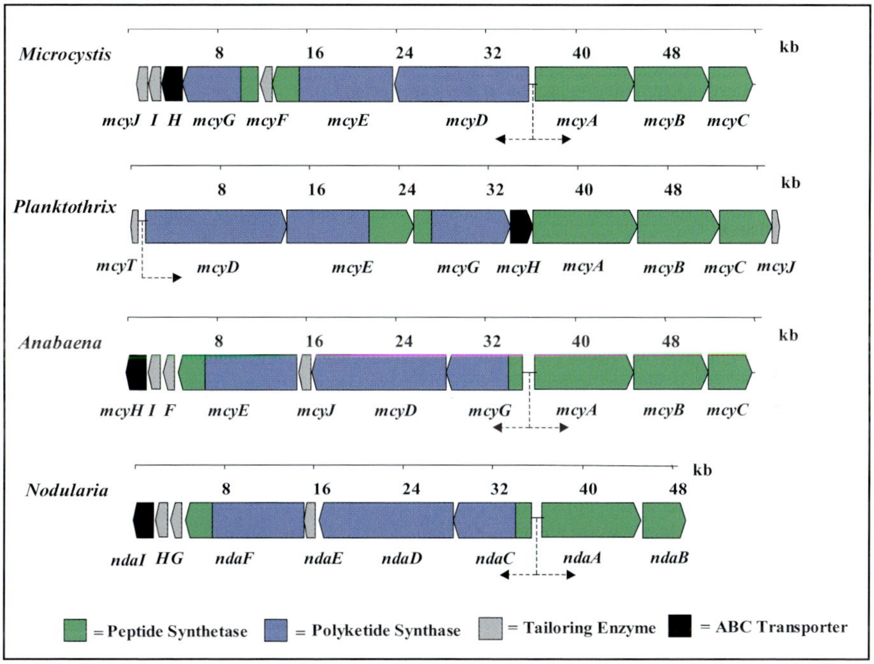

Figure 5. Gene clusters coding for the biosynthesis of microcystin in Microcystis *(Tillett et al., 2000; Nishizawa et al., 2000),* Planktothrix *(Christiansen et al., 2003), and* Anabaena *(Rouhiainen et al., 2004) and of nodularin in* Nodularia *(M.C. Moffitt and B.A. Neilan; accession no. AY210783). Arrows indicate the transcriptional start sites from the putative promoter regions.*

There is a further difference between *Microcystis/Planktothrix* and *Anabaena/Nodularia.* NRPS and PKS genes are usually found to follow a "colinearity rule", i.e. the order of genes is the same as the order of the single enzymatic steps of the respective biosynthesis (Marahiel et al., 1997; von Döhren et al., 1997). As described above, microcystin synthesis starts with the step catalysed by McyG, followed by the steps involving McyD, E, A, B and C. An identical arrangement of genes is found in *Anabaena* (Fig. 5). The order of *nda* genes in

Nodularia follows also the "colinearity rule" (Fig. 5). In *Microcystis* and *Planktothrix,* however, the order of the first three genes is *mcyD*, *E* and *G,* i.e. there is a clear deviation from the "colinearity rule". Similarly, the position of tailoring genes and the gene encoding a component of a putative ABC transporter within the cluster is different in the individual genera, except in *Anabaena* and *Nodularia* where, remarkably, it is identical. *Microcystis* still exhibits the same complement of tailoring genes as *Anabaena* and *Nodularia*, but arranged in a different order. *Planktothrix* is most distinctive: two of the tailoring genes (*mcyF, I*) are completely missing, whereas an additional gene, encoding a thioesterase (*mcyT*), was found in close proximity to the biosynthesis gene cluster (Christiansen et al., 2003).

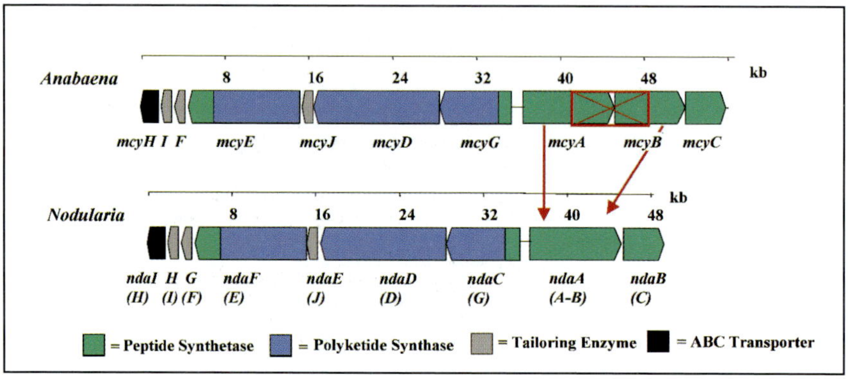

*Figure 6. The microcystin biosynthesis gene (*mcy*) cluster of* Anabaena *90 and the nodularin biosynthesis gene cluster (*nda*) of* Nodularia spumigena *NSOR10 show the same order of genes. The* nda *cluster is proposed to have evolved from an* Anabaena-*type* mcy *cluster by deletion of two NRPS modules from* mcyA *and* B *and fusion of the remaining sequences of* mcyA *and* B *resulting in gene* ndaA. *Letters in brackets indicate the* mcy *gene with homology to the respective* nda *gene.*

Although the overall similarity of the Mcy and Nda multi-enzymes is very high, there is a striking difference between the *mcy* clusters and the *nda* cluster: the *nda* cluster lacks two NRPS modules that are present in all *mcy* clusters. In *Nodularia,* the genes *mcyA* and *B* found separately in the other genera, have been fused together with a simultaneous loss of two NRPS modules resulting in the *ndaA* gene (Fig. 6). The loss of two NRPS modules is consistent with the fewer amino acid residues (five *vs.* seven) in nodularin compared to microcystin (Fig. 7; Annila et al., 1996). Despite the loss of two modules, microcystin/nodularin biosynthesis gene clusters in *Anabaena* and *Nodularia* are the most similar, as expected from the position of these genera in phylogenetic trees (Iteman et al., 2002; Janson and Graneli, 2002; Rantala et al., 2004). It is, therefore, proposed that the nodularin gene cluster evolved from a microcystin biosynthesis gene cluster of the *Anabaena*-type by one major step, the deletion of two NRPS modules (cf. Rantala et al., 2004).

In the case of *Microcystis*, *mcy* clusters of two strains have been sequenced and found to have the same gene order and a high similarity in their nucleotide sequences (Dittmann et al., 1997; Nishizawa et al., 1999 and 2000; Tillett et al., 2000). In contrast, several recombination and many mutation events must have occurred to lead to the differences in structural organisation and nucleotide sequence between the *mcy* gene clusters in *Anabaena, Microcystis* and *Planktothrix* (Fig. 6; Nishizawa et al., 2000; Tillett et al., 2000; Christiansen et al., 2003; Rouhiainen et al., 2004). These differences make a lateral transfer of *mcy* gene clusters between genera unlikely and rather suggest an independent evolution over a long time span. Important data concerning the evolution of *mcy* genes were obtained by analysing the phylogenetic relationship of conserved *mcy* sequences and of sequences of housekeeping genes. The results of this study are not in agreement with lateral gene transfer between the genera, but strongly support the idea of a common ancestor of the *mcy* genes in *Anabaena, Microcystis, Nostoc* and *Planktothrix* (Rantala et al., 2004). Microcystin biosynthesis, therefore, appears to be a very old pathway of secondary metabolism. The lack of *mcy* (or *nda*) genes in many cyanobacterial taxa must consequently be due to losses of these genes during evolution (Rantala et al., 2004). The loss of *mcy* genes seem to be an ongoing process that might start with a loss of function by mutation. There are reports of *Microcystis* and *Planktothrix* genotypes (strains) that contain *mcy* genes, but do not produce microcystin. Whereas such genotypes seem to be rare in *Microcystis* (Mikalsen et al., 2003; Nishizawa et al., 1999; Tillett et al., 2001), they have been more frequently observed in some Austrian (Kurmayer et al., 2004) and German lakes (S. Mbedi and C. Wiedner, pers. commun.). Preliminary data suggest that their inability to produce microcystin is due to mutation of *mcy* gene(s) (Kurmayer et al., 2004).

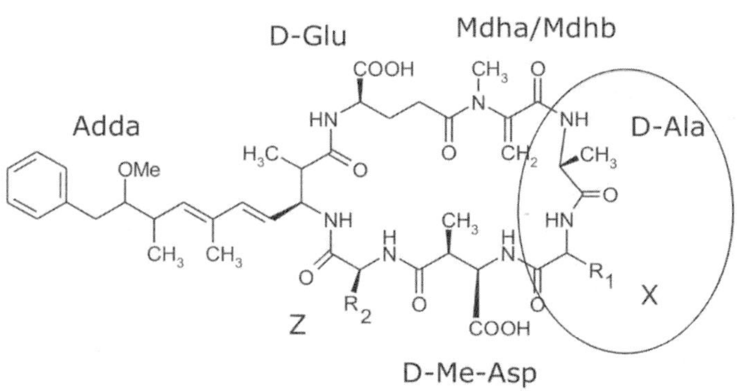

Figure 7. The molecular structure of microcystins and nodularins is similar. Nodularins lack the two amino acids shown in the oval and have Mdhb instead of Mdha.

6 Structure and function of adenylation domains

Further insight into the evolution of microcystin biosynthesis can be expected from analyses of the sequences of individual *mcy* genes and their modules and domains (Christiansen et al., 2003; Kurmayer et al., 2001; Mikalsen et al., 2003; Rouhiainen et al., 2004). A fast evolution should occur of those enzymes and domains that contribute to the observed large variation in microcystin structure. Most of this variation is due to differences in individual amino acid positions, whereas the polyketide side chain of the Adda moiety is highly conserved. The specificity of the so-called adenylation (A) domain determines which amino acid(s) are activated by an individual NRPS module.

A-domains specifically recognise their substrate amino acids and form an amino-acyl-AMP intermediate. Analysis of the crystal structure of the phenylalanine activating domain of the tyrocidine synthetase from *Bacillus subtilis* (PheAT), as well as comparison of the growing number of NRPS adenylation domains, have led to the development of a "specificity-conferring code of the amino acids" (Stachelhaus et al., 1999; Challis et al., 2000). Eight single residues of the amino acid binding pocket were proposed to discriminate between the different amino acid substrates. This code can be used for the prediction of the specificity of NRPSs that have not been characterised.

The four individual strains that have been used for sequencing microcystin biosynthesis gene clusters show different complements of microcystin variants. Another two strains have been investigated that produce different variants of nodularin. Differences occur at four amino acid positions, namely the Mdha/Mdhb position, the L-Leu/L-Arg position, the D-Asp/D-MeAsp position and the L-Arg/L-Har position (Table 1). The different specificities can at least in part be recognised by comparing the "amino acid specificity codes" of the corresponding A-domains. The first A-domain in McyA is supposed to activate serine as the precursor of the Mdha moiety. All microcystin producing strains conform to the code, i.e. show high similarity to serine-activating domains in other bacteria. The analogous NdaAA1 domain, however, shows a clear threonine code, which is expected to be the precursor of the Mdhb moiety. The McyBA1 domain seems to either recognise only L-leucine (as seen for the two *Microcystis* examples) or to be multispecific for L-leucine and L-arginine. In the first case a "leucine code" can be observed, whereas the *Planktothrix* CYA 126 and *Anabaena* 90 McyBA1 domains are not homologous to other NRPS A-domains, but even show some variation between each other. McyBA2 activates D-aspartate and D-methyl-aspartate in all strains investigated, except *Planktothrix* CYA 126, which seems to accept only D-aspartate as a substrate. It is not yet clear, whether this is related to the single amino acid difference in the corresponding codes (Table 1). Alternatively, the enzyme providing D-methyl-aspartate as a substrate might be missing in *Planktothrix*. Strikingly, McyCA1 activates L-arginine in all microcystin-producing strains, even though in all four cases no clear "arginine code" can be observed. The *Nodularia* strains investigated incorporate L-Arg (*Nodularia spumigena* NSOR10) or L-homoarginine (Har, *Nodularia harveyana* PCC7804) in the nodularin structure. There is a single amino acid difference in the specificity code between the two *Nodularia* strains that could account for the deviating substrate specificity (Table 1), but biochemical evidence is needed to verify this assumption.

Differences in the specificities of A-domains can be deduced not only from the few amino acids determining the 'specificity codes', but also by comparing complete sequences of binding pockets. In contrast to the major part of the Mcy complex, these domains are highly variable in their sequence. Recombination

Table 1. Adenylation domains of microcystin synthetase and nodularin synthetase gene clusters: Comparison of specificity conferring amino acids in the substrate binding pocket, predicted substrates and real substrate. (Org.: organism; Ma = Microcystis aeruginosa *PCC7806, Pa =* Planktothrix agardhii, *A90 =* Anabaena *90, Ns =* Nodularia spumigena *NSOR10, Nh =* N. harveyana *PCC7804. Ref.: reference; 1 = Tillett et al., 2000, 2 = Christiansen et al., 2003, 3 = Rouhiainen et al., 2004, 4 = accession no. AY210783)*

A domain	*Org*	*Binding pocket*	*Predicted substrate*	*Substrate*	*Ref*
McyAA1	*Ma*	DVWHFSLI	Ser	L-Ser	1
McyAA1	*Pa*	DVWHISLI	Ser	L-Ser	2
McyAA1	*A90*	DVWHISLI	Ser	L-Ser	3
NdaAA1	*Ns*	DFWNIGMV	Thr	L-Thr	4
McyAA2	*Ma*	DLFNNALT	Gly	L-Ala	1
McyAA2	*Pa*	DLFNNALS	Gly	L-Ala	2
McyAA2	*A90*	DLFNNALT	Gly	L-Ala	3
---	*Ns*	--	--	--	
McyBA1	*Ma*	DVWFLGNV	Leu	L-Leu	1
McyBA1	*Pa*	DALFFGLV	--	L-Arg/L-Leu	2
McyBA1	*A90*	DVWFFGLV	--	L-Arg/L-Leu	3
---	*Ns*	--	--	--	
McyBA2	*Ma*	DARHVGIV	--	D-MeAsp/D-Asp	1
McyBA2	*Pa*	DPRHVGIF	--	D-Asp	2
McyBA2	*A90*	DARHVGIF	--	D-MeAsp/D-Asp	3
NdaAA2	*Ns*	DARHVGIF	--	D-MeAsp	4
McyCA1	*Ma*	DVWTIGAV	--	L-Arg	1
McyCA1	*Pa*	DPWVFGLV	--	L-Arg	2
McyCA1	*A90*	DVWVFGLV	--	L-Arg	3
NdaBA1	*Ns*	DVWNFGFV	--	L-Arg	4
NdaBA1	*Nh*	DVWSFGFV	--	L-Har	4
McyEA1	*Ma*	DPRHSGVV	--	D-Glu	1
McyEA1	*Pa*	DPRHSGVV	--	D-Glu	2
McyEA1	*A90*	DPRHSGVV	--	D-Glu	3
NdaFA1	*Ns*	DPRHSGVV	--	D-Glu	4
McyGA1	*Ma*	VGIWVAAS	--	Phenylacetate	1
McyGA1	*Pa*	VGIWVAAS	--	Phenylacetate	2
McyGA1	*A90*	VGIWVAAS	--	Phenylacetate	3
NdaCA1	*Ns*	VGIWVAAS	--	Phenylacetate	4

between different modules of the *mcy* gene cluster seems to contribute to the observed sequence variability (Christiansen et al., 2003; Mikalsen et al., 2003; Rouhiainen et al., 2004). Since individual strains usually produce several different peptides and polyketides, they should also harbour several NRPS and PKS gene clusters. It remains to be determined to what extent an exchange of DNA sequences between the different clusters has contributed to their evolution.

The evolution of *mcy* gene clusters is a dynamic and continuing process. Differences in the general structure of the *mcy* clusters and co-evolution of conserved *mcy* sequences with house-keeping genes do not favour lateral gene transfer between the genera and clearly point to the early appearance of microcystin biosynthesis in the evolution of cyanobacteria (see above). There has been rapid evolution of certain regions of the *mcy* cluster that bear the genetic information for the variability of microcystins. This evolution has occurred 'conventionally' by mutation and by recombination between modules and domains, though a contribution of intraspecific lateral transfer to the evolution of these sequences has not been ruled out.

7 Acknowledgements

We thank Michael Hisbergues for providing the data shown in Figure 4 and Tony Walsby for critically reading the manuscript. The work of the authors was supported by grants from the European Union (CYANOTOX-ENV4-CT98-0802 and TOPIC-FMRX-CT98-0246), the Deutsche Forschungsgemeinschaft (Bo 1045/13-3) and the Fond der Chemischen Industrie.

8 References

Annila, A., Lehtimäki, J., Mattila, K., Eriksson, J.E., Sivonen, K., Rantala, T.T. and Drakenberg, T. (1996) Solution structure of nodularin: an inhibitor of serine/threonine-specific protein phosphatases, *Journal of Biological Chemistry* **27**, 16695-16702.

Arment, A.R. and Carmichael, W.W. (1996) Evidence that microcystin is a thio-template product, *Journal of Phycology* **32**, 591-597.

Bishop, C.T., Anet, E.F.L.J. and Gorham, P.R. (1959) Isolation and identification of the fast-dead factor in *Microcystis aeruginosa* NRC-1, *Canadian Journal of Biochemistry and Physiology* **37**, 453-458.

Botes, D.P., Tuiman, A.A., Wessels, P.L., Viljoen, C.C., Kruger, H., Williams, D.H., Santikarn, S., Smith, R.J. and Hammond, S.J. (1984) The structure of cyanoginosin-LA, a cyclic peptide from the cyanobacterium *Microcystis aeruginosa*, *Journal of the Chemical Society - Perkin Transactions* **I**, 2311-2319.

Challis, G.L., Ravel, J. and Townsend, C.A. (2000) Predictive, structure-based model of amino acid recognition by nonribosomal peptide synthetase adenylation domains, *Chemistry and Biology* **7**, 211-224.

Christiansen, G., Dittmann, E., Ordorika, L.V., Rippka, R., Herdman, M. and Börner, T. (2001) Nonribosomal peptide synthetase genes occur in most cyanobacterial genera as evidenced by their distribution in axenic strains of the PCC, *Archives of Microbiology* **76**, 452-458.

Christiansen, G., Fastner, J., Erhard, M., Börner, T. and Dittmann, E. (2003) Microcystin biosynthesis in *Planktothrix*: genes, evolution. and manipulation, *Journal of Bacteriology* **185**, 564-572.

Dittmann, E., Meissner, K. and Börner, T. (1996) Conserved sequences of peptide synthetase genes in the cyanobacterium *Microcystis aeruginosa, Phycologia* **35**, 62-67.

Dittmann, E., Neilan, B.A., Erhard, M., von Döhren, H. and Börner, T. (1997) Insertional mutagenesis of a peptide synthetase gene that is responsible for hepatotoxin production in the cyanobacterium *Microcystis aeruginosa* PCC 7806, *Molecular Microbiology* **26**, 779-787.

Dittmann, E., Erhard, E., Kaebernick, M., Scheler, C., Neilan, B.A., von Döhren, H. and Börner, T. (2001) Altered expression of two light-dependent genes in a microcystin-lacking mutant of *Microcystis aeruginosa* PCC 7806, *Microbiology* **147**, 3113-3119.

Golden, J.W. and Yoon, H.S. (2003) Heterocyst development in *Anabaena*, *Current Opinion in Microbiology* **185**, 6995-7000.

Hisbergues, M., Christiansen, G., Rouhiainen, L., Sivonen, K. and Börner, T. (2003) PCR-based identification of microcystin-producing genotypes of different cyanobacterial genera, *Archive for Microbiology* **180,** 402-410.

Hopwood, D.A. (1997) Genetic contributions to understanding polyketide synthases, *Chemical Reviews* **97**, 2465-2497.

Iteman, I., Rippka, R., Tandeau de Marsac, N. and Herdman, M. (2002) rDNA analyses of planktonic heterocystous cyanobacteria, including members of the genera *Anabaenopsis* and *Cyanospira, Microbiology* **148**, 481-496.

Janson, S. and Graneli, E. (2002) Phylogenetic analyses of nitrogen-fixing cyanobacteria from the Baltic Sea reveal sequence anomalies in the phycocyanin operon, *International Journal of Systematic and Evolutionary Microbiology* **52**, 1397-1404.

Kaebernick, M., Neilan, B.A., Börner, T. and Dittmann, E. (2000) Light and the transcriptional response of the microcystin gene cluster, *Applied and Environmental Microbiology* **66**, 3387-3392.

Kaebernick, M., Dittmann, E., Börner, T. and Neilan, B.A. (2002) Multiple alternate transcripts direct the biosynthesis of microcystin, a cyanobacterial nonribosomal peptide, *Applied and Environmental Microbiology* **68**, 449-455.

Khosla, C. (1997) Harnessing the biosynthetic potential of modular polyketide synthases, *Chemical Reviews* **97**, 2577-2590.

Kurmayer, R., Dittmann, E., Fastner, J. and Chorus, I. (2001) Diversity of microcystin genes in lake Wannsee (Germany), *Microbial Ecology* **43**, 107-118.

Kurmayer, R. and Kutzenberger, T. (2003) Application of real-time PCR for quantitation of microcystin genotypes in a population of the toxic cyanobacterium *Microcystis sp.*, *Applied and Environmental Microbiology* **69**, 6723-6730.

Kurmayer, R., Christiansen, G., Fastner, J. and Börner, T. (2004) Abundance of active and inactive microcystin genotypes in populations of the toxic cyanobacterium *Planktothrix* spp., *Environmental Microbiology*, in press.

Lehtimäki, J., Lyra, C., Suomalainen, S., Sundman, P., Rouhiainen, L., Paulin, L., Salkinoja-Salonen, M. and Sivonen, K. (2000) Characterization of *Nodularia* strains, cyanobacteria from brackish waters, by genotypic and phenotypic methods, *International Journal of Systematic and Evolutionary Microbiology* **50**, 1043-1053.

Lyra, C., Suomalainen, L.C., Gugger, M., Vezie, C., Sundman, P., Paulin, L. and Sivonen, K. (2001) Molecular characterization of planktic cyanobacteria of *Anabaena*, *Aphanizomenon*, *Microcystis* and *Planktothrix* genera, *International Journal of Systematic and Evolutionary Microbiology* **51**, 513-526.

Marahiel, M.A., Stachelhaus, T. and Mootz, H.D. (1997) Modular peptide synthetases involved in nonribosomal peptide synthesis, *Chemical Reviews* **97**, 2651-2673.

Meissner, K., Dittmann, E. and Börner, T. (1996) Toxic and non-toxic strains of the cyanobacterium *Microcystis aeruginosa* contain sequences homologous to peptide synthetase genes, *FEMS Microbiology Letters* **135**, 295-303.

Mikalsen, B., Boison, G., Skulberg, O.M., Fastner, J., Davies, W., Gabrielsen, T.M., Rudi, K. and Jacobsen, K.S. (2003) Natural variation in the microcystin synthetase operon *mcyABC* and impact on microcystin production in *Microcystis* strains, *Journal of Bacteriology* **185**, 2774-2785.

Moore, R.E., Chen, J.L., Moore, B.S., Patterson, G.M.L. and Carmichael, W.W. (1991) Biosynthesis of microcystin-LR: origin of carbons in the Adda and Masp units, *Journal of the American Chemical Society* **113**, 5083-5084.

Neilan, B.A., Dittmann, E., Rouhiainen, L., Bass, R.A., Schaub, V., Sivonen, K. and Börner, T. (1999) Nonribosomal peptide synthesis and toxigenicity of cyanobacteria, *Journal of Bacteriology* **181**, 4089-4097.

Nishizawa, T., Asayama, M., Fujii, K., Harada, K. and Shirai, M. (1999) Genetic analysis of the peptide synthetase genes for a cyclic heptapeptide microcystin in *Microcystis* ssp., *Journal of Biochemistry* **126**, 520-526.

Nishizawa, T., Ueda, A., Asayama, M., Fujii, K., Harada, K.-I., Ochi, K. and Shirai, M. (2000) Polyketide synthase gene coupled to the peptide synthetase module involved in the biosynthesis of the cyclic heptapeptide microcystin, *Journal of Biochemistry* **127**, 779-789.

Otsuka, S., Suda, S., Matsumoto, S. and Watanabe, W.W. (2000) Morphological variability of colonies of *Microcystis* morphospecies in culture, *Journal of General and Applied Microbiology* **46**, 39-50.

Ouellete A.J.A. and Wilhelm, S.W. (2003) Toxic cyanobacteria: the evolving molecular toolbox, *Frontiers for Ecology and Environment* **1**, 359-366.

Pearson, L., Hisbergues, M., Dittmann, E., Börner, T. and Neilan, B.A. (2004) Inactivation of an ABC transporter, McyH, results in loss of microcystin production in the cyanobacterium *Microcystis aeruginosa* PCC 7806, *Applied and Environmental Microbiology*, in press.

Rantala, A., Fewer, D.P., Hisbergues, M., Rouhiainen, L., Vaitomaa, J., Börner, T. and Sivonen, K. (2004) Phylogenetic evidence for the early evolution of microcystin synthesis, *Proceedings of the National Academy of Sciences (USA)* **101**, 568-573.

Rinehart, K.L., Harada, K.I., Namikoshi, M., Chen, C., Harvis, C.A., Munroe, M.H.G., Blunt, J.W., Mulligan, P.E., Beasley, V.R., Dahlem, A.M. and Carmichael, W.W. (1988) Nodularin, microcystin, and the configuration of Adda, *Journal of the American Chemical Society* **110**, 8557-8558.

Rouhiainen, L., Paulin, L., Suomalainen, S., Hyytiainen, H., Buikema, W., Haselkorn, R. and Sivonen, K. (2000) Genes encoding synthetases of cyclic depsipeptides, anabaenopeptilides, in *Anabaena* strain 90, *Molecular Microbiology* **37**, 156-167.

Rouhiainen, L., Vakkilainen, T., Lumbye Siemer, B., Buikema, W., Haselkorn, R. and Sivonen, K. (2004) Genes coding for the synthesis of hepatotoxic heptapeptides (microcystins) in the cyanobacterium *Anabaena* strain 90, *Applied and Environmental Microbiology* **70**, 686-692.

Schembri, M.A., Neilan, B.A. and Saint, C.P. (2001) Identification of genes implicated in the toxin production in the cyanobacterium *Cylindrospermopsis raciborskii*, *Environmental Toxicology* **16**, 413-421.

Sielaff, H., Dittmann, E., Tandeau de Marsac, N., Bouchier, C., von Döhren, H., Börner, T. and Schwecke, T. (2003) The *mcyF* gene of the microcystin biosynthetic gene cluster from *Microcystis aeruginosa* encodes an aspartate racemase, *Biochemical Journal* **373**, 903-916.

Sivonen, K. and Jones, G.J. (1999) Cyanobacterial toxins, in I. Chorus and J. Bartram (eds.), *Toxic Cyanobacteria in Water: A Guide to their Public Health Consequences, Monitoring and Management*, Spon, London, pp. 41-111.

Stachelhaus, T., Mootz, H.D. and Marahiel, M.A. (1999) The specificity-conferring code of adenylation domains in nonribosomal peptide synthetases, *Chemistry and Biology* **6**, 493-505.

Tillett, D., Dittmann, E., Erhard, M., von Döhren, H., Börner, T. and Neilan, B.A. (2000) Structural organisation of microcystin biosynthesis in *Microcystis aeruginosa* PCC 7806: an integrated peptide-polyketide synthetase system, *Chemistry and Biology* **7**, 753-764.

Tillett, D., Parker, D.L. and Neilan, B.A. (2001) Detection of toxigenicity by a probe for the microcystin synthetase A gene (*mcyA*) of the cyanobacterial genus *Microcystis*: comparison of toxicities with 16S rRNA and phycocyanin operon (phycocyanin intergenic spacer), *Applied and Environmental Microbiology* **67**, 2810-2818.

Turgay, K. and Marahiel, M.A. (1994) A general approach for identifying and cloning peptide synthetase genes, *Peptide Research* **7**, 753-764.

Vaitomaa, J., Rantala, A., Halinen, K., Rouhiainen, L., Tallberg, P., Mokelke, L. and Sivonen, K. (2003) Quantitative real-time PCR for determination of microcystin synthetase E copy numbers for *Microcystis* and *Anabaena* in lakes, *Applied and Environmental Microbiology* **69**, 7289-7297.

Van Liere, L. and Mur, L.R. (1980) Occurrence of *Oscillatoria agardhii* and some related species, a survey, in J. Barica and L.R. Mur (eds.), *Hypertrophic Ecosystems*; *Developments in Hydrobiology* **2**, 67-77.

Von Döhren, H., Keller, U., Vater, J. and Zocher, R. (1997) Multifunctional peptide synthetases, *Chemical Reviews* **97**, 2675-2705.

CHAPTER 3

DYNAMICS OF CYANOBACTERIAL TOXINS

Sources of variability in microcystin concentrations

W. Edwin A. KARDINAAL & Petra M. VISSER

1 Introduction

Cyanobacteria can form dense blooms that hamper recreation by diminished water clarity, bad odour and taste. Moreover, some cyanobacteria are toxic and cause incidental illness or death of cattle, dogs and humans, or may even affect entire ecosystems with disastrous impact on fish, birds and eukaryotic algae (Codd, 1995; Jochimsen et al., 1998; Singh et al., 2001; Sipiä et al., 2002; Vardi et al., 2002; Codd et al., 2003; see Chapter 1 of Codd et al., in this volume). Dense blooms dominated by toxic cyanobacteria can generate acute hazardous situations. Proper water management requires knowledge about the environmental factors that control cyanobacterial biomass and toxin production.

The key cyanobacterial genera known for their potential ability to produce toxic substances include *Anabaena*, *Aphanizomenon*, *Cylindrospermopsis, Lyngbya, Microcystis, Nodularia*, *Nostoc* and *Planktothrix* (Carmichael, 2001). These genera can produce a wide variety of different toxic compounds (Sivonen and Jones, 1999). The toxins can roughly be classified in hepatotoxins, neurotoxins and dermatotoxins. Hepatotoxins are toxins that can damage the liver. The hepatotoxins produced by cyanobacteria consist of microcystins, nodularins and cylindrospermopsins. In this review emphasis will be on microcystins, produced by among others *Microcystis*, *Anabaena* and *Planktothrix*. These microcystin-producing cyanobacteria belong to the most abundant cyanobacteria in freshwaters and brackish waters. Consequently, microcystins can be found throughout the world, in rivers, lakes, reservoirs, and also in brackish seas like the Baltic Sea. To date, at least 71 different structural variants of microcystin have been found (Codd et al., Chapter 1 in this volume). The variant microcystin-LR is considered to be most widespread, and is regarded as one of the most toxic microcystins (its LC_{50} is estimated at 50 µg microcystin per kg bodyweight; Sivonen and Jones, 1999).

Monitoring of microcystin concentrations in various lakes has revealed a high variability in space and time. Table 1 illustrates some of the variability in both microcystin concentrations (µg/L) and microcystins per unit dry weight (µg/g DW)

J. Huisman, H.C.P. Matthijs and P.M. Visser (eds.), Harmful Cyanobacteria, 41–63.

in several lakes. This shows that microcystin concentrations may range over more than 5 orders of magnitude. Essentially four sources of variation may be responsible for this tremendous variability in microcystin concentrations:

Variability in cyanobacterial biomass. A higher biomass of microcystin-producing cyanobacteria will result in higher microcystin concentrations. Several environmental factors affect the growth of cyanobacteria in lakes.

Methodological variability. Microcystin concentrations can be expressed in various ways. For instance, microcystin can be expressed per volume of water, per unit of dry weight, per unit of chlorophyll *a*, or per biovolume. Part of the observed variation in microcystin concentrations might stem from the variability within the biomass estimates themselves.

Table 1. Overview of recently published microcystin concentrations from various countries. Concentrations are presented in μg/g dry weight (DW) or else in μg/L as indicated.

Country	***No. of samples (% toxic)***	***Concentration (μg/g DW or μg/L)***	***Reference***
Argentina	35 (97)	5.8 – 2400	Amé et al. (2003)
Brazil	7	0.08 – 3.70	Domingos et al. (1999)
Brazil	50 (18)	0.003 – 10.0 (μg/L)	Hirooka et al. (1999)
Chile	2 (100)	8 – 130	Campos et al. (1999)
Chile	1 (100)	20	Neumann et al. (2000)
China	6 (100)	24.5 – 97.3	Shen et al. (2003)
Germany	11 (100)	500 – 6500	Ernst et al. (2001)
Germany	129 (100)	0.14 – 119 (μg/L)	Fromme et al. (2000)
Germany	1 (100)	14700	Jungmann et al. (1996)
Hungary	2 (100)	1710-6600	Reskóné and Törökné (2000)
France	25 (72)	0 - 5.2 (μg/L)	Briand et al. (2002)
Kenya	4 (100)	310 – 19822	Ballot et al. (2003)
Korea	25 (100)	0 – 0.2 (μg/L)	Oh et al. (2001)
Morocco	9 (89)	700 – 8800	Oudra et al. (2001)
Morocco	8 (100)	0 – 8800	Loudiki et al. (2002)
Netherlands	48 (96)	0-2420	STOWA (2000)
Ohio, USA	5 (?)	1200 (max.)	Brittain et al. (2000)
Philippines	3 (100)	649 – 4019	Cuvin-Aralar et al. (2002)
Philippines	11 (55)	0 – 1344	Baldia et al. (2003)
Uruguay	9 (100)	101 – 1074	De Leon and Yunez (2001)

Physiological variability. Microcystin production by cyanobacteria is affected by several environmental factors like nutrient availability, light conditions, and temperature. Physiological studies with isolated cyanobacterial strains under controlled laboratory conditions have shed light on this source of variability.

Variability in cyanobacterial species and genotype composition. The composition of structural microcystin variants and the microcystin content can be quite variable among species and even among different genotypes within the same species (Fastner et al., 1999b; Rohrlack et al., 2001). As a result, changes in the species composition of cyanobacteria, and also changes in genotype composition within the same species, may lead to considerable changes in microcystin concentration.

In this chapter, we review the impact of these sources of variability on microcystin dynamics. In particular, we will focus on environmental factors that control cyanobacterial growth and microcystin production, on different methods to express microcystin concentrations, on changes in species and genotype composition, and on recent methodological advances for the discrimination between toxic and non-toxic genotypes of cyanobacteria.

2 Variability in cyanobacterial biomass

To a large extent, the natural variability of the microcystin concentration in lakes is tied to dynamic changes in the biomass of microcystin-producing cyanobacteria. In eutrophic lakes in the temperate zone, cyanobacterial biomass typically increases after the clear-water phase in late spring or early summer. Depending on latitude and the nutritional status of the lake, cyanobacterial dominance may be sustained until late summer or early autumn (Wetzel, 2001).

Hypotheses to explain cyanobacterial dominance in eutrophic lakes are rather diverse (see reviews by Hyenstrand et al., 1998; Dokulil and Teubner, 2000; Huisman and Hulot, Chapter 7 in this volume) and will be summarized here:

Elevated water temperature. Maximum growth rates of cyanobacteria are often found at 25°C or higher (Robarts and Zohary, 1987). This temperature optimum is, in general, higher than the optimum temperature found for green algae and diatoms.

Low photon irradiance. Dense phytoplankton blooms or high background turbidity may result in light limitation. Some cyanobacteria, in particular filamentous species of the Oscillatoria group like *Planktothrix agardhii*, have very low light requirements (Mur et al., 1977; Scheffer et al., 1997). This enables them to become dominant over green algae and diatoms under light-limited conditions. Other cyanobacteria, like *Microcystis*, have higher light requirements, however, and hence will generally not become dominant under light-limited conditions (Huisman et al., 1999).

Water-column stability. Many cyanobacteria possess gas vesicles, which provide them with buoyancy (see Visser et al., Chapter 6 in this volume). In a stable water column, buoyant cyanobacteria like *Microcystis, Anabaena* and *Aphanizomenon* may float upwards, and thus increase their daily light dose (Ibelings et al., 1991). This gives buoyant cyanobacteria a competitive advantage over non-buoyant phytoplankton species during periods of stable weather with little turbulent mixing (Visser et al., 1996; Walsby et al., 1997; Huisman et al., 2004). Furthermore, buoyancy regulation may enable these cyanobacteria to migrate between surface layers with high light availability and deeper layers with high nutrient availability (e.g. Ganf and Oliver, 1982). Stratifying cyanobacteria like *Planktothrix rubescens* can accumulate at a depth where the combination of photon irradiance, light quality and nutrient availability is most favourable for their growth (Konopka, 1989).

Zooplankton grazing. Zooplankton species like *Daphnia* are important phytoplankton grazers in freshwater ecosystems. *Daphnia* can graze efficiently on phytoplankton not exceeding a size of around 50 μm in diameter (Burns, 1968). Many colony-forming and filamentous cyanobacteria are much larger than 50 μm, however, and for that reason may experience a much lower grazing pressure than smaller planktonic algae. In addition, the toxins produced by cyanobacteria may prevent high grazing pressures (e.g. Haney, 1987).

Carbon dioxide/pH. Many micro-algae and cyanobacteria possess active transport systems for carbon dioxide and bicarbonate (Marcus et al., 1982; Espie et al., 1990; Miller et al., 1990; Badger et al., 2002; Price and Badger, 2002). There is evidence that these carbon-concentrating mechanisms are more efficient in cyanobacteria (Shapiro, 1973, 1990; Raven, 1985), which makes them good competitors at high pH values, a general characteristic of eutrophic lakes.

Phosphorus storage. Several cyanobacteria have the ability to store excess phosphorus as polyphosphates. This internal storage allows growth at low external phosphorus concentrations and favours these cyanobacteria in the competition with other phytoplankton when phosphorus availability is low (Pettersson et al., 1993; Ducobu et al., 1998; see Dignum et al., Chapter 4 in this volume).

Nitrogen source. Based on enclosure experiments, Blomqvist et al. (1994) proposed the hypothesis that the nitrogen source may explain cyanobacterial dominance. Non-nitrogen fixing cyanobacteria were favoured by ammonium-nitrogen, as also shown for *Oscillatoria* (*Planktothrix*) by Klemer (1976) in enclosures. Nitrate-nitrogen favours the development of eukaryotic phytoplankton. Nitrogen scarcity favours the development of nitrogen-fixing cyanobacteria.

N/P ratio. Low ratios of the N to P-source favour cyanobacteria. Smith (1983) postulated this hypothesis after analysing data from 17 lakes. He concluded that cyanobacteria are generally better competitors for nitrogen than for phosphorus, and are thus favoured in lakes with low total N/total P ratios.

All these mechanisms have been demonstrated to affect cyanobacterial dominance. This implies that the commonly observed dominance of harmful cyanobacteria in eutrophic lakes cannot be attributed to a single master factor. Rather, different cyanobacterial species have different physiological traits and requirements, and hence are favoured by different environmental conditions. Furthermore, many cyanobacteria exploit combinations of several traits to become dominant. *Aphanizomenon*, for instance, combines at least three different strategies that may contribute to its competitive success, as it profits from its buoyancy, is capable of nitrogen fixation, and its filamentous morphology and colony size offers protection against grazing. Further aspects of cyanobacterial dominance are discussed by Huisman and Hulot (Chapter 7 in this volume).

3 Methodological variability: how to express microcystin concentrations?

Given that microcystin concentrations can be expressed in various ways, what would be the most suitable unit of measurement for microcystin concentrations?

For water management purposes, expression of the microcystin concentration per volume of water (e.g. in μg microcystin/L) seems most relevant. This allows a straightforward evaluation of the health risks associated with cyanobacterial blooms, and it also allows comparison of measured microcystin concentrations with the guideline levels set by national authorities and the World Health Organization. Unfortunately, for studies on the dynamics of microcystin production this measure is less useful because changes in the population densities of cyanobacterial blooms will cause concomitant changes in the microcystin concentration per volume of water, as most of the microcystin is cell bound.

In studies on microcystin dynamics, microcystins are therefore often expressed per unit of biomass. Here, biomass is usually measured in terms of dry weight or chlorophyll *a*. These approaches also have their pitfalls, however. For instance, estimates of dry weight are often based on total seston, which may include toxic cyanobacteria but also other phytoplankton, zooplankton, and particulate organic matter. Consequently, the microcystin concentration expressed per unit dry weight may be quite variable owing to changes in, for instance, particulate organic matter, which may have little to do with shifts in the abundances of toxic cyanobacteria. Accordingly, Fastner et al. (1999b) showed that the variability of microcystin concentrations between lakes was smaller when normalised to chlorophyll *a* than when normalised per unit seston dry weight. Good correlations of microcystin concentrations with phytoplankton numbers and chlorophyll *a* concentrations have been reported (Kotak et al., 1995; Oh et al., 2001). A problem that applies when microcystin concentrations are normalised per unit chlorophyll *a* is that all phytoplankton species contain chlorophyll *a*, not only toxic cyanobacteria. Thus, changes in phytoplankton species composition may lead to changes in microcystin concentrations per unit chlorophyll *a*, even if abundances of toxic cyanobacteria remain constant. Furthermore, the chlorophyll contents of phytoplankton species are quite sensitive to a wide range of environmental factors, including light and nutrient conditions (Collier and Grossman, 1992; Burnap et al., 1993; Falkowski and Raven, 1997; Bibby et al., 2001; Jeanjean et al., 2003; Yeremenko et al., 2004), such that chlorophyll is not a very robust proxy of phytoplankton biomass.

Quantification of the abundances of potentially toxic cyanobacteria and the expression of microcystin concentrations per cell or per unit biovolume of these cyanobacteria seems the most relevant approach to monitor microcystin dynamics. This approach is quite laborious, as it requires identification of potentially toxic cyanobacterial species and the measurement of their biovolumes (e.g., by microscope, image analysis techniques, or automated cell counters). Expression of microcystins per unit biovolume of toxic cyanobacteria has the advantage, though, that differences in cell dimensions among species are cancelled out, and that the microcystin dynamics are directly linked to the species that may produce these microcystins. The use of biovolume has been reported over the last years to be a good biomass estimate. Wiedner et al. (2002) showed that the dynamics of microcystin concentrations closely followed the dynamics of biovolumes of *Planktothrix agardhii* in a German lake. Likewise, Fastner et al. (1999a) noted that observed changes in microcystin concentration reflected the contributions of different cyanobacterial species to the phytoplankton biomass.

4 Physiological variability: the impact of environmental factors

Dynamics of microcystin concentrations cannot always be explained by changes in the abundances of cyanobacteria. In many cases, variation in toxin concentration occurs independent of changes in the size of the cyanobacterial population. Welker et al. (2003) found that the microcystin content of *Microcystis* (expressed per biovolume) in Lake Müggelsee ranged from 0.1 $\mu g/mm^3$ to more than 5 $\mu g/mm^3$. For another German lake, Jähnichen et al. (2001) reported that a relative low biovolume of cyanobacteria coincided with a high concentration of microcystin at the onset of cyanobacterial growth, while later in the season samples with high biovolumes but low microcystin concentrations were observed. In addition to changes in cyanobacterial abundance, other processes regulating the microcystin concentration per biomass must be involved to explain this variation.

4.1 FIELD STUDIES

In search for explanations of observed microcystin variability, several studies have been initiated that perform year-round monitoring of environmental factors and microcystin dynamics in lakes. In a study by Kotak et al. (1995), positive correlations were found between the microcystin content (expressed per *Microcystis* dry weight) and the concentrations of total phosphorus (TP) and soluble reactive phosphorus (SRP). In the same study, negative relations were found between the microcystin content of *Microcystis* and dissolved nitrogen (nitrate and ammonium), temperature and Secchi depth, but none were significant. In contrast, Wicks and Thiel (1990) found negative relations for the microcystin concentration in *Microcystis* scums (mainly *M. aeruginosa*) and the orthophosphate concentration. Positive correlations were found between microcystin concentrations and solar radiation, oxygen saturation, water temperature, and primary production. Jungmann et al. (1996) monitored various chemical and physical parameters as well as the microcystin content of *Microcystis*, but found no significant relations between environmental parameters and microcystin concentrations. Kotak et al. (2000) compared environmental data to toxin concentrations of *Microcystis* in multi-species assemblages, and found negative correlations of both microcystin-LR concentrations and *Microcystis* biomass with Secchi depth and positive correlations with chlorophyll *a*, TP and pH.

In conclusion, these field studies indicate that microcystin dynamics (expressed per cyanobacterial biomass) correlate with light, nutrients or the ratio between nutrient pools, i.e., the same factors that influence cyanobacterial growth.

4.2 LABORATORY STUDIES

4.2.1 Effect of environmental factors on microcystin production

In the field, several factors may vary at the same time. This makes it difficult to assess causal relationships between environmental factors and physiological variables (e.g. microcystin production). In laboratory cultures, conditions can be controlled and the effect of single environmental factors on microcystin production can be investigated. During the past few years, research has focused on the effects of phosphorus, nitrogen, light, iron and temperature on the microcystin content in laboratory cultures. Sivonen and Jones (1999) reviewed this work, and conclude that the majority of studies indicate that cyanobacteria produce the highest amount of toxins under conditions that are most favourable for growth. This is consistent with results of the field studies described in the previous paragraph. In particular, Sivonen and Jones conclude that cyanotoxins are produced in highest amounts under optimal light conditions. These optimal light conditions may differ among various cyanobacterial taxa, depending on their light requirements (e.g., *Planktothrix* prefers low light, whereas *Aphanizomenon* prefers high light). Also, nutrient-rich conditions generally result in a higher microcystin production.

Sivonen and Jones (1999) further note that differences in the experimental set-up of laboratory cultures as well as the use of different biomass estimates probably cause inconsistencies in the results of different studies. Laboratory cultures that are typically used for this type of research include batch cultures, semi-continuous cultures, and continuous cultures. In batch cultures, growth conditions cannot be controlled at a constant value. That is, population densities in a batch culture change, with concomitant changes in nutrient depletion and shading. If, say, population density or light availability would affect microcystin production, this might interfere with the actual factor under investigation in a batch culture, such as a nutrient limitation. In steady-state continuous cultures, all factors except the variable of interest can be kept constant for a prolonged time. Thus, although continuous-culture experiments are time-consuming and rather expensive, continuous cultures generally provide more controlled laboratory conditions allowing more detailed study of cyanobacterial toxin production.

Laboratory studies with isolated strains of different species have revealed that microcystin production and contents of cyanobacteria can be affected by a wide variety of factors, including temperature (van der Westhuizen and Eloff, 1985; Rapala et al., 1997), pH (Song et al., 1998), light (Sivonen, 1990; Utkilen and Gjølme, 1992; Wiedner et al., 2003), nitrogen (Sivonen, 1990; Orr and Jones, 1998; Long et al., 2001), phosphorus (Sivonen, 1990; Rapala et al., 1997; Oh et al., 2000), and iron (Utkilen and Gjølme, 1995). Orr and Jones (1998) and Long et al. (2001) hypothesised that the impact of so many environmental factors on microcystin production might be explained by the growth rate forcing the microcystin production. This hypothesis was supported by the observation of a positive relation between the microcystin production rate and cell specific growth rate in N-limited *Microcystis* batch cultures. Oh et al. (2000) found a similar relation in P-limited continuous cultures of *Microcystis* (strain UTEX 2388).

Cellular microcystin contents are also affected by growth rate. For instance, Long et al. (2001) found that microcystin contents of *Microcystis* under stringent nitrogen limitation were about three times lower than microcystin contents under nitrogen-saturated conditions. Similarly, Wiedner et al. (2003) found that, under light-limited conditions, the specific growth rate and cellular microcystin content both increased about twofold with increasing light intensity. However, this relation did not apply to light-saturated conditions. At high light levels the specific growth rate remained constant while the cellular microcystin content declined with increasing light intensity.

The studies of Long et al. (2001) and Wiedner et al. (2003) used different *Microcystis* strains. How robust are their findings? Would experiments with other strains yield similar results? Hesse and Kohl (2001) compared the microcystin production of six *Microcystis* strains under different conditions of nitrogen, phosphorus, and light. This revealed that the microcystin contents of these strains varied by a factor of 2-3 in response to changes in nutrient and light conditions, consistent with the findings of Long et al. (2001) and Wiedner et al. (2003). However, the strains differed in the direction of their response. For instance, strain W334 increased its microcystin content under light limitation, whereas strain W368 decreased its microcystin content. Analogously, strain W368 increased its microcystin content under phosphorus limitation, while the strains HUB 5-2-4 and W334 decreased their microcystin content (Hesse and Kohl, 2001).

In conclusion, laboratory studies indicate that the microcystin contents of cyanobacteria may vary by a factor of 2-3 in response to different environmental conditions. However, the observed changes in microcystin content seem to differ between different strains and different environmental conditions. This variability among strains complicates any further generalisations on the adaptive response of microcystin contents to environmental factors.

4.2.2 Are microcystins primary or secondary metabolites?

Cyanobacteria can synthesize a broad range of secondary metabolites, including peptides, polyketides and alkaloids. Microcystins are peptides produced by the non-ribosomal peptide synthetase pathway (see Börner and Dittmann, Chapter 2 in this volume) and are considered to be secondary metabolites (Carmichael, 1992). However, Orr and Jones (1998) objected to microcystins being secondary metabolites. Their argument is based on the observation that most secondary metabolites are normally triggered when growth ceases, whereas microcystin production may increase with increasing growth rate. A strong indication that microcystin is a true secondary metabolite came from the observation that the growth rate of a wild type *Microcystis* strain with microcystin-synthetase gene was retained in a derived mutant without functional microcystin-synthetase (Hesse et al., 2002). Similarly, different wild-type strains of *Microcystis* with and without microcystin production lack substantial differences in growth rate (Hesse and Kohl, 2001; Böttcher et al., 2001). Yet, the functional role of microcystin in intracellular or extracellular metabolism is still not elucidated. A further search for conditions that trigger transcription and functional activity of the assembled microcystin synthetase would be very informative. Based on ecological and molecular investigations, Kaebernick and Neilan (2001) have suggested putative extracellular and intracellular functions for microcystin. Extracellular functions can be the reduction of zooplankton grazing mediated by microcystin, or the use of microcystin in allelopathic interactions or cell-cell communication (Dittmann et al., 2001). Intracellular functions of microcystin have been related to iron-ion balance (Utkilen and Gjølme, 1995) or functions in regulation of light harvesting (Hesse et al., 2002).

4.2.3 Genetic regulation of microcystin production

Plant toxins may be produced constitutively (e.g. precursors of antimicrobial compounds) or may be synthesized in response to environmental triggers (Wittstock and Gershenzon, 2002; Osbourn et al., 2003). In microcystin-producing cyanobacteria, genes encoding for the microcystin synthetase enzyme complexes are always present and are apparently expressed constitutively. Only a few examples exist of non-microcystin-producing species with microcystin synthetase genes present (Neilan et al., 1999; Pan et al., 2002). Variation in cellular microcystin contents, by a factor of 2-3, results from changes in transcription of the genes (i.e. the number of mRNA copies produced), the subsequent translation of mRNA and assembly of the synthetase complex, the specific activity of the synthetase complex to produce microcystin, and/or the potential turnover of microcystin. Questions on the extent to which microcystin production is genetically controlled are under current investigation (see Börner and Dittmann, Chapter 2 in this volume).

5 Variability in cyanobacterial species and genotype composition

The variability in microcystin concentration per unit cyanobacterial biomass observed in lakes (Table 1) greatly exceeds the two- to three-fold variability in cellular microcystin content typically observed in isolated laboratory strains. Changes in species and genotype composition of the cyanobacterial community may to a large extent explain the dynamics of the microcystin concentration in lakes. In this section, therefore, the succession of various cyanobacterial species, morphotypes, and genotypes will be discussed.

5.1 DYNAMICS WITHIN CYANOBACTERIAL COMMUNITIES

Harmful cyanobacteria comprise a range of different genera, species and genotypes. Fastner et al. (1999b) showed that different cyanobacterial genera may produce different amounts and structural variants of microcystin. Lakes dominated by *Planktothrix* species had significantly higher concentrations of microcystin per biomass than lakes dominated by *Microcystis* species. Changes in cyanobacterial species composition may thus contribute to the variability of the microcystin dynamics often observed in lakes dominated by cyanobacteria.

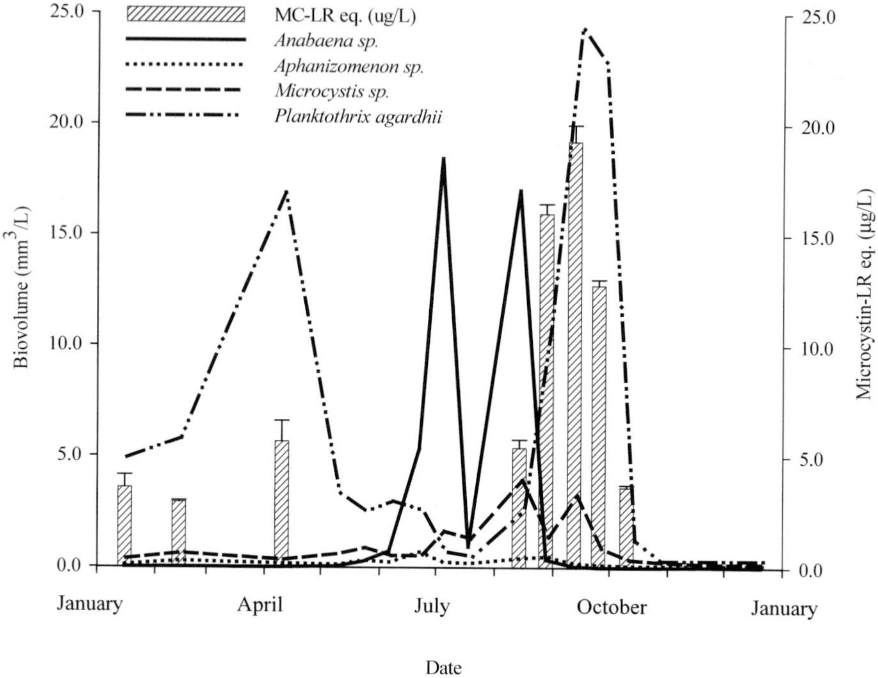

Figure 1. Seasonal changes in the abundance of Anabaena, Aphanizomenon, Microcystis *and* Planktothrix *in Lake Kinselmeer, The Netherlands. Bars indicate microcystin concentrations (in μg/L), as measured by HPLC.*

In shallow polymictic eutrophic lakes, cyanobacterial biomass is often dominated by filamentous species of the Oscillatoriales group. Examples include the perennial blooms of e.g. *Planktothrix agardhii* observed in shallow lakes in The Netherlands (Berger, 1989; Scheffer et al., 1997), Germany (Wiedner et al., 2002) and France (Briand et al., 2002). Figure 1 shows the seasonal changes in cyanobacterial dominance and microcystin concentration in shallow Lake Kinselmeer, The Netherlands. The peaks in microcystin concentration in this lake coincided with the dominance of *Planktothrix* and the less abundant *Microcystis* during spring and fall, while microcystin concentrations remained low during the two summer blooms of *Anabaena*. Wiedner et al. (2002) studied several shallow polymictic lakes where either *Planktothrix* or *Limnothrix* dominated cyanobacterial biomass during late summer. Prior to the dominance of these Oscillatoriales, the Nostocales (mainly *Anabaena*, *Aphanizomenon* and *Anabaenopsis* species) reached their maximum population density in early summer. In a *Planktothrix*-dominated lake, toxin concentrations closely followed the biovolume of *Planktothrix*. In lakes dominated by *Limnothrix* species, no toxin production was detected. In contrast, in the study of Briand et al. (2002) microcystin dynamics did not track the dynamics in *Planktothrix* biomass. Briand et al. suggest that the microcystin dynamics they observed might result from a succession of closely related *Planktothrix* genotypes, each producing different amounts of microcystin. From cultured strains, it is known that toxic and non-toxic *Planktothrix* strains indeed exist (Lyra et al., 2001).

Several deep stratified lakes are characterized by a seasonal succession of buoyant cyanobacteria, from *Anabaena* and *Aphanizomenon* dominance in spring towards *Microcystis*-dominated communities in summer (Watanabe et al., 1992; Kotak et al., 1995; Welker et al., 2003). Isolation and cultivation of cyanobacterial strains from deep stratified lakes in The Netherlands showed that microcystin production could not be detected in the *Anabaena* and *Aphanizomenon* strains. Here microcystin production was governed by *Microcystis* (Kardinaal, unpublished data).

For a long time, taxonomic identification of different *Microcystis* species has been based on morphological criteria like colony morphology, colony size, cell size, and mucilage envelope (Komárek and Anagnostidis, 1999). Fastner et al. (2001) showed that the morphotypes of *Microcystis* may differ in composition of microcystin structural variants. It appears that *M. aeruginosa* is the most widespread morphotype worldwide. In an extensive study, Kurmayer et al. (2002) found that 73% of the *M. aeruginosa* colonies contained the *mcyB* gene (part of the total gene cluster encoding for the microcystin synthesis; see Chapter 2 of Börner and Dittmann in this volume). In contrast, this gene could be detected in only 17% of the *M. ichtyoblabe* and in none of the *M. wesenbergii* colonies, illustrating the different abilities for microcystin production of these morphotypes.

Succession of different *Microcystis* morphotypes might therefore explain the microcystin dynamics in a lake. In Lake Suwa, Japan, succession of different *Microcystis* morphotypes could indeed be related to microcystin concentrations (Park et al., 1998). In three out of four seasons, the microcystin concentration was closely related to the exponential growth phase of the dominant *M. aeruginosa* and *M. viridis*, whereas *M. wesenbergii* was found to be non-toxic. In Danish lakes, however, blooms of toxic *M. wesenbergii* have been described and in several experimental studies toxic *M. wesenbergii* strains have been reported (Neilan et al., 1997; Otsuka et al., 1999; Henriksen, 2001; Pan et al., 2002). In another Japanese lake, Lake Kasumigaura, the temporal variation in microcystin concentration could not be related to *Microcystis* biomass or *Microcystis* species composition (Watanabe et al., 1992). Jähnichen et al. (2001) showed that in Bautzen Reservoir, Germany, four potential microcystin producers, *M. aeruginosa*, *M. viridis*, *M. wesenbergii* and *Planktothrix agardhii*, alternated in dominance. Within a single season subsequent peaks in cyanobacterial biomass were dominated by different cyanobacterial species. Peaks in microcystin concentration, however, did not coincide with peaks in cyanobacterial biomass. In laboratory studies several *Microcystis* strains appear to show so much morphological plasticity that they can display characteristics of various different morphospecies (Otsuka et al., 2000). Hence, identification of the different morphotypes of *Microcystis* is insufficient for reliable prediction of microcystin dynamics.

Within a single morphotype of *Microcystis*, different toxic and non-toxic genotypes may coexist. Rohrlack et al. (2001) isolated 22 *Microcystis* strains from Lake Wannsee and Lake Pehlitzsee, Germany. They found much dissimilarity in microcystin concentration per unit biomass among these strains, ranging from strains without microcystin to strains containing 4000 μg microcystin/g DW. Carrillo et al. (2003) confirmed these findings for 26 strains of *M. aeruginosa* isolated from three reservoirs in Spain. Thus, microcystin contents can vary considerably among different *Microcystis* strains that are morphologically indistinguishable and co-occur in the same lake.

To complicate things further, Jungmann et al. (1996) suggested that colony size could play a role in the microcystin content of *Microcystis*. They found that in colonies larger than 100 μm diameter the microcystin content was higher than in the smaller size classes (30-66 μm). This impact of colony size has been confirmed by recent studies. Kurmayer et al. (2003) isolated *Microcystis* colonies, and each colony was tested for the presence of the *mcyB* gene. It was shown that larger colonies (> 100 μm) more often contained the *mcyB* gene than small colonies (< 100μm). Microcystin dynamics could be related to the population development of the larger colonies (> 100 μm). Another study with colonies isolated from all over Europe showed that 88% of the colonies larger than 1100 μm produced microcystin. The proportion of microcystin-producing colonies dropped to around 50% for colonies of 400–1100 μm and to around 20% for colonies smaller than 400 μm (Via-Ordorika et al., 2004).

5.2 PATTERNS IN MICROCYSTIN DYNAMICS

To recapitulate, there is increasing evidence that the microcystin dynamics in freshwater lakes largely result from successional replacement of a multitude of cyanobacterial genotypes that differ in microcystin production (Sivonen and Jones, 1999; Chorus et al., 2001; Rohrlack et al., 2001; Briand et al., 2002; Wiedner et al., 2002; Carillo et al., 2003; Welker et al., 2003). These various microcystin producing and non-producing genotypes cannot be differentiated using standard microscopy. Despite this apparent complexity, however, some striking patterns in seasonal microcystin dynamics may emerge. Welker et al. (2003) recently pointed at a significant negative relation between the microcystin content per unit biovolume of *Microcystis* (μg/mm^3) and the actual *Microcystis* biomass (mm^3) in Lake Müggelsee, Germany. They suggest that toxigenic cells are relatively more abundant or more toxigenic at the onset of the *Microcystis* bloom (end of June/early July) than at the height of the *Microcystis* bloom (August). Reanalysis of data from four Dutch lakes dominated by *Microcystis* revealed a similar negative relation between the microcystin content of *Microcystis* and the total *Microcystis* biomass (Fig. 2). Also in these Dutch lakes, the microcystin content per unit biomass was highest in early summer, at the onset of the *Microcystis* blooms.

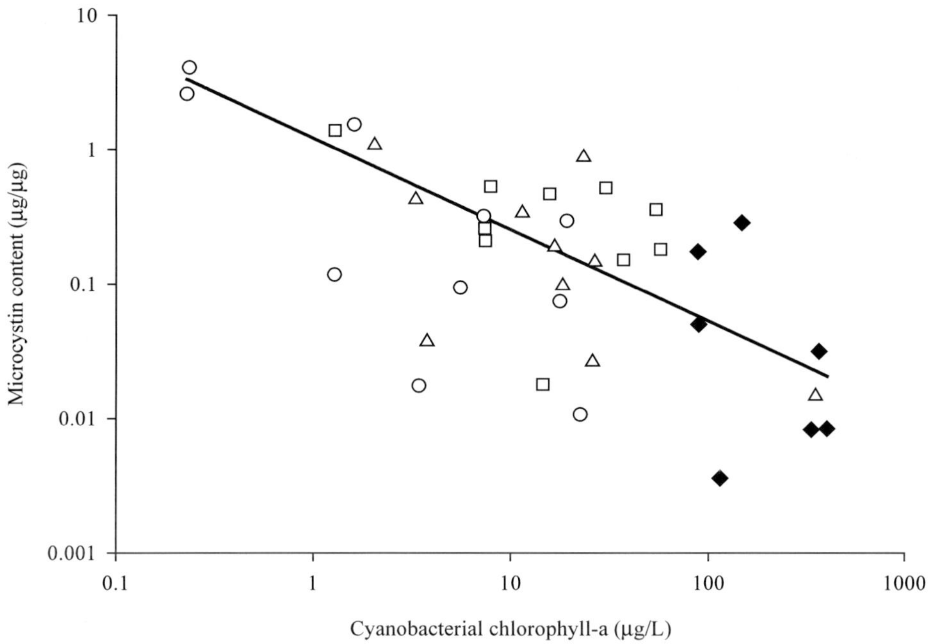

Figure 2. Relation between the microcystin content of cyanobacteria (expressed as microcystin-LR equivalents per cyanobacterial chlorophyll a*) and cyanobacterial abundance (expressed as cyanobacterial chlorophyll* a*, based on flow-cytometric data) in four Dutch lakes: Lake Gouden Ham (○), Lake 't Joppe (□), Lake Sloterplas (Δ) and Lake de Eend (♦) (Linear regression: $R^2 = 0.42$; $P < 0.0001$; $n = 36$). Data were collected by STOWA (2000).*

To investigate the generality of this emergent pattern, we screened the scientific literature for other studies on seasonal microcystin dynamics in *Microcystis*-dominated lakes. In Hartbeespoort Dam, South Africa, a similar negative relation between the microcystin content per unit of cyanobacterial biomass and *Microcystis* abundance has been observed (Wicks and Thiel, 1990). There, the peak in microcystin content occurred in February, during midsummer in the Southern hemisphere. In studies of lakes in Japan (Watanabe, 1992), USA (Jacoby, 1999) and Sweden (Cronberg, 1999), microcystin contents were expressed per unit of total dry weight. On the basis of their reported plankton composition, we partitioned the total dry weight into dry weight originating from cyanobacteria and dry weight from other plankton species. This again revealed a trend in which the highest microcystin levels per unit cyanobacterial biomass occurred at the start of the cyanobacterial bloom; microcystin contents declined during further bloom development. In the Bautzen Reservoir, Germany, Jungmann et al. (1996) monitored temporal dynamics of microcystin: at the end of June the microcystin content per *Microcystis* biomass (μg/mg) was highest. Assuming that the *Microcystis* bloom in this lake increased further towards the end of August, the data are in line with the other observations in this paragraph. A decline of microcystin content with increasing *Microcystis* biomass was also observed in three lakes in Canada (Kotak et al., 1995). Peak abundance of *M. aeruginosa* occurred in August and September, while in five out of the nine lake years investigated the highest toxin concentration per *Microcystis* biomass was found at the end of June. In a wastewater treatment plant in Portugal the cellular microcystin contents were highest in June, approximately 320 fg/cell. Later in summer, microcystin contents dropped to around 10 fg/cell (Vasconcelos and Pereira, 2001). In the Daechung Reservoir, Korea, cellular microcystin contents peaked at the end of June, but were still higher after the blooming of *Microcystis* at the end of August, probably because of the increasing abundance of *Planktothrix* species (Oh et al., 2001).

All in all, 24 of the 31 lake years that we found in the literature clearly showed a negative density dependence of the microcystin content. That is, the microcystin content per unit of *Microcystis* biomass declined while the *Microcystis* population increased. In 7 of the 31 lake-years, this relation was either less evident or not present at all. These findings suggest that in the majority of *Microcystis*-dominated lakes the seasonal succession starts with one or more toxic genotypes that are gradually (but not completely) displaced by non-toxic genotypes during the course of summer. Which factors might drive this seasonal pattern? Experimental studies of Hesse and Kohl (2001) and Vezie et al. (2002) suggest that high nutrient levels favour the growth of toxic *Microcystis* genotypes over non-toxic genotypes. In many lakes, nutrient availability is gradually depleted during summer, thereby possibly favouring a shift from toxic to non-toxic genotypes. An alternative explanation might be that in dilute populations natural selection favours a high toxicity per cell to deter potential grazers, whereas with increasing population densities less microcystin per individual cell is required to generate a similar deterrent effect. Additional research is needed to investigate these hypotheses, and to establish whether similar patterns can be found for other cyanobacterial species as well.

6 Molecular tools for the identification of toxic genotypes

The studies reviewed in the preceding sections show that variability in genotype composition within cyanobacterial communities has a great impact on the variability in microcystin concentrations in lakes. However, different genotypes within the same genus generally cannot be distinguished by microscope. Recent advances in molecular biology offer promising tools to differentiate between toxic and non-toxic genotypes. Two different molecular approaches have thus far been applied. One approach makes use of the presence of the microcystin synthetase genes, whereas the other approach focuses on the genetic relatedness of the cyanobacteria. Both approaches and their possible applications will be discussed here.

6.1 MICROCYSTIN SYNTHETASE GENES

Genes encoding for microcystin synthetases have been largely identified for microcystin-producing strains of *Microcystis*, *Anabaena* and *Planktothrix* (Dittmann et al., 1997; Tillet et al., 2000; Christiansen et al., 2003; Hisbergues et al., 2003; Rouhiainen et al., 2004; see Börner and Dittmann, Chapter 2 in this volume). Based on sequence information of these genes, specific genetic primers have been developed for the amplification of the *mcyA* and *mcyB* regions, both encoding a step in microcystin biosynthesis. The high sensitivity and specificity of these primers has been demonstrated in laboratory cultures as well as in field samples (Tillet et al., 2001; Pan et al., 2002; Hisbergues et al., 2003), allowing identification of microcystin-producing cyanobacteria. During bloom development the presence of cyanobacteria with the *mcyA* synthetase gene was monitored and their toxicity was confirmed by HPLC measurements (Baker et al., 2002).

An approach to quantify the abundances of microcystin-producing genotypes was recently developed by Kurmayer and Kutzenberger (2003). They developed two independent quantitative real-time PCR assays. One of the assays was used to quantify the total *Microcystis* cell number, and the other to quantify the number of microcystin-producing cells. For the first assay the ITS region of the phycocyanin operon was amplified, whereas for the second assay a part of the *mcyB* gene was amplified. The assays were tested in laboratory samples as well as in field samples from Lake Wannsee, Germany. A good relation was found between microscopic cell counts and the quantitative PCR results of the phycocanin quantification assay. However, although all tested laboratory strains showed equal amplification efficiencies, the test rendered a systematic overestimation of cell numbers by a factor of two to six, probably because of the presence of multiple genome copies per cell. The field samples indicated that the proportion of *mcyB*-containing cells in the *Microcystis* population was relatively constant, at about 11% during all seasons.

Vaitomaa et al. (2003) developed a comparable method to identify and quantify genotypes responsible for microcystin production. To this end, genus-specific *mcyE* primers were used to estimate the number of *mcyE* copies by quantitative real-time PCR. The method was tested in lakes in Finland. In Lake Tuusulanjärvi the results show clearly that the *Microcystis* genotypes were mainly responsible for producing microcystin within a cyanobacterial population dominated by *Microcystis*, *Anabaena* and *Aphanizomenon*. No correlation was found between the number of *mcyE* copies and the microcystin concentration. The number of *mcyE* copies was high compared to microscopic cell counts, analogous to the findings of Kurmayer and Kutzenberger (2003) for phycocyanin genes. Again, this overestimation is likely a consequence of multiple genome copies per cell, or results from underestimation of the genome sizes of the external standard strains used for comparison.

6.2 PHYLOGENY-BASED RECOGNITION OF STRAINS

Another approach to distinguish between toxic and non-toxic strains is to search for a phylogenetic relation with microcystin production. Lyra et al. (2001) characterised cyanobacterial phylogeny by comparison of 16S rRNA gene sequences of a large number of laboratory isolates of freshwater cyanobacteria. Four clades became apparent: clade 1 contained filamentous heterocyst-forming cyanobacteria (like *Anabaena* and *Aphanizomenon*); clade 2 contained *Planktothrix* species; clade 3 revealed colony-forming *Microcystis* strains, and clade 4 contained *Synechococcus* strains. However, it was not possible to make a straightforward differentiation between toxic and non-toxic strains within those clades.

In a more detailed follow-up study, analyses of sequences for 16S rRNA, 16S-23S rRNA internal transcribed spacer (rRNA-ITS) and a RubisCO gene spacer (*rbcLX*) showed high similarity between the genera *Anabaena* and *Aphanizomenon* (Gugger et al., 2002). A first cluster contained anatoxin-producing *Anabaena*, a second cluster contained only non-toxic strains of *Anabaena* and *Aphanizomenon* and a third cluster contained microcystin-producing *Anabaena* strains. However, cluster 1 also contained non-toxic strains from both genera, and cluster 3 also contained non-toxic *Anabaena* strains.

Likewise, other studies revealed that the microcystin-producing genera *Microcystis* and *Planktothrix* could not be divided into distinct clusters of toxic and non-toxic strains on the basis of their 16S rRNA sequences (Neilan et al., 1997; Otsuka et al., 1998; Lyra et al., 2001). Otsuka et al. (1999) investigated the rRNA internal transcribed spacer of 47 *Microcystis* strains, and distinguished three major clusters. The first cluster contained a mix of microcystin-producing *M. aeruginosa* and various non-toxic morphotypes; the second cluster contained solely microcystin producing-strains of three different morphotypes, and the third cluster mainly consisted of the non-toxic morphotype *M. wesenbergii*. Thus, generally speaking, direct sequencing and comparison of 16S rRNA and rRNA-ITS genes is still too coarse to distinguish between toxic and non-toxic strains.

Janse et al. (2003) developed a high-resolution method based on the application of denaturing gradient gel electrophoresis (DGGE) to rRNA-ITS sequences. DGGE is a powerful method for analysis of the biodiversity and dynamics of microbial populations. Primers and PCR protocols were designed to target the cyanobacterial rRNA-ITS. DGGE profiles of rRNA-ITS yielded high-resolution differentiation of closely related organisms. The method proved to differentiate between cyanobacterial genera. Moreover, most of the 20 *Microcystis* cultures tested could be differentiated when using this technique. As a next step, Janse et al. (2004) studied 107 *Microcystis* colonies from 16 different lakes in Europe and Morocco. The presence of microcystins in each colony was examined by MALDI-TOF mass spectrometry, and the colonies were differentiated on the basis of their rRNA-ITS PCR amplified products via DGGE. This approach rendered 59 different classes of microcystin-producing and non-producing genotypes. Sequences of the ITS fragment from representative genotypes were congruent with the classification based on rRNA-ITS DGGE. Alignment of the colony sequences and published rRNA-ITS sequences from known toxigenic *Microcystis* cultures confirmed the unique recognition of microcystin-producing genotypes based on their rRNA-ITS sequences. This high-resolution method can be useful to monitor the population dynamics of specific genotypes of microcystin-producing cyanobacteria within a mixture of species.

7 Conclusions

The studies reviewed in this chapter reveal a tremendous variability in the microcystin contents of cyanobacterial blooms, spanning at least 5 orders of magnitude. Some blooms dominated by potentially microcystin-producing genera are not toxic at all, whereas other blooms dominated by the same genera are highly toxic and can thus pose a serious health threat. Part of the reported variability can be attributed to methodological differences in microcystin analysis. International standardisation of methodology is therefore recommended, preferably by quantification of microcystin contents per unit of cyanobacterial biovolume. Part of the variability can also be attributed to physiological variation in microcystin contents induced by environmental factors. Changes in nutrient and light conditions do affect the microcystin contents of isolated cyanobacterial strains to some extent, but the two- to threefold physiological variation commonly observed in laboratory strains does not account for the microcystin variability over one or more orders of magnitude commonly observed in the field. By and large, the major source of variation in microcystin contents seems to be related to the successive replacement of toxic and non-toxic genotypes within cyanobacterial genera. This variability is a major problem for water management authorities, because it implies that the toxicity of cyanobacterial blooms cannot be assessed by standard microscopic enumeration of cell numbers of cyanobacterial species. Recognition of toxic genotypes is therefore of key importance to forecast the microcystin dynamics in any given situation. The recently developed molecular approaches seem promising tools to monitor the population dynamics of toxic and non-toxic genotypes and, thereby, to elucidate the factors driving microcystin dynamics within cyanobacterial communities.

8 Acknowledgements

We like to thank Jef Huisman for restructuring the manuscript, which greatly improved it. We also like to thank Thomas Börner, Hans C.P. Matthijs and Martin Welker for their helpful comments on the manuscript. WEAK was supported by the DYNATOX project funded by the Dutch Technology Foundation (STW). PMV was supported by the Earth and Life Sciences Foundation (ALW), which is subsidized by the Netherlands Organization for Scientific Research (NWO).

9 References

Amé, M.V., Díaz, M.D. and Wunderlin, D.A. (2003) Occurrence of toxic cyanobacterial blooms in San Roque Reservoir (Cordoba, Argentina): a field and chemometric study, *Environmental Toxicology* **18**, 192-201.

Badger, M.R., Hanson, D. and Price G.D. (2002) Evolution and diversity of CO_2 concentrating mechanisms in cyanobacteria, *Functional Plant Biology* **29**, 161-173.

Baker, J.A., Entsch, B., Neilan, B.A. and McKay, D.B. (2002) Monitoring changing toxigenicity of a cyanobacterial bloom by molecular methods, *Applied and Environmental Microbiology* **68**, 6070-6076.

Baldia, S.F., Conaco, M.C.G., Nishijima, T., Imanishi, S. and Harada, K.I. (2003) Microcystin production during algal bloom occurrence in Laguna de Bay, the Philippines, *Fisheries Science* **69**, 110-116.

Ballot, A., Pflugmacher, S., Wiegand, C., Kotut, K. and Krienitz, L. (2003) Cyanobacterial toxins in Lake Baringo, Kenya, *Limnologica* **33**, 2-9.

Berger, C. (1989) In situ primary production, biomass and light regime in the Wolderwijd, the most stable *Oscillatoria-agardhii* lake in The Netherlands, *Hydrobiologia* **185**, 233-244.

Bibby, T.S., Nield, J. and Barber, J. (2001) Iron deficiency induces the formation of an antenna ring around trimeric photosystem I in cyanobacteria, *Nature* **412**, 743-745.

Blomqvist, P., Pettersson, A. and Hyenstrand, P. (1994) Ammonium-nitrogen: A key regulatory factor causing dominance of non-nitrogen-fixing cyanobacteria in aquatic systems, *Archiv für Hydrobiologie* **132**, 141-164.

Böttcher, G., Chorus, I., Ewald, S., Hintze, T. and Walz, N. (2001) Light-limited growth and microcystin content of *Planktothrix agardhii* and *Microcystis aeruginosa* in turbidostats, in I. Chorus (ed.), *Cyanotoxins - Occurrence, Causes, Consequences*, Springer-Verlag, Berlin, Germany, pp. 115-132.

Briand, J.F., Robillot, C., Quiblier-Lloberas, C. and Bernard, C. (2002) A perennial bloom of *Planktothrix agardhii* (Cyanobacteria) in a shallow eutrophic French lake: limnological and microcystin production studies, *Archiv für Hydrobiologie* **153**, 605-622.

Brittain, S.M., Wang, J., Babcock-Jackson, L., Carmichael, W.W., Rinehart, K.L. and Culver, D.A. (2000) Isolation and characterization of microcystins, cyclic heptapeptide hepatotoxins from a Lake Erie strain of *Microcystis aeruginosa*, *Journal of Great Lakes Research* **26**, 241-249.

Burnap, R., Troyan, T. and Sherman, L.A. (1993) The highly abundant chlorophyll-protein complex of iron-deficient *Synechococcus* sp. PCC7942 (CP43') is encoded by the *isiA* gene, *Plant Physiology* **103**, 893-902.

Burns, C.W. (1968) The relationship between body size of filter-feeding Cladocera and the maximum size of particle ingested, *Limnology and Oceanography* **13**, 675-678.

Campos, V., Cantarero, S., Urrutia, H., Heinze, R., Wirsing, B., Neumann, U. and Weckesser, J. (1999) Microcystin in cyanobacterial blooms in a Chilean lake, *Systematic and Applied Microbiology* **22**, 169-173.

Carmichael, W.W. (1992) Cyanobacteria secondary metabolites: the cyanotoxins, *Journal of Applied Bacteriology* **72**, 445-459.

Carmichael, W.W. (2001) Health effects of toxin-producing cyanobacteria: the CyanoHABs, *Human and Ecological Risk Assessment* **7**, 1393-1407.

Carrillo, E., Ferrero, L.M., Alonso-Andicoberry, C., Basanta, A., Martin, A., Lopez-Rodas, V. and Costas, E. (2003) Interstrain variability in toxin production in populations of the cyanobacterium *Microcystis aeruginosa* from water-supply reservoirs of Andalusia and lagoons of Doñana National Park (southern Spain), *Phycologia* **42**, 269-274.

Chorus, I., Niesel, V., Fastner, J., Wiedner, C., Nixdorf, B. and Lindenschmidt, K.E. (2001) Environmental factors and microcystin levels in waterbodies, in I. Chorus (ed.), *Cyanotoxins - Occurrence, Causes, Consequences*, Springer-Verlag, Berlin, Germany, pp. 159-177.

Christiansen, G., Fastner, J., Erhard, M., Börner, T. and Dittmann, E. (2003) Microcystin biosynthesis in *Planktothrix*: genes, evolution, and manipulation, *Journal of Bacteriology* **185**, 564-572.

Codd, G.A. (1995) Cyanobacterial toxins: occurrence, properties and biological significance, *Water Science and Technology* **32**, 149-156.

Codd, G.A., Metcalf, J.S., Morrison, L.F., Krienitz, L., Ballot, A., Pflugmacher, S., Wiegand, C. and Kotut, K. (2003) Susceptibility of flamingos to cyanobacterial toxins via feeding, *Veterinary Record* **152**, 722-723.

Collier, J.L. and Grossman, A.M. (1992) Chlorosis induced by nutrient deprivation in *Synechococcus* sp. strain PCC 7942: not all bleaching is the same, *Journal of Bacteriology* **174,** 4718-4726.

Cronberg, G., Annadotter, H. and Lawton, L. A. (1999) The occurrence of toxic blue-green algae in Lake Ringsjon, southern Sweden, despite nutrient reduction and fish biomanipulation, *Hydrobiologia* **404**, 123-129.

Cuvin-Aralar, M.L., Fastner, J., Focken, U., Becker, K. and Aralar, E.V. (2002) Microcystins in natural blooms and laboratory cultured *Microcystis aeruginosa* from Laguna de Bay, Philippines, *Systematic and Applied Microbiology* **25**, 179-182.

De Leon, L. and Yunes, J.S. (2001) First report of a microcystin-containing bloom of the cyanobacterium *Microcystis aeruginosa* in the La Plata River, South America, *Environmental Toxicology* **16**, 110-112.

Dittmann, E., Neilan, B.A., Erhard, M., von Döhren, H. and Börner, T. (1997) Insertional mutagenesis of a peptide synthetase gene that is responsible for hepatotoxin production in the cyanobacterium *Microcystis aeruginosa* PCC 7806, *Molecular Microbiology* **26**, 779-787.

Dittmann, E., Erhard, M., Kaebernick, M., Scheler, C., Neilan, B.A., von Döhren, H. and Börner, T. (2001) Altered expression of two light-dependent genes in a microcystin-lacking mutant of *Microcystis aeruginosa* PCC 7806, *Microbiology* **147,** 3113-3119.

Dokulil, M.T. and Teubner, K. (2000) Cyanobacterial dominance in lakes, *Hydrobiologia* **438**, 1-12.

Domingos, P., Rubim, T.K., Molica, R.J.R., Azevedo, S. and Carmichael, W.W. (1999) First report of microcystin production by picoplanktonic cyanobacteria isolated from a northeast Brazilian drinking water supply, *Environmental Toxicology* **14**, 31-35.

Ducobu, H., Huisman, J., Jonker, R.R. and Mur, L.R. (1998) Competition between a prochlorophyte and a cyanobacterium under various phosphorus regimes: comparison with the Droop model, *Journal of Phycology* **34**, 467-476.

Ernst, B., Hitzfeld, B. and Dietrich, D. (2001) Presence of *Planktothrix* sp and cyanobacterial toxins in Lake Ammersee, Germany and their impact on whitefish (*Coregonus lavaretus* L.), *Environmental Toxicology* **16**, 483-488.

Espie, G.S., Miller, A.G., Kandasamy, R.A. and Canvin, D.T. (1990) Active HCO_3-transport in cyanobacteria, *Canadian Journal of Botany* **69**, 936-944.

Falkowski, P.G. and Raven, J.A. (1997) *Aquatic Photosynthesis*, Blackwell Science, Oxford, UK.

Fastner, J., Neumann, U., Wirsing, B., Weckesser, J., Wiedner, C., Nixdorf, B. and Chorus, I. (1999a) Microcystins (hepatotoxic heptapeptides) in German fresh water bodies, *Environmental Toxicology* **14**, 13-22.

Fastner, J., Erhard, M., Carmichael, W.W., Sun, F., Rinehart, K.L., Ronicke, H. and Chorus, I. (1999b) Characterization and diversity of microcystins in natural blooms and strains of the genera *Microcystis* and *Planktothrix* from German freshwaters, *Archiv für Hydrobiologie* **145**, 147-163.

Fastner, J., Erhard, M. and von Döhren, H. (2001) Determination of oligopeptide diversity within a natural population of *Microcystis* spp. (Cyanobacteria) by typing single colonies by matrix-assisted laser desorption ionisation - time of flight mass spectrometry, *Applied and Environmental Microbiology* **67**, 5069-5076.

Fromme, H., Kohler, A., Krause, R. and Fuhrling, D. (2000) Occurrence of cyanobacterial toxins: microcystins and anatoxin-a in Berlin water bodies with implications to human health and regulations, *Environmental Toxicology* **15**, 120-130.

Ganf, G.G. and Oliver, R.L. (1982) Vertical separation of light and available nutrients as a factor causing replacement of green algae by blue-green algae in the plankton of a stratified lake, *Journal of Ecology* **70**, 829-844.

Gugger, M., Lyra, C., Henriksen, P., Coute, A., Humbert, J.F. and Sivonen, K. (2002) Phylogenetic comparison of the cyanobacterial genera *Anabaena* and *Aphanizomenon*, *International Journal of Systematic and Evolutionary Microbiology* **52**, 1867-1880.

Haney, J.F. (1987) Field studies on zooplankton-cyanobacteria interactions, *New Zealand Journal of Marine and Freshwater Research* **21**, 467-477.

Henriksen, P. (2001) Toxic freshwater cyanobacteria in Denmark, in I. Chorus (ed.), *Cyanotoxins - Occurrence, Causes, Consequences*, Springer-Verlag, Berlin, Germany, pp. 49-55.

Hesse, K. and Kohl, J.G. (2001) Effects of light and nutrient supply on growth and microcystin content of differerent strains of *Microcystis aeruginosa*, in I. Chorus (ed.), *Cyanotoxins - occurence, causes, consequences*, Springer, Berlin, Germany, pp. 104 -114.

Hesse, K., Dittmann, E. and Börner T. (2002) Consequences of impaired microcystin production for light-dependent growth and pigmentation of *Microcystis aeruginosa* PCC 7806, *FEMS Microbiology Ecology* **37**, 39-43.

Hirooka, E.Y., Pinotti, M.H.P., Tsutsumi, T., Yoshida, F. and Ueno, Y. (1999) Survey of microcystins in water between 1995 and 1996 in Parana, Brazil using ELISA, *Natural Toxins* **7**, 103-109.

Hisbergues, M., Christiansen, G., Rouhiainen, L., Sivonen, K. and Börner, T. (2003) PCR-based identification of microcystin-producing genotypes of different cyanobacterial genera, *Archives of Microbiology* **18**, 402-410.

Huisman, J., Jonker, R.R., Zonneveld, C. and Weissing, F.J. (1999) Competition for light between phytoplankton species: experimental tests of mechanistic theory, *Ecology* **80**, 211-222.

Huisman, J., Sharples, J., Stroom, J.M., Visser, P.M., Kardinaal, W.E.A., Verspagen, J.M.H. and Sommeijer, B. (2004) Changes in turbulent mixing shift competition for light between phytoplankton species, *Ecology* **85**, 2960-2970.

Hyenstrand, P., Blomqvist, P. and Petterson, A. (1998) Factors determining cyanobacterial success in aquatic systems: a literature review, *Archiv für Hydrobiologie Special Issues Advances in Limnology* **51**, 41 - 62.

Ibelings, B.W., Mur, L.R. and Walsby, A.E. (1991) Diurnal changes in buoyancy and vertical distribution in populations of *Microcystis* in two shallow lakes, *Journal of Plankton Research* **13**, 419-436.

Jacoby, J.M., Collier, D.C., Welch, E.B., Hardy, F.J. and Crayton, M. (2000) Environmental factors associated with a toxic bloom of *Microcystis aeruginosa*, *Canadian Journal of Fisheries and Aquatic Sciences* **57**, 231-240.

Jähnichen, S., Petzoldt, T. and Benndorf, J. (2001) Evidence for control of microcystin dynamics in Bautzen reservoir (Germany) by cyanobacterial population growth rates and dissolved inorganic carbon, *Archiv für Hydrobiologie* **150**, 177-196.

Janse, I., Meima, M., Kardinaal, W.E.A. and Zwart, G. (2003) High-resolution differentiation of cyanobacteria by using rRNA-internal transcribed spacer denaturing gradient gel electrophoresis, *Applied and Environmental Microbiology* **69**, 6634 - 6643.

Janse, I., Kardinaal, W.E.A., Meima, M., Fastner, J., Visser, P.M. and Zwart, G. (2004) Toxic and non-toxic *Microcystis* colonies in natural populations can be differentiated on the basis of rRNA internal transcribed spacer gene diversity, *Applied and Environmental Microbiology* **70**, 3979 - 3987.

Jeanjean, R., Zuther, E., Yeremenko, N., Havaux, M., Matthijs, H.C.P. and Hagemann, M. (2003) A photosystem 1 psaFJ-null mutant of the cyanobacterium *Synechocystis* PCC 6803 expresses the isiAB operon under iron replete conditions, *FEBS Letters* **549**, 52-56.

Jochimsen, E.M., Carmichael, W.W., An, J.S., Cardo, D.M., Cookson, S.T., Holmes, C.E.M., Antunes, M.B.D., de Melo, D.A., Lyra, T.M., Barreto, V.S.T., Azevedo, S. and Jarvis, W.R. (1998) Liver failure and death after exposure to microcystins at a hemodialysis center in Brazil, *New England Journal of Medicine* **338**, 873-878.

Jungmann, D., Ludwichowski, K.U., Faltin, V. and Benndorf, J. (1996) Field study to investigate environmental factors that could effect microcystin synthesis of a *Microcystis* population in the Bautzen reservoir, *Internationale Revue der Gesamten Hydrobiologie* **81**, 493-501.

Kaebernick, M. and Neilan, B.A. (2001) Ecological and molecular investigations of cyanotoxin production, *FEMS Microbiology Ecology* **35**, 1-9.

Klemer, A.R. (1976) The vertical distribution of *Oscillatoria agardhii* var. *isothrix*, *Archiv für Hydrobiologie* **78**, 343-362.

Komárek, J. and Anagnostides, K. (1999) Cyanoprokaryota 1. Teil Chroococcales, Gustav Fischer Verlag, Jena, Germany.

Konopka, A. (1989) Metalimnetic cyanobacteria in hard-water lakes: buoyancy regulation and physiological state, *Limnology and Oceanography* **34**, 1174-1184.

Kotak, B.G., Lam, A.K.Y., Prepas, E.E., Kenefick, S.L. and Hrudey, S.E. (1995) Variability of the hepatotoxin microcystin-LR in hypereutrophic drinking-water lakes, *Journal of Phycology* **31**, 248-263.

Kotak, B.G., Lam, A.K.Y., Prepas, E.E. and Hrudey, S.E. (2000) Role of chemical and physical variables in regulating microcystin-LR concentration in phytoplankton of eutrophic lakes, *Canadian Journal of Fisheries and Aquatic Sciences* **57**, 1584-1593.

Kurmayer, R., Dittmann, E., Fastner, J. and Chorus, I. (2002) Diversity of microcystin genes within a population of the toxic cyanobacterium *Microcystis* spp. in Lake Wannsee (Berlin, Germany), *Microbial Ecology* **43**, 107-118.

Kurmayer, R., Christiansen, G. and Chorus, I. (2003) The abundance of microcystin-producing genotypes correlates positively with colony size in *Microcystis* sp and determines its microcystin net production in Lake Wannsee, *Applied and Environmental Microbiology* **69**, 787-795.

Kurmayer, R. and Kutzenberger, T. (2003) Application of real-time PCR for quantification of microcystin genotypes in a population of the toxic cyanobacterium *Microcystis* sp., *Applied and Environmental Microbiology* **69**, 6723-6730.

Long, B.M., Jones, G.J. and Orr, P.T. (2001) Cellular microcystin content in N-limited *Microcystis aeruginosa* can be predicted from growth rate, *Applied and Environmental Microbiology* **67**, 278-283.

Loudiki, M., Oudra, B., Sabour, B., Sbiyyaa, B. and Vasconcelos, V. (2002) Taxonomy and geographic distribution of potential toxic cyanobacterial strains in Morocco, *International Journal of Limnology* **38**, 101-108.

Lyra, C., Suomalainen, S., Gugger, M., Vezie, C., Sundman, P., Paulin, L. and Sivonen, K. (2001) Molecular characterization of planktic cyanobacteria of *Anabaena, Aphanizomenon, Microcystis* and *Planktothrix* genera, *International Journal of Systematic and Evolutionary Microbiology* **51**, 513-526.

Marcus, Y., Zenvirth, D., Harel, E. and Kaplan, A. (1982) Induction of HCO_3^- transporting capability and high photosynthetic affinity to inorganic carbon by low concentration of CO_2 in *Anabaena variabilis, Plant Physiology* **69**, 1008-1012.

Miller, A.G., Espie, G.S. and Canvin, D.T. (1990) Active CO_2 transport in cyanobacteria, *Canadian Journal of Botany* **69**, 925-935.

Mur, L.R., Gons, H.J. and Van Liere, L. (1977) Some experiments on the competition between green algae and blue-green bacteria in light-limited environments, *FEMS Microbiology Letters* **1**, 335-338.

Neilan, B.A., Jacobs, D., DelDot, T., Blackall, L.L., Hawkins, P.R., Cox, P.T. and Goodman, A.E. (1997) rRNA sequences and evolutionary relationships among toxic and nontoxic cyanobacteria of the genus *Microcystis*, *International Journal of Systematic Bacteriology* **47**, 693-697.

Neilan, B.A., Dittmann, E., Rouhiainen, L., Bass, R.A., Schaub, V., Sivonen, K. and Börner, T. (1999) Nonribosomal peptide synthesis and toxigenicity of cyanobacteria, *Journal of Bacteriology* **181**, 4089-4097.

Neumann, U., Campos, V., Cantarero, S., Urrutia, H., Heinze, R., Weckesser, J. and Erhard, M. (2000) Co-occurrence of non-toxic (cyanopeptolin) and toxic (microcystin) peptides in a bloom of *Microcystis* sp from a Chilean Lake, *Systematic and Applied Microbiology* **23**, 191-197.

Oh, H.M., Lee, S.J., Jang, M.H. and Yoon, B.D. (2000) Microcystin production by *Microcystis aeruginosa* in a phosphorus-limited chemostat, *Applied and Environmental Microbiology* **66**, 176-179.

Oh, H.M., Lee, S.J., Kim, J.H., Kim, H.S. and Yoon, B.D. (2001) Seasonal variation and indirect monitoring of microcystin concentrations in Daechung Reservoir, Korea, *Applied and Environmental Microbiology* **67**, 1484-1489.

Orr, P.T. and Jones, G.J. (1998) Relationship between microcystin production and cell division rates in nitrogen-limited *Microcystis aeruginosa* cultures, *Limnology and Oceanography* **43**, 1604-1614.

Osbourn, A.E., Xiaoquan, Q., Townsend, B. and Qin, B. (2003) Dissecting plant secondary metabolism: constitutive chemical defences in cereals, *New Phytologist* **159**, 101-108.

Otsuka, S., Suda, S., Li, R.H., Watanabe, M., Oyaizu, H., Matsumoto, S. and Watanabe, M.M. (1998) 16S rDNA sequences and phylogenetic analyses of *Microcystis* strains with and without phycoerythrin, *FEMS Microbiology Letters* **164**, 119-124.

Otsuka, S., Suda, S., Li, R.H., Watanabe, M., Oyaizu, H., Matsumoto, S. and Watanabe, M.M. (1999) Phylogenetic relationships between toxic and non-toxic strains of the genus *Microcystis* based on 16S to 23S internal transcribed spacer sequence, *FEMS Microbiology Letters* **172**, 15-21.

Otsuka, S., Suda, S., Li, R.H., Matsumoto, S. and Watanabe, M.M. (2000) Morphological variability of colonies of *Microcystis* morphospecies in culture, *Journal of General and Applied Microbiology* **46**, 39-50.

Oudra, B., Loudiki, M., Sbiyyaa, B., Martins, R., Vasconcelos, V. and Namikoshi, N. (2001) Isolation, characterization and quantification of microcystins (heptapeptides hepatotoxins) in *Microcystis aeruginosa* dominated bloom of Lalla Takerkoust lake-reservoir (Morocco), *Toxicon* **39**, 1375-1381.

Pan, H., Song, L.R., Liu, Y.D. and Börner, T. (2002) Detection of hepatotoxic *Microcystis* strains by PCR with intact cells from both culture and environmental samples, *Archives of Microbiology* **178**, 421-427.

Park, H.D., Iwami, C., Watanabe, M.F., Harada, K., Okino, T. and Hayashi, H. (1998) Temporal variabilities of the concentrations of intra- and extracellular microcystin and toxic *Microcystis* species in a hypertrophic lake, Lake Suwa, Japan (1991-1994), *Environmental Toxicology and Water Quality* **13**, 61-72.

Pettersson, K., Herlitz, E. and Istvanovics, V. (1993) The role of *Gloeotrichia echinulata* in the transfer of phosphorus from sediments to water in Lake Erken, *Hydrobiologia* **253**, 123-129.

Price, G.D. and Badger, M.R. (2002) Advances in understanding how aquatic photosynthetic organisms utilize sources of dissolved inorganic carbon for CO_2 fixation, *Functional Plant Biology* **29,** 117-121.

Rapala, J., Sivonen, K., Lyra, C. and Niemelä, S.I. (1997) Variation of microcystins, cyanobacterial hepatotoxins, in *Anabaena* spp. as a function of growth stimuli, *Applied and Environmental Microbiology* **63**, 2206-2212.

Raven, J.A. (1985) The CO_2 concentrating mechanism, in W.J. Lucas and J.A. Berry (eds.), *Inorganic Carbon Uptake by Aquatic Photosynthetic Organisms*, Waverly Press, Baltimore, USA, pp. 67-83.

Reskóné, M.N. and Törökné, A.K. (2000) Toxic *Microcystis aeruginosa* in Lake Velencei, *Environmental Toxicology* **15**, 554-557.

Robarts, R.D. and Zohary, T. (1987) Temperature effects on photosynthetic capacity, respiration, and growth rates of bloom-forming cyanobacteria, *New Zealand Journal of Marine and Freshwater Research* **21**, 391-399.

Rohrlack, T., Henning, M. and Kohl, J.G. (2001) Isolation and characterization of colony-forming *Microcystis aeruginosa* strains, in I. Chorus (ed.), *Cyanotoxins - Occurrence, Causes, Consequences*, Springer-Verlag, Berlin, Germany, pp. 152-158.

Rouhiainen, L., Vakkilainen, T., Siemer, B.L., Buikema, W., Haselkorn, R. and Sivonen, K. (2004) Genes coding for hepatotoxic heptapeptides (microcystins) in the cyanobacterium *Anabaena* strain 90, *Applied and Environmental Microbiology* **70**, 686-692.

Scheffer, M., Rinaldi, S., Gragnani, A., Mur, L.R. and Van Nes, E.H. (1997) On the dominance of filamentous cyanobacteria in shallow, turbid lakes, *Ecology* **78**, 272-282.

Shapiro, J. (1973) Blue-green algae: why they become dominant, *Science* **179**, 382-384.

Shapiro, J. (1990) Current beliefs regarding dominance by blue-greens: the case for the importance of CO_2 and pH, *Verhandlungen der Internationalen Vereinigung für Theoretische und Angewandte Limnologie* **24**, 38-54.

Shen, P.P., Shi, Q., Hua, Z.C., Kong, F.X., Wang, Z.G., Zhuang, S.X. and Chen, D.C. (2003) Analysis of microcystins in cyanobacteria blooms and surface water samples from Meiliang Bay, Taihu Lake, China, *Environment International* **29**, 641-647.

Singh, D.P., Tyagi, M.B., Kumar, A. and Thakur, J.K. (2001) Antialgal activity of a hepatotoxin-producing cyanobacterium, *Microcystis aeruginosa*, *World Journal of Microbiology and Biotechnology* **17**, 15-22.

Sipiä, V.O., Kankaanpaa, H.T., Pflugmacher, S., Flinkman, J., Furey, A. and James, K.J. (2002) Bioaccumulation and detoxication of nodularin in tissues of flounder (*Platichthys flesus*), mussels (*Mytilus edulis, Dreissena polymorpha*), and clams (*Macoma balthica*) from the northern Baltic Sea, *Ecotoxicology and Environmental Safety* **53**, 305-311.

Sivonen, K. (1990) Effects of light, temperature, nitrate, orthophosphate, and bacteria on growth of and hepatotoxin production by *Oscillatoria agardhii* strains, *Applied and Environmental Microbiology* **56**, 2658-2666.

Sivonen, K. and Jones, G.J. (1999) Cyanobacterial toxins, in I. Chorus and J. Bartram (eds.), *Toxic Cyanobacteria in Water: A Guide to their Public Health Consequences, Monitoring and Management*, Spon, London, United Kingdom, pp. 41–111.

Smith, V.H. (1983) Light and nutrient dependence of photosynthesis by algae, *Journal of Phycology* **19**, 306-313.

Song, L., Sano, T., Li, R., Watanabe, M.M., Liu, Y. and Kaya, K. (1998) Microcystin production of *Microcystis viridis* (cyanobacteria) under different culture conditions, *Phycological Research* **46**, 19-23.

STOWA (2000) Toxische blauwalgen in recreatiewateren, STOWA report 2000–20, Utrecht, The Netherlands.

Tillett, D., Dittmann, E., Erhard, M., von Döhren, H., Börner, T. and Neilan, B.A. (2000) Structural organization of microcystin biosynthesis in *Microcystis aeruginosa* PCC7806: an integrated peptide-polyketide synthetase system, *Chemistry and Biology* **7**, 753-764.

Tillett, D., Parker, D.L. and Neilan, B.A. (2001) Detection of toxigenicity by a probe for the microcystin synthetase A gene (*mcyA*) of the cyanobacterial genus *Microcystis*, comparison of toxicities with 16S rRNA and phycocyanin operon (phycocyanin intergenic spacer) phylogenies, *Applied and Environmental Microbiology* **67**, 2810-2818.

Utkilen, H. and Gjølme, N. (1992) Toxin production by *Microcystis aeruginosa* as a function of light in continuous cultures and its ecological significance, *Applied and Environmental Microbiology* **58**, 1321-1325.

Utkilen, H. and Gjølme, N. (1995) Iron-stimulated toxin production in *Microcystis aeruginosa*, *Applied and Environmental Microbiology* **61**, 797-800.

Vaitomaa, J., Rantala, A., Halinen, K., Rouhiainen, L., Tallberg, P., Mokelke, L. and Sivonen, K. (2003) Quantitative real-time PCR for determination of microcystin synthetase E copy numbers for *Microcystis* and *Anabaena* in lakes, *Applied and Environmental Microbiology* **69**, 7289-7297.

Van der Westhuizen, A.J. and Eloff, J.N. (1985) Effect of temperature and light on the toxicity and growth of the blue-green alga *Microcystis aeruginosa* (UV-006), *Planta* **163**, 55-59.

Vardi, A., Schatz, D., Beeri, K., Motro, U., Sukenik, A., Levine, A. and Kaplan, A. (2002) Dinoflagellate-cyanobacterium communication may determine the composition of phytoplankton assemblage in a mesotrophic lake, *Current Biology* **12**, 1767-1772.

Vasconcelos, V.M. and Pereira, E. (2001) Cyanobacteria diversity and toxicity in a wastewater treatment plant (Portugal), *Water Research* **35**, 1354-1357.

Vezie, C., Rapala, J., Vaitomaa, J., Seitsonen, J. and Sivonen, K. (2002) Effect of nitrogen and phosphorus on growth of toxic and nontoxic *Microcystis* strains and on intracellular microcystin concentrations, *Microbial Ecology* **43**, 443-454.

Via-Ordorika, L., Fastner, J., Hisbergues, M., Dittmann, E., Erhard, M., Komarèk, J., Kurmayer, R. and Chorus, I. (2004) Distribution of microcystin-producing and non-microcystin producing *Microcystis* in European freshwater bodies: detection of microcystins and *mcy* genes in single colonies, *Systematic and Applied Microbiology* **27**, in press.

Visser, P.M., Ibelings, B.W., Van der Veer, B., Koedood, J. and Mur, L.R. (1996) Artificial mixing prevents nuisance blooms of the cyanobacterium *Microcystis* in Lake Nieuwe Meer, the Netherlands, *Freshwater Biology* **36**, 435-450.

Walsby, A.E., Hayes, P.K., Boje, R. and Stal, L.J. (1997) The selective advantage of buoyancy provided by gas vesicles for planktonic cyanobacteria in the Baltic Sea, *New Phytologist* **136**, 407-417.

Watanabe, M.M., Kaya, K. and Takamura, N. (1992) Fate of the toxic cyclic heptapeptides, the microcystins, from blooms of *Microcystis* (Cyanobacteria) in a hypertrophic lake, *Journal of Phycology* **28**, 761-767.

Welker, M., von Döhren, H., Tauscher, H., Steinberg, C.E.W. and Erhard, M. (2003) Toxic *Microcystis* in shallow lake Müggelsee (Germany): dynamics, distribution, diversity, *Archiv für Hydrobiologie* **157**, 227-248.

Wetzel, R.G. (2001) *Limnology: Lake and River Ecosystems*, third edition, Academic Press, San Diego, USA.

Wicks, R.J. and Thiel, P.G. (1990) Environmental factors affecting the production of peptide toxins in floating scums of the cyanobacterium *Microcystis aeruginosa* in a hypertrophic African reservoir, *Environmental Science and Technology* **24**, 1413-1418.

Wiedner, C., Nixdorf, B., Heinz, R., Wirsing, B., Neumann, U. and Weckesser, J. (2002) Regulation of cyanobacteria and microcystin dynamics in polymictic shallow lakes, *Archiv für Hydrobiologie* **155**, 383-400.

Wiedner, C., Visser, P.M., Fastner, J., Metcalf, J.S., Codd, G.A. and Mur, L.R. (2003) Effects of light on the microcystin content of *Microcystis* strain PCC 7806, *Applied and Environmental Microbiology* **69**, 1475-1481.

Wittstock, U. and Gershenzon, J. (2002) Constitutive plant toxins and their role in defense against herbivores and pathogens, *Current Opinion in Plant Biology* **5**, 300-307.

Yeremenko, N., Kouril, R., Ihalainen, J.A., D'Haene, S., Oosterwijk, N., Andrizhiyevskaya, E.G., Keegstra, W., Dekker, H.L., Hagemann, M., Boekema, E.J., Matthijs, H.C.P. and Dekker, J.P. (2004) Supramolecular organization and dual function of the IsiA chlorophyll-binding protein in cyanobacteria, *Biochemistry* **43**, 10308-10313.

CHAPTER 4

NUTRIENT LIMITATION OF FRESHWATER CYANOBACTERIA

Tools to monitor phosphorus limitation at the individual level

Marco DIGNUM, Hans C.P. MATTHIJS, Roel PEL, Hendrikus J. LAANBROEK & Luuc R. MUR

1 Introduction

Changes in species composition that are often observed in phytoplankton communities result from a complex interplay between physical and chemical properties of the aquatic environment on the one hand and the responsiveness of the individual species on the other hand. The potential role of nutrient limitation in accommodating these changes in species composition has since long been recognized. However, the interpretation of results, to the benefit of understanding nutrient competition between species in phytoplankton communities, is very much hampered by the limited knowledge that is gained by chemical assay methods for measurements of actual nutrient concentrations in surface waters. Release of limiting nutrients in lake water may even simply escape detection, because phytoplankton is able to sequester occasional nutrient pulses with high affinity. Although biological assays, which define the limiting nutrient from an increase in biomass following addition of that particular nutrient, are in common use, their reporting value is generally limited to the total phytoplankton community only. Furthermore, these bioassays require relatively long incubation times before results are known. Rapid detection of early changes in the phytoplankton community is required to combat the development of unwanted blooms of harmful cyanobacteria. For this purpose, we have invested in the development of fluorescent staining procedures that enable the recognition of nutrient limitation within individual phytoplankters. That way, more detailed information on the contribution of different species to phytoplankton community development is gained.

To detect nutrient limitation at the level of the individual cell, we have searched for marker proteins that indicate nutrient limitation. The choice of such marker proteins requires a sound knowledge of metabolic pathways and gene expression

J. Huisman, H.C.P. Matthijs and P.M. Visser (eds.), Harmful Cyanobacteria, 65–86.

patterns involved. Furthermore, proteins that will be used as a marker should preferably be exposed on the periphery of the cell to facilitate fluorescent staining. At the same time they must be tightly connected to the cell to prevent excretion or loss. Cell wall proteins are the first choice; these are ideally located at the outer membrane surface, or else in the periplasmic space. Possibly, integral cytoplasmic membrane proteins would still provide an adequate target. Monitoring the expression levels of marker proteins in individual cells permits a characterization of nutrient availability to different phytoplankton species during a growth season. In this contribution the focus is on phosphorus, because this nutrient is often the limiting factor in freshwater communities. We provide an overview of what is known about phosphorus metabolism in cyanobacteria, with emphasis on the cell wall proteins. Initial results from combined application of flow cytometry and fluorescent staining of phosphatase activity are presented.

2 The nutrient status of phytoplankton

2.1 NUTRIENT LIMITATION VERSUS EUTROPHICATION

The presence and growth of phytoplankton in surface water is not only related to the availability of sunlight and carbon dioxide, but also depends on the availability of certain key nutrients. Of these nutrients, phosphorus (P) and nitrogen (N) have attracted the most attention. Low nutrient levels (oligotrophic conditions) allow biodiversity to be high because a variable flow of materials and energy through the ecosystem creates complex nutritional interrelationships and specialised niches. Competition processes in a complex community may generate non-equilibrium dynamics, which sustain a large number of coexisting phytoplankton species (Huisman and Weissing, 1999). This situation, of high biodiversity and low biomass, is associated with a high water quality. At higher levels of nutrients, however, cyanobacteria may become dominant, and may form dense surface blooms. These organisms can be toxic, and generally their over-abundance has a deteriorating effect on the aquatic ecosystem, resulting in low water quality. The increased availability of nutrients through eutrophication is mostly caused by our agricultural and industrial habits (Reynolds, 1998, and references therein). The reverse situation, in which measures are taken to reduce nutrient inputs into aquatic ecosystems (remediation), may eventually lead to nutrient limitation and water quality improvement.

2.2 HOW TO MONITOR THE NUTRIENT STATUS ?

Attempts to control algal problems have focused primarily on managing the reduction of phytoplankton biomass by, for example, reduced phosphorus loading or biomanipulation (e.g. Hosper, 1997). This approach was often appropriate to reduce the overall biomass production, yet in many cases failed to render complete restoration. A more balanced approach appears to be required, taking into account the interplay of various nutrients and the effect on community complexity. The ideas of Hecky and Kilham (1988) are still relevant in this respect. According to these authors, we need to refine our understanding of nutrient limitation to be able to manage species composition. Therefore, indicators for the phytoplankton nutrient status are required. The term 'diagnostic tool' has been used to denote a signal or analytical procedure that empirically identifies the symptoms of an environmental constraint on phytoplankton growth rates (Falkowski et al., 1992; Mann et al., 1993). Assessment of the nutrient status down to the individual level is desirable to account for all aspects of heterogeneity in the phytoplankton community. A direct approach to address this challenge is to monitor the presence of those cell-constituents, which are specifically linked to the nutrient status of interest by fluorescent detection in a flow cytometer (Scanlan et al., 1997; Palenik and Wood, 1998). Obvious requirements for suitable markers of the nutrient status are that composition, accumulation, or activity of these markers (mRNA, proteins, lipids, carbohydrates, metabolites) depends on the physiological state of the cell and especially on the availability of specific nutrients. The results from whole-genome analysis and protein characterisation studies in model organisms help identify the genes involved in nutrient uptake. In the case of freshwater cyanobacteria, the complete genome of two model organisms is available: *Synechocystis* PCC 6803 and *Synechococcus elongatus* PCC 7942. Obviously, the general applicability of techniques that work for specific model organisms needs to be checked for the species actually present in ecosystems of interest. Fortunately, however, biochemical knowledge gained from model organisms can often be applied to a wider range of organisms in ecosystems as biochemical traits usually follow a common blue print.

3 Nutrient uptake through the cell envelope

The periphery of the cell is of interest for two reasons. Firstly, this is where contact with the environment takes place. The process of adaptation to changing conditions in the environment is likely to be well pronounced in this cell fraction. Secondly, the use of non-intrusive diagnostics for the nutrient status requires the target to be exposed at the cell surface, to minimise the extent of manipulation of the sample. Protection against osmosis and specific transport of nutrients are both function of the cell envelope. The thick multi-layered envelopes of cyanobacteria form a considerable mechanical and permeability barrier for mostly larger molecules (Hoiczyk and Hansel, 2000). The outermost surface of cyanobacteria is usually covered with a carbohydrate sheath and a two-dimensional crystalline, single-protein surface-layer (S-layer; Šmarda et al., 2002). The various sections of the cell wall are discussed next.

3.1 THE OUTER MEMBRANE

The outer membrane (OM) protects the cell from harmful agents such as proteases, bile salts, antibiotics, toxins and phage, and against drastic changes in osmotic pressure (Cowan et al., 1992). The OM is an asymmetric lipid bilayer. The outer monolayer contains lipopolysaccharide (LPS) as major lipid, whereas the inner monolayer contains phospholipids. In addition to lipids and proteins, carotenoids are components of the isolated OM fractions of cyanobacteria (Resch and Gibson, 1983; Jürgens and Weckesser, 1985). The function of these pigments is to protect the cells from oxidative stress, by shielding them from excessive light (Hirschberg and Chamovitz, 1994; Miller et al., 2002). The asymmetry of the outer membrane relies on an intact peptidoglycan layer in the periplasmic space. The outer membrane contains structural lipoprotein, which links to the underlying peptidoglycan either covalently, or via ion-bridges. Pore-forming proteins, called porins, mainly determine the permeability of the outer membrane. The *Synechocystis* PCC 6803 genome encodes six homologues to the outer membrane porins SomA and SomB with highly conserved surface layer homologous domains (Hoiczyk and Hansel, 2000). The predicted structure of the translated proteins shows all characteristics of porins, and furthermore, the expression pattern of several of these proteins is modified by nutrient availability (Dignum, 2003). The fact that these porins are responsive to nutrient availability and have regions that are exposed at the periphery of the cell makes this class of proteins of keen interest for further research.

3.2 THE PERIPLASMIC SPACE AND THE PEPTIDOGLYCAN LAYER

The layer between the inner and outer membrane of Gram-negative bacteria is defined as the periplasmic space, and is estimated to contribute about 7% of the total volume of the cell. The osmotic pressure in the water-filled periplasm is only slightly higher than that of the medium (Koch, 1998). Close contact with the exterior through the semi-permeable outer membrane allows (small-molecule) tags to be used to visualise periplasmic proteins as diagnostic for the nutrient status. The periplasmic space is much more viscous and dense with proteins than the cytoplasm (Raivio and Silhavy, 2001). Soluble proteins in the periplasm are released by applying cold osmotic shock (Heppel, 1967; Fulda et al., 1999). The periplasm contains other proteins than the cytoplasm, including binding proteins and hydrolytic enzymes that degrade substances for nutrition (Koch, 1998). Most of the proteins identified in the periplasm of *Synechocystis* represent 'hypothetical proteins' with unknown function (Fulda et al., 2000). The majority of the assigned proteins are involved in the generation and modification of the external cell layers. About 10% of the assigned proteins belong to the family of proteases (Fulda et al., 2000). The periplasmic space also contains the peptidoglycan layer, which constitutes the actual cell wall in mechanical terms. The peptidoglycan (synonym: murein) layer primarily functions in maintaining the cell shape and withstands the very high internal osmotic pressure that the cytoplasm exerts in dilute environments. It consists of a relatively open network of amino sugars and amino acids. Long strands of amino sugars are covalently linked by pentapeptides, forming a giant, hollow, net-like molecule. The high degree of cross-linking found in cyanobacteria is similar to that in Gram-positive bacteria. However, teichoic acid, a typical constituent of Gram-positive bacteria is missing in cyanobacterial cell walls. But then again, the cyanobacterial peptidoglycan forms complexes with polysaccharides, similar to Gram-positive peptidoglycan.

3.3 THE CYTOPLASMIC MEMBRANE

The cytoplasmic membrane acts as a real diffusion barrier, and contains a large number of uptake systems for solutes. An interesting approach to assess the transport capability of the cytoplasmic membrane of *Synechocystis* was conducted by microbial genome analysis (Paulsen et al., 1998). The *Synechocystis* genome encodes 92 transporters. The number of ATP-dependent transporters (see below) was ten times the number of proton-motive-force dependent transporters. This in contrast with *E. coli*, which has equal numbers of the two types. The special status of the thylakoid membranes that are embedded in the cytoplasm and add ATP through photophosphorylation for use at the cytoplasmic membrane can explain this difference. The transport proteins may render a focus for our interest in characterisation of proteins with a reporting value for nutrient limitation. Another relevant feature of the cytoplasmic membrane is its role in registration of fluctuations in the environment. Two-component regulatory systems with sensor proteins located in the cytoplasmic membrane function to sense specific changes in the surroundings (Ronson et al., 1987; Parkinson and Kofoid, 1992; Bijisma and Groisman, 2003). The signal is transferred over the membrane and passed on to the transcription apparatus by corresponding regulatory proteins.

4 Importance of phosphate

The important role that phosphate plays in cells is apparent on three levels.

Structure. Phosphate forms covalent ester links between carbohydrate monomers, creating a rigid structure. Organically bound phosphate is found in the peptidoglycan fraction of the cell wall. Its probable function is to bind polysaccharide covalently to the peptidoglycan matrix (Jürgens et al., 1983). The role of these compounds in giving the cell its shape and strength was discussed above. Phosphate is also a structure determining part of nucleic acids: it connects the ribonucleotide monomers that form the DNA or RNA backbone. Another way in which phosphate gives structure to cell components is by acting as an intermediate for ionic binding. The inner leaflet of the outer membrane consists of phospholipids, which stick together by binding to intermediate divalent cations. The lipopolysaccharides of cyanobacteria also contain small amounts of bound P (Hoiczek and Hansel, 2000).

Energy carrier. Energy-rich phosphates are involved in the cells' metabolism as universal free energy carriers. The most important energy carrier is adenosine triphosphate (ATP). Energy can be directly stored inside the cell by conversion to poly-phosphate (polyP), or indirectly by generating carbohydrates. The biosynthesis of many macromolecules is accompanied by liberation of pyrophosphate (PP_i) as waste product (Lahti, 1983). More than one PP_i molecule is liberated for every monomer in protein, nucleic acid and polysaccharide (Klemme, 1976). Both PP_i and ATP can be synthesised phototrophically.

Information carrier. Environmental and internal signals provide important information for the adaptive responses of bacteria. Prokaryotic signalling systems are complex, with multiple components, connections to other regulatory circuits, and feedback loops (Parkinson and Kofoid, 1992). These networks commonly contain transmitter and receiver modules. The communication between transmitter and receiver usually involves activation by phosphorylation (kinase activity), deactivation by dephosphorylation (phosphatase activity; Ronson et al., 1987). Well-conserved two-component regulatory systems (TCRS) function to sense specific changes in the environment (sensory component) and transduce that information to the transcriptional apparatus (regulatory component). The sensor is usually a transmembrane protein, which binds to ligands with a variable periplasmic domain, and transmits the signal to a conserved cytoplasmic domain (transmitter), through allosteric alteration. The activated sensor interacts with the N-terminal part of the regulator (receiver). The response regulator is usually a DNA binding protein, facilitating transcription by activating the promoter. The phosphoryl transfer pathways may diverge (more than one regulator phosphorylated by one kinase), or converge (more than one kinase phosphorylates one response regulator) (Hellingwerf et al., 1995, 1998). Accordingly, signal transduction pathways by phosphoryl transfer meet all the criteria of a neural network. In *Synechocystis*, at least 80 TCRS pairs have been found in the genome (Mizuno et al., 1996).

5 Phosphate incorporation

Orthophosphate ($H_2PO_4^-$, HPO_4^{2-}, or PO_4^{3-}), synonym inorganic phosphate (abbreviated P_i), some phosphorylated sugars and phosphonate are the only directly available P sources for phytoplankton (Palenik et al., 2003). Bio-available P can be defined as the sum of immediately available P (P_i), and P that can be transformed by naturally occurring physical (e.g. desorption), chemical (e.g. dissolution) and biological processes (e.g. enzymatic degradation; Boström et al., 1988). Three factors determine the efficiency of P_i-uptake: The permeability of the cell membranes, the relative concentrations of P_i inside and outside the cells, and the capacity to use a variety of phosphorylated compounds. P_i incorporation takes place in several steps (Falkner et al., 1989), which are schematically indicated in Figure 1. Organic or inorganic phospho-esters are converted into P_i by the activity of alkaline phosphatase enzymes outside the cell, at the cell-surface or in the periplasm. Transport through the cytoplasmic membrane is the rate-limiting step of P_i incorporation (e.g. Falkner et al., 1989). At low concentrations many organisms can induce a high-affinity uptake system, which transports P_i at the expense of ATP. Uptake of P_i ceases at a threshold concentration because the energy available for the uptake process is then insufficient. P_i is subsequently converted into energy-rich ATP. The energy required to drive the process of ATP formation from ADP and sequestered P_i is provided by photo-phosphorylation, which is coupled to the proton flux across the thylakoid membranes in the light reactions of photosynthesis (Simonis and Urbach, 1973). Excess P_i is stored inside the cells as polyphosphate granules (Poly-P). PolyP-bodies are also referred to as metachromatic granules or volutin (Kornberg 1995), and are similar to acidocalcisomes (Marchesini et al., 2002). PolyP is present in these granules in microcrystalline aggregates, consistent with their very high electron density (Ruiz et al., 2001). Many functions have been attributed to polyP. It is obviously a reservoir for P_i, and connected to the polyPoverplus phenomenon: the rapid and extensive polyP accumulation, when P_i is added to cells previously subjected to P_i-depletion. As expected for a polyanion, polyP is a strong chelator of metal ions. Chelation of various metals (e.g. Zn, Fe, Cu, and Cd) may reduce their toxicity or affect their functions (Kornberg, 1995). Other functions reported are as a buffer against alkali ions (Pick and Weiss, 1991), and as a channel for DNA entry (Kornberg, 1995). Furthermore, polyP has a function in physiological adjustments to growth, development, and deprivations. A function as phosphagen, a store of P-bound energy is disputable. Although high amounts of polyP can be stored in the cells, even highly elevated levels of polyP could supply the cell with ATP for only a few seconds (Rao et al., 1998). PolyP is coupled to the energy charge of the cell, though, because it is reversibly synthesized from ATP. PolyP formation is an equilibrium reaction that does not require anything but the energy gained from the hydrolysis of ATP into ADP and P_i.

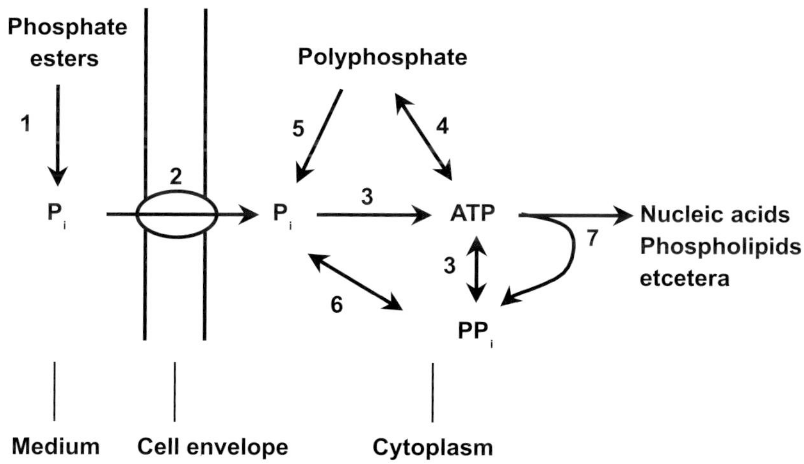

Figure 1. Cellular phosphate incorporation. The numbers indicate the following enzymes: 1: Alkaline phosphatase, 2: P_i-transport system, 3: ATP synthase, 4: Polyphosphate kinase, 5: Polyphosphatase, 6: Pyrophosphatase, and 7: Anabolic enzymes.

5.1 IS P_i THE LIMITING FACTOR FOR PHYTOPLANKTON IN LAKES ?

The notion that phosphate may be a limiting factor for algal growth goes back more than a century (see Shapiro, 1988; Correll, 1999). Since the beginning of eutrophication management in the 1970s there has been general agreement that phytoplankton biomass production is determined by availability of phosphate in a wide variety of lakes. In lake surveys as well as experimental lakes the concentration of chlorophyll and the concentration of carbon in the seston fraction were proportional to the concentration of total phosphorus in the water column (Vollenweider, 1975; Schindler, 1977). The plausible explanation for this general correlational pattern is that organisms can draw on the massive atmospheric sources and sinks for carbon and nitrogen to maintain, on average, the C:N:P ratios that reflect their requirements for growth. However, phosphorus cannot be replenished in a similar way from atmospheric resources. It follows that phosphorus ultimately controls phytoplankton abundance in many aquatic environments. Even while a sudden increase in the phosphate input (pulse) may cause algae to enter N- or C-limitation, there are long-term processes at work in the environment, which cause these deficiencies to be corrected (Schindler, 1977; Hecky and Kilham, 1988). In freshwater situations phosphate is thus the most likely macronutrient to become limiting to phytoplankton production. It would be simplistic, though, to assume that all the species in a community are limited by the same factor. Different species within the same phytoplankton community may have different resource requirements, and consequently may be limited by different factors. In fact, multiple resource limitation of phytoplankton communities has often been demonstrated (Hecky and Kilham, 1988; Elser et al., 1990; Huisman and Weissing, 2001).

5.2 ACQUIRING EVIDENCE FOR NUTRIENT LIMITATION

Several classical approaches for detection of nutrient limitation have been described (e.g. reviewed in Hecky and Kilham, 1988; Beardall et al., 2001). ***Dissolved nutrient concentrations*** present negative evidence: when a nutrient concentration becomes very low, it is likely to limit algal production. However, phytoplankton species and communities have such high affinity for N and P that nutrient limitations occur at concentrations that are generally not analytically detectable, or otherwise, in case of pulsed supply, may be missed due to low sampling frequency. ***Elemental ratios*** have often been used to indicate which nutrient will eventually limit biomass production ('Liebig' limitation). They are less useful, however, to ascertain whether a certain phytoplankton population is nutrient limited at a given point in time (Beardall et al., 2001). ***Nutrient enrichment bioassays*** are operational tests in which one or more nutrients are added to a volume of water to determine to what extent algal growth is stimulated. The test result of these bioassays is likely to be dependent on the level of complexity of the assay system used (whole lake nutrient enrichment; laboratory scale enclosure; continuous culture). Especially natural systems are capable of much more complex responses at longer time scales. The only concern to aquatic resource managers is the highest level, the natural system, whereas inferences are often made based on evidence from lower level test systems. For instance, the artificial system in which the nutrient enrichment bioassay is carried out may seriously underestimate the role of tight nutrient recycling in the aquatic ecosystem that it is presumed to represent. If ***calculated fluxes of nutrients*** can show that growth is dependent on one nutrient rather than any other, then that nutrient may limit algal growth. The problem is that the system must be very well defined and all nutrient inputs must be measured with significant accuracy, which is difficult to apply in a natural system. ***Physiological responses to nutrient limitation*** can be used instantaneously as indicators for the nutrient status. Prolonged incubations are not required, but repetitive sampling to characterize the general state of an ecosystem is. Examples of nutrient status indicators are: variation in cell contents, maximum uptake rate, maximal growth yield estimations based on variable fluorescence (Healey and Hendzel, 1980), bio-availability to reporter strains (e.g. Pat et al., 2001), immunofluorescence detection of specific markers (Dyhrman and Palenik, 1999), and enzymatic assays (Rengefors et al., 2003; Nedoma et al., 2003). The search in our present work focuses on the identification of biomolecules or processes that are useful as physiological indicator for nutrient limitation.

6 Adaptation strategies to P-deficiency

6.1 CLASSIFICATION OF ADAPTATION STRATEGIES

As P_i is essential for the growth of phytoplankton, they have developed strategies to cope with (temporary) P_i-depletion. These growth strategies underlie the ecological patterns that allow us to interpret and understand the processes of community assembly. They are pre-adaptations that are permanent, quantifiable features of the organism, which can be invoked experimentally (Reynolds, 1998). It is important to distinguish general from specific responses. P-depleted cyanobacteria display several phenomenological changes as compared to nutrient-replete cultures. The colour changes from intense blue-green to yellowish (bleaching, chlorosis), the cells appear smaller, and sometimes foam appears on the culture (cell lysis). At the physiological level, there are three prerequisites to survive P_i-deficiency (Falkner et al., 1998). Firstly, the cells must possess uptake systems that operate efficiently at very low and fluctuating P_i concentrations. Secondly, P_i must be stored inside the cells to secure the availability in times when P_i-uptake ceases. Thirdly, the cellular processes must be largely independent of external P_i-concentrations, but proportional to the amount of P_i stored in the cells. According to La Roche et al. (1999), phytoplankton cells exhibit three major categories of responses to nutrient limitation: retrenchment, compensation and acquisition. Retrenchment, or down-regulation of physiological rates is a progressive and reversible response, resulting in a modulation of the overall growth rate and changes in biochemical composition of the cells (proper growth rate limitation). Examples are: changes in the relative amounts of photo-pigments, a marked decrease in RNA content in P_i-depleted cells (N. Yeremenko, personal communication). This last observation is in line with results from a study with *Synechococcus* cells that are growing with a P-limitation, in which the RNA contents varies directly with the growth rate (Grillo and Gibson, 1979). Compensation includes all cellular responses that alleviate the effects that the lack of nutrients imposes (La Roche et al., 1999), exemplifying a general response. For example, lack of P_i in the cells obstructs photophosphorylation. This may lead to photo-inhibitory stress, for which the cells have to compensate as if they were experiencing excessive light conditions, e.g. by increased synthesis of carotenoids in the outer membrane. Acquisition is the development of more efficient uptake systems, which is a specific response. This involves synthesis of both high-affinity P_i-uptake system and alkaline phosphatase enzymes that convert alternate chemical forms of P.

Similarly, in phytoplankton ecology three different strategies are usually distinguished to cope with temporary nutrient limitations (Sommer, 1989; Ducobu et al., 1998). These strategies may operate separately or in co-operation. Cells with a

growth strategy use transient nutrient enrichments to achieve a high growth rate by optimising their specific yield (Y_x). A well-known example of the growth strategy to cope with shortage of phosphate is the replacement of the P-containing cell wall component teichoic acid by teichuronic acid in Gram-positive bacteria. In phosphate-rich conditions 15% of the cellular phosphate of *Bacillus subtilis* is stored in the cell wall polymer teichoic acid. During P-deficiency this compound is replaced by teichuronic acid, which contains no phosphate (e.g. Lahooti and Harwood, 1999). The result is a higher amount of biomass on the same amount of P, and also a higher growth rate with a lower P-concentration. Although the cyanobacterial peptidoglycan has phosphate links, similar use of this potential transient nutrient enrichment has as far as the authors know never been described for cyanobacteria. Alternatively, cells can increase the uptake rate, by making more uptake proteins or changing the conformation of uptake proteins. This is connected to the ***affinity strategy***, in which new synthesis of an uptake system with higher affinity for the nutrient causes a decrease of the saturation constant (K_m). Affinity strategists are efficient users of low external nutrient concentrations. A potential response is the induced synthesis of a high-affinity uptake system for P_i. Cells with a ***storage strategy*** secure the P_i-availability in times when uptake ceases, by build-up of internal supplies. This implies synthesis of enzymes that transform P_i into insoluble macromolecules (polyphosphates) inside the cell. This internal storage may enable the cells to produce several generations when external nutrient concentrations are low. To these three strategies, a fourth can be added that may be referred to as the ***scavenging strategy***. This strategy involves the induced synthesis of enzymes that transform generally inaccessible phosphate-containing compounds into a form that the cell can use, or the excretion of compounds that liberate inaccessible forms of phosphate. Organisms employing this strategy are able to use organic molecules that contain the required nutrient (e.g., P-esters), or they may use alternative inorganic forms of the nutrient.

6.2 P_i-UPTAKE NEAR THE THERMODYNAMIC LIMIT

Microorganisms cannot completely deplete their environment of phosphate; a threshold concentration exists, below which incorporation is thermodynamically impossible (typically 3-10 nM: Falkner et al., 1989). The threshold concentration, however, is influenced by the affinity of the uptake system. P_i-uptake of *Synechococcus elongatus* PCC 7942 efficiently takes place near the threshold concentration (Wagner et al., 2000, and references therein). The P_i-uptake rate by *Synechococcus* near the thermodynamic equilibrium was found to be linearly dependent on the free energy of polyphosphate formation and the pH-gradient at the thylakoid membrane (Wagner et al., 1995). Cells were able to independently change their kinetic and energetic properties, and seemed to optimise the efficiency of P_i-uptake, rather than absorb as much as possible. In this situation, P_i-uptake and growth were not directly coupled; the growth rate was rather related to the amount of stored phosphate, as intuited by Droop (1973). Wagner and co-authors suggest that cyanobacteria have a capacity to "memorise" nutrient fluctuations for several hours, which provides a means to adjust growth rate to P_i-availability. The nature of this adaptive response to changes in the external concentrations appears to be an index of previous growth conditions, and reflect the whole concentration range the population has previously experienced (Falkner et al., 1993; Wagner et al., 1995). The information about previous P-supply is lost, however, after complete growth arrest (Falkner et al., 1995). Analysis of the uptake activity of phytoplankton by the methods developed in these studies may also serve as a proper tool for monitoring P_i inflow into lakes, and to establish threshold concentrations below the normal measurable range (Falkner et al., 1993; Aubriot et al., 2000). The observed adaptive flexibility towards P_i concentrations fluctuating around the thermodynamic threshold may be provided by staggered interplay of two parallel uptake systems with distinct high affinities (see below).

7 The *pho*-operon: a high-affinity P_i uptake system

In a P_i-depleted environment, P_i is taken up along a steep concentration gradient in an energy-dependent process. In these conditions, the efficiency of P_i-uptake from environments with a low P_i concentration is considerably enhanced by channels through the outer membrane that exhibit strong anion selectivity or bear specific P_i-binding sites, and by a periplasmic space that possesses high affinity P_i-binding proteins. The high-affinity P_i-uptake system, Pst, known from early research on *Escherichia coli* (Enterobacteriaceae), has all these characteristics. The Pst-system is carefully regulated, and is directly ATP-driven (Horiuchi et al., 1959; Rosenberg et al., 1977). The pathway for P_i-uptake into cells grown under P_i-depleting conditions is summarised in Figure 2. It starts with the passage of P_i or phosphorylated compounds through an outer membrane porin protein channel (phosphate-specific PhoE or a non-specific porin) into the periplasmic space. In the periplasm, the phosphorylated compounds are hydrolysed by alkaline phosphatases; P_i is captured by the P_i-binding protein PstS, and directed to the ATP-binding cassette-dependent (ABC) transporter in the cytoplasmic membrane. This transporter consists of two integral membrane-bound proteins, PstA and PstC, and a cytoplasmic peripheral membrane protein, PstB (Silver and Walderhaug, 1992). Homologous high-affinity P_i-uptake systems have been found in cyanobacteria, and have most extensively been studied for *Synechococcus*. For this cyanobacterium, the P_i-binding protein SphX (Scanlan et al., 1993; Mann and Scanlan, 1994), two alkaline phosphatases PhoA (Ray et al., 1991) and PhoV (Wagner et al., 1995), and the sensor/regulator pair SphS/SphR (Aiba et al., 1993) have been described. A similar P_i-uptake system has been identified in *Synechocystis*. The genome of *Synechocystis* contains two operons (sll0679-0684 and slr1247-1250) that are homologous to the Pst-system (Kaneko et al., 1996; Dignum, 2003; Suzuki et al., 2004; Yeremenko, 2004). The first system is involved in an initial, fast response toward P_i-depleting conditions; the second system is the major high-affinity P_i-uptake system, but takes longer to be derepressed (Suzuki et al., 2004). The presence of two high-affinity P_i-uptake systems with different sensitivity for P_i-concentrations may be important in providing a flexible response near the thermodynamic threshold concentration, as described above for *Synechococcus* (Falkner et al., 1998). Regulation of several porins by P_i-availability also occurs in *Synechocystis* (Dignum, 2003). It is presently unknown whether these porins are selective towards P_i. The genes encoding alkaline phosphatase *pho*A (sll0654), P_i-sensing histidine kinase *pho*R (sll0337) and response regulator *pho*B (slr0081) have been identified (Hirani et al., 2001). The regulation of the Pst-system is highly specific for P_i-availability, and the proteins are located in the cell periphery. Therefore, the Pst-system is of major interest as potential markers for the nutrient status of cyanobacteria.

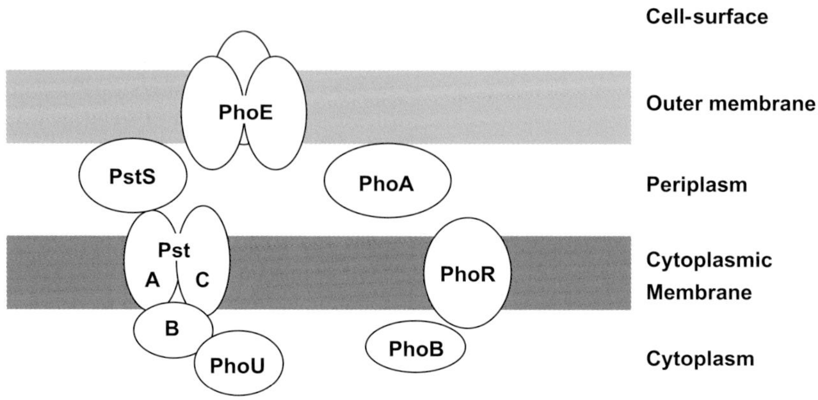

Figure 2. High-affinity phosphate uptake by the Escherichia coli pho-*operon (after Silver and Walderhaug, 1992), PhoA: Alkaline phosphatase, PhoE: phosphate-selective outer membrane porin, PstS: phosphate-binding protein, PstABC: ATP-binding cassette transporter, PhoR: sensory kinase, PhoB: transcription regulator, PhoU: tentative inhibitor of pho-expression.*

8 P-availability and the importance of phosphatases

Although P_i and some small organic phosphate esters are readily available for phytoplankton, these compounds are usually present in only minute amounts in lakes. Dissolved organic phosphates are predominantly of large molecular weight or colloidal material (Bentzen et al., 1992). These are not ubiquitous (Jansson et al., 1988; Rijkeboer et al., 1991), and the colloidal-P is very stable and resistant against dephosphorylation (Olsson and Jansson, 1984). Nonetheless, the available fraction is actively utilised by bacteria and algae (Bentzen et al., 1992; Cotner and Wetzel, 1992). Particulate organic P is also found to be available to phytoplankton to a certain extent (Boström et al., 1988). In other words, when P_i is released from dead cells (detritus), or from dissolved substrates, it is quickly reabsorbed by living cells (Jansson et al., 1988). Most naturally produced organic phosphorus compounds, except phosphonate, are esters of orthophosphoric acid. The main forms of P in detrital matter are probably adsorbed P_i, sugar phosphates, glycerophosphate, polynucleotides and phospholipids (e.g. Pant and Reddy, 2001). The P-moiety in these compounds is only available after hydrolysis of the phosphate ester bonds. This reaction is carried out mainly by phosphohydrolases. These can be classified as phosphomonoesterases, phosphodiesterases, triphosphoric monoester hydrolases, hydrolases splitting anhydride bonds in phosphoryl-containing anhydrides, and hydrolases splitting P-N bonds (Siuda, 1984). The term 'phosphatases' is mostly used synonymously with non-specific phosphomonoesterases, which have broad specificity towards different substrates. In addition to phosphomonoesters, nucleic acids may become available as P-source, although at a slower rate than phosphomonoesters (Boström et al., 1988). The phosphate forms a diester bond between nucleotide monomers, which can be hydrolysed by either alkaline phosphatases or the more specific 5'-nucleotidase (Bentzen et al., 1992). Other potential substrates for phosphatases are inorganic pyrophosphate, polyP, and short chain metapolyP (Siuda, 1984).

Acid phosphatases, on the one hand, have mostly been found inside cells, and are generally not repressed by orthophosphate (Jansson et al., 1988). Alkaline phosphatases, on the other hand, usually have an extracellular function, and their expression is sensitive to phosphate availability. The fact that alkaline phosphatases are synthesized under P-deficiency, in combination with the notion that the product of phosphatase activity, P_i, is readily assimilated by phytoplankton, provides the basis for the hypothesis that phosphatases have an essential function in the P dynamics of lakes (Jansson et al., 1988). According to Currie and Kalff (1984), a consequence of the hypothesis that algal P comes predominantly from organic P, is that there is no theoretical reason to expect algal growth to depend on ambient P_i-concentrations, nor to expect that resource competition among freshwater phytoplankton should depend on interspecific differences in P_i-uptake kinetics alone. Rather, resource competition depends on the combination of uptake-, storage-, and conversion-capacities in relation to the availability of P. In P_i-depleted conditions, the factor that limits P-availability is probably the rate at which organic P becomes available as substrate for alkaline phosphatases. Ubiquity of inducible alkaline phosphatases among phytoplankton supports these conclusions for subsaturating P-loads. In general, alkaline phosphatase activity reduces the residence time of P in the water phase.

9 Alkaline phosphatases as indicator for P-deficiency

The presence and activity of alkaline phosphatases can be used as an indicator for P-deficiency, whereas direct measurements of the P_i-concentration do not give the relevant information (Dignum, 2003). Theoretically, the production rate of derepressible phosphatases should give the best measurement of P-deficiency. In practice, the potential phosphatase activity has been used as an indicator of P-deficiency. Potential phosphatase activity is assayed with a suitable artificial substrate (e.g., a P-ester giving a coloured product upon dephosphorylation), at substrate concentrations near the saturation concentration to allow the reaction to proceed at maximum rate. The phosphatase activity measured in routine assays cannot be used for predictions of in situ hydrolytic activity, for the following reasons (Jansson et al., 1988). Firstly, natural substrate concentrations, and thus conversion rates may be different from those used in the assays. Secondly, pH and temperature used in the assay are not comparable to those in lake waters. Thirdly, the artificial substrates may not be representative of the natural substrates. Although bulk phosphatase activity has been widely used as a means of diagnosing P-deficiency, an important flaw in its use as a P-deficiency indicator is the uncertainty about the origin of the enzymes (Jansson et al., 1988). Cells may actively excrete dissolved enzymes, and another important but non-specific contribution is delivered from the cytoplasm of dying and disintegrating cells (Jansson et al., 1988). Also, there is uncertainty whether the activity comes from (constitutive) acid phosphatases, or (derepressible) alkaline phosphatases. It is therefore crucial to trace phosphatase activity back to the cells that produced the enzyme. The ELF-97^{TM} phosphate (ELF-P) substrate yields intensely green fluorescent precipitates of ELF-97^{TM} alcohol (ELF-A) upon enzymatic dephosphorylation. Applications of the ELF substrate to study the phosphatase activity of freshwater phytoplankton were recently published (Rengefors et al., 2003; Nedoma et al., 2003; Dignum et al., 2004). Figure 3 shows some examples. The cells show intense fluorescence of ELF-A only under P_i-limited conditions (Fig. 3B,D).

10 Flow cytometry to detect P-deficiency

Flow cytometry offers several advantages over traditional methods used in aquatic microbial ecology: phytoplankton data on large numbers of samples can be obtained in a short time, on a scale relevant for individual cells or colonies. Various parameters are analysed simultaneously, and gated amplification allows separate investigation of species groups. Furthermore, the sorting capacity of some flow cytometers offers the possibility to select (sub)populations for further analysis elsewhere (Vrieling and Anderson, 1996; Collier and Campbell, 1999; Vives-Rego et al., 2000). Spectral characteristics of endogenous pigments (autofluorescence) provide the means to discriminate between phytoplankton groups. The emission spectra of phytoplankton show red fluorescence of the photosynthetic pigment chlorophyll *a*, with a maximum at about 685 nm. In addition, cyanobacteria have accessory pigments, the phycobiliproteins, which emit in the orange and red regions of the spectrum. Phycoerythrin has an emission maximum in the 560-590 nm or in the 620-650 nm range, depending on the kind of chromophore it carries. Phycocyanin has an emission maximum at about 652 nm. The orange-red autofluorescence is therefore a highly selective property for the detection of cyanobacteria (Hofstraat et al., 1991). An example of the flow-cytometric separation of several phytoplankton groups from Lake Loosdrecht (The Netherlands) is shown in Figure 4A. The lake's phytoplankton community consists for more than 90% of filamentous cyanobacteria with very similar morphological characteristics. Due to differences in pigmentation, clusters of the dominant species *Limnothrix* sp. (with phycocyanin; cluster III in Fig. 4A) and the second-most abundant species in the lake, *Prochlorothrix hollandica* (without phycocyanin; cluster II in Fig. 4A), are separated. Application of the ELF-P staining method to phytoplankton extends the use of flow cytometry (González-Gil et al., 1998; Rengefors et al., 2001; Nedoma et al., 2003; Rengefors et al., 2003; Dignum et al., 2004). The green ELF-A fluorescence is maximal at 520-530 nm. In P-depleted batch-cultures of *Limnothrix* sp. isolate MR1 about 50% of the trichomes showed ELF-A fluorescence, whereas only a few trichomes fluoresced in nutrient-replete cultures (Dignum et al., 2004). In a laboratory scale enclosure initially started up with natural lake water in spring

2001, but with continuously added P_i in excess, about 45% of the *Limnothrix*-trichomes initially showed ELF-A fluorescence. While the number of *Limnothrix*-trichomes increased quickly, about 15% had ELF-A fluorescence after one week, and only 5.5% after two weeks (Dignum, 2003). These results show that growth of the *Limnothrix*-population was P_i-limited, and that the percentage of trichomes with ELF-A fluorescence reflects the extent of P_i-limitation for a natural population of filamentous cyanobacteria.

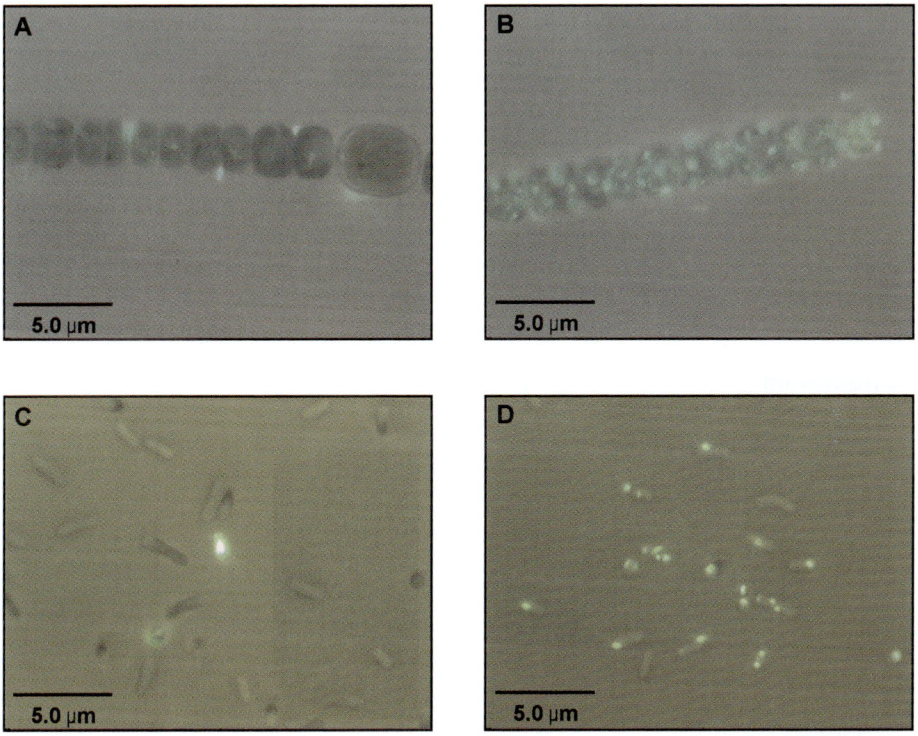

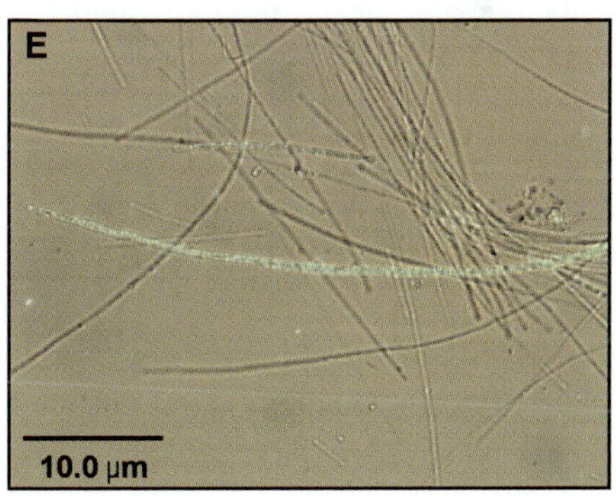

Figure 3. Fluorescence microscopic images of ELF-A stained cyanobacteria. (A) Nutrient replete, and (B) P_i-depleted batch cultures of Anabaena variabilis. *(C) Nutrient replete, and (D) P_i-depleted batch cultures of* Synechococcus elongatus. *(E) A sample from Lake Loosdrecht. Reprinted from Dignum et al. (2004), with permission from Elsevier.*

Figure 4 shows flow-cytometric analysis of a sample from Lake Loosdrecht with and without ELF-staining. The additional cluster in Figure 4B is the cyanobacterium *Aphanizomenon*, which was not visible without ELF-staining (Fig. 4A). The ELF-A versus phycocyanin fluorescence histogram of the ELF-stained sample (Fig. 4D) shows that about 25% of the *Limnothrix*-trichomes had ELF-A fluorescence (intense fluorescence at 525 nm) at that moment. Further ELF-measurements with lake samples revealed that the growth of *Limnothrix* sp. in Lake Loosdrecht was P_i-limited during part of 2001, that there was distinct seasonal variation in the extent of P_i-limitation, and that there was heterogeneity in P_i-limitation within the *Limnothrix* population (Dignum, 2003). Furthermore, about 15% of the *Prochlorothrix hollandica*-trichomes in Figure 4 show ELF-A fluorescence. Batch cultures of *P. hollandica* PCC 9006 were shown to have low, non-inducible phosphatase activity (Dignum et al., 2004). The two species thus seem to employ different strategies towards low P_i-availability. *Limnothrix* can be described as a scavenging and affinity strategists, whilst *Prochlorothrix* can be described as an affinity and storage strategist (Ducobu et al., 1998). In conclusion, the ELF-method is a useful indicator for rapid monitoring of alkaline phosphatase activity at the level of the individual cell by means of flow cytometry.

11 Discussion

Diagnostic tools for the phytoplankton nutrient status should comprise fluorescent stains that are optimised for broad applicability, nutrient status specificity, and non-interference with the autofluorescence. Cell surface marker proteins are adequate indicators of nutrient limitation because their expression is highly sensitive to changes in nutrient concentration; their expression is specific for each nutrient, can potentially be quantified per cell, and is possibly taxon specific (Scanlan and Wilson, 1999). Compared to the immunological approach (Scanlan et al., 1997; La Roche et al., 1999), enzymatic assays are applicable to a wide variety of species and may have an enormous amplification factor. The ELF-method satisfies the criteria for diagnostic tools proposed by Falkowski et al. (1992): the tool is broadly applicable in the field across phylogenetic lines, it identifies a process that imposes a truly physiological limitation, and it is uniquely affected by a specific limiting factor. The ELF-method also has some important drawbacks, however: Not all species are responsive, even when they have inducible phosphatase activity. Distinction between inducible, extracellular alkaline phosphatases and constitutive, intracellular acid phosphatases is not possible. Furthermore, ELF-A fluorescence cannot be quantified in terms of the production rate of derepressible phosphatases, which would provide the best measurement of P-deficiency. It is not sure if a decrease in ELF-A fluorescence is caused by an alleviation of the P-limitation, or conversely, by an extreme P-limitation hampering de novo synthesis or accessibility of alkaline phosphatase. Other aspects of the physiological status should be included in order to get a complete picture of the present conditions and the nutrient history. The affinity, scavenging, and storage status give complementary information, reflecting the present growth conditions and the cells' nutrient history.

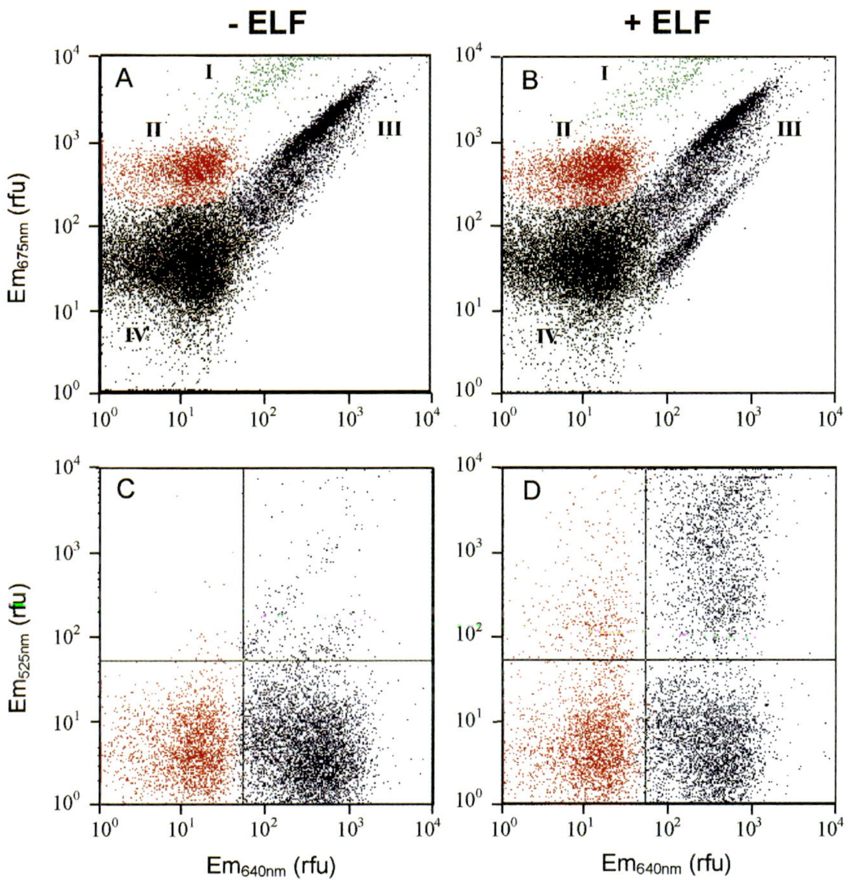

Figure 4. Flow-cytometric separation and ELF-staining of major phytoplankton groups from Lake Loosdrecht (The Netherlands). (A,B) Two-dimensional histograms showing chlorophyll fluorescence (675 nm) versus phycocyanin fluorescence (640 nm), in (A) samples without ELF-staining, and (B) samples with ELF-staining. Species clusters are: I, green algae and diatoms; II, prochlorophytes (here the filamentous Prochlorothrix hollandica*); III, cyanobacteria (predominantly filamentous* Limnothrix *sp.); IV, detritus (dead cells). (C,D) ELF-fluorescence (525 nm) versus phycocyanin fluorescence (640 nm) histograms of species clusters II and III, in (C) samples without ELF-staining, and (D) samples with ELF staining.*

Future work should involve development of indicators for the P-affinity status and the P-storage status, for example by creating antibodies against PstS or porins, and by optimising quantitative staining of internal polyP-stores. Furthermore, to study growth limitation, detailed knowledge about population-specific growth rates is required. By linking fluorescence-activated cell sorting (FACS) and isotope-ratio mass spectrometry through in-line pyrolytic methylation, phytoplankton can be probed for their population-specific $\delta^{13}C$ signature. This novel method allows assessment of population-specific growth rates from phytoplankton cells that are labelled with ^{13}C-CO_2 (Pel et al., 2003). Alternatively, the cell division cycle can be monitored by fluorogenic staining of DNA as a measure for the growth rate. Parpais et al. (1996) have reported a remarkable, irreversible arrest of the marine cyanobacterium *Prochlorococcus* sp. in the DNA replication (S) phase, when subjected to P-starvation. Moreover, these authors suggested that examination of the cell cycle of natural populations could be used to determine whether they are limited by P (Parpais et al., 1996; Veldhuis et al., 2001).

To make these fluorescent stains successful tools in water management, affordable and amenable flow cytometers are required (e.g. Dubelaar et al., 1999). In conjunction with the possibility to separate populations according to their contents of endogenous fluorescent pigments, these tools provide insight in factors that constrain productivity, and also affect community structure and species succession (e.g. La Roche et al., 1999). This approach allows rapid assessment of the limiting factors that determine the growth of harmful cyanobacteria.

12 Acknowledgements

We thank the reviewers for their critical reading of our manuscript. This work was supported by the Earth and Life Sciences Foundation (ALW) which is subsidized by the Netherlands Organisation for Scientific Research NWO.

13 References

Aiba, H., Nagaya, M. and Mizuno, T. (1993) Sensor and regulator proteins from the cyanobacterium *Synechococcus* species PCC7942 that belong to the bacterial signal-transduction protein families: implication in the adaptive response to phosphate limitation, *Molecular Microbiology* **8**, 81-91.

Aubriot, L., Wagner, F. and Falkner, G. (2000) The phosphate uptake behaviour of phytoplankton communities in eutrophic lakes reflects alterations in the phosphate supply, *European Journal of Phycology* **35**, 255-262.

Beardall, J., Young, E. and Roberts, S. (2001) Approaches for determining phytoplankton nutrient limitation, *Aquatic Sciences* **63**, 44-69

Bentzen, E., Taylor, W.D. and Millard, E.S. (1992) The importance of dissolved organic phosphorus to phosphorus uptake by limnetic plankton, *Limnology and Oceanography* **37**, 217-231.

Bijisma, J.J.E. and Groisman, E.A. (2003) Making informed decisions: regulatory interactions between two-component systems, *Trends in Microbiology* **11**, 359-366.

Boström, B., Persson, G. and Broberg, B. (1988) Bioavailability of different phosphorus forms in freshwater systems, *Hydrobiologia* **170**, 133-155.

Collier, J.L. and Campbell, L. (1999) Flow cytometry in molecular aquatic ecology, *Hydrobiologia* **401**, 33-53.

Correll, D.L. (1999) Phosphorus: a rate-limiting nutrient in surface waters, *Poultry Sciences* **78**, 674-682.

Cotner Jr., J.B. and Wetzel, R.G. (1992) Uptake of dissolved inorganic and organic phosphorus compounds by phytoplankton and bacterioplankton, *Limnology and Oceanography* **37**, 232-243.

Cowan, S.W., Schirmer, T., Rummel, G., Steiert, M., Ghosh, R., Pauptit, R.A., Jansonius, J.N. and Rosenbusch, J.P. (1992) Crystal structures explain functional properties of two *E. coli* porins, *Nature* **358**, 727-733.

Currie, D.J. and Kalff, J. (1984) The relative importance of bacterioplankton and phytoplankton in phosphorus uptake in freshwater, *Limnology and Oceanography* **29**, 311-321.

Dignum, M. (2003) Phosphate uptake proteins as markers for the nutrient status of freshwater cyanobacteria. PhD Thesis, University of Amsterdam, The Netherlands.

Dignum, M., Hoogveld, H.L., Matthijs, H.C.P., Laanbroek, H.J. and Pel R. (2004) Detecting the phosphate status of phytoplankton by enzyme labelled fluorescence and flow cytometry, *FEMS Microbiology Ecology* **48**, 29-38.

Droop, M.R. (1973) Some thoughts on nutrient limitation in algae, *Journal of Phycology* **9**, 264-272.

Dubelaar, G.B.J., Gerritzen, P.L., Beeker, A.E.R., Jonker, R.R. and Tangen, K. (1999) Design and first results of the CytoBuoy: A wireless flow cytometer for in situ analysis of marine and fresh waters, *Cytometry* **37**, 247-254.

Ducobu, H., Huisman, J., Jonker, R.R. and Mur, L.R. (1998) Competition between a prochlorophyte and a cyanobacterium under various phosphorus regimes: comparison with the Droop model, *Journal of Phycology* **34**, 467-476.

Dyhrman, S.T. and Palenik, B. (1999) Phosphate stress in cultures and field populations of the dinoflagellate *Prorocentrum minimum* detected by a single-cell alkaline phosphatase assay, *Applied and Environmental Microbiology* **65**, 3205-3212.

Elser, J.J., Marzolf, E.R. and Goldman, C.R. (1990) Phosphorus and nitrogen limitation of phytoplankton growth in the freshwaters of North America: a review and critique of experimental enrichments, *Canadian Journal of Fisheries and Aquatic Sciences* **47**, 1468-1477.

Falkner, G., Falkner R. and Schwab, A.J. (1989) Bioenergetic characterization of transient state phosphate uptake by the cyanobacterium *Anacystis nidulans*, *Archives of Microbiology* **152**, 353-361.

Falkner, G., Falkner, R. and Wagner, F. (1993) Adaptive phosphate uptake behaviour of the cyanobacterium *Anacystis nidulans*: analysis by a proportional flow-force relation, *Critical Reviews of the Academy of Sciences in Paris* **316**, 784-787.

Falkner, G., Wagner, F., Small, J.V. and Falkner, R. (1995) Influence of fluctuating phosphate supply on the regulation of phosphate uptake by the blue-green alga *Anacystis nidulans*, *Journal of Phycology* **31**, 745-753.

Falkner, R., Wagner, F., Aiba, H. and Falkner, G. (1998) Phosphate-uptake behaviour of a mutant of *Synechococcus* sp. PCC 7942 lacking one protein of the high-affinity phosphate-uptake system, *Planta* **206**, 461-465.

Falkowski, P.G., Greene, R.M. and Geider, R.J. (1992) Physiological limitations on phytoplankton productivity in the ocean, *Oceanography* **5**, 84-91.

Fulda, S., Mikkat, S., Schröder, W. and Hagemann, M. (1999) Isolation of salt-induced periplasmic proteins from *Synechocystis* sp. strain PCC 6803, *Archives of Microbiology* **171**, 214-217.

Fulda, S., Huang, F., Nilsson, F., Hagemann, M. and Norling, B. (2000) Proteomics of *Synechocystis* sp. strain PCC 6803, *European Journal of Biochemistry* **267**, 5900-5906.

González-Gil, S., Keafer, B.A., Jovine, R.V.M., Aguilera, A., Liu, S. and Anderson, D.M. (1998) Detection and quantification of alkaline phosphatase in single cells of phosphorus-starved marine phytoplankton, *Marine Ecology Progress Series* **164**, 21-35.

Grillo, J.F. and Gibson, J. (1979) Regulation of phosphate accumulation in the unicellular cyanobacterium *Synechococcus*, *Journal of Bacteriology* **140**, 508-517.

Healey, F.P. and Hendzel, L.L. (1980) Physiological indicators of nutrient deficiency in lake phytoplankton, *Canadian Journal of Fisheries and Aquatic Sciences* **37**, 442-453.

Hecky, R.E., and Kilham, P. (1988) Nutrient limitation of phytoplankton in freshwater and marine environments: a review of recent evidence on the effects of enrichment, *Limnology and Oceanography* **33**, 796-822.

Hellingwerf, K.J., Postma, P.W., Tommassen, J. and Westerhoff, H.V. (1995) Signal transduction in bacteria: phospho-neural network(s) in *Escherichia coli*? *FEMS Microbiology Reviews* **16**, 309-21.

Hellingwerf, K.J., Crielaard, W.C., Teixeira de Mattos, M.J., Hoff, W.D., Kort, R., Verhamme, D.T. and Avignone-Rossa, C. (1998) Current topics in signal transduction in bacteria, *Antonie Van Leeuwenhoek* **74**, 211-227.

Heppel, L.A. (1967) Selective release of enzymes from bacteria, *Science* **156**, 1451-1455.

Hirani, T.A., Suzuki, I., Murata, N., Hayashi, H. and Eaton-Rye, J.J. (2001) Characterization of a two-component signal transduction system involved in the induction of alkaline phosphatase under phosphate-limiting conditions in *Synechocystis* sp. PCC 6803, *Plant Molecular Biology* **45**, 13-144.

Hirschberg, J. and Chamovitz, D. (1994) Carotenoids in cyanobacteria, in D.A. Bryant, (ed.), *The Molecular Biology of Cyanobacteria*, Kluwer Academic Publishers, Dordrecht, pp. 559-579.

Hofstraat, J.W., de Vreeze, M.E.J., van Zeijl, W.J.M., Peperzak, L., Peters, J.C.H. and Balfoort, H.W. (1991) Flow cytometric discrimination of phytoplankton classes by fluorescence emission and excitation properties, *Journal of Fluorescence* **1**, 249-265.

Hoiczyk, E. and Hansel, A. (2000) Cyanobacterial cell walls: news from an unusual prokaryotic envelope, *Journal of Bacteriology* **182**, 1191-1199.

Horiuchi, T., Horiuchi, S. and Mizuno, D. (1959) A possible negative feedback phenomenon controlling formation of alkaline phosphomonoesterase in *Escherichia coli*, *Nature* **183**, 1529-1530.

Hosper, H. (1997) Clearing lakes: an ecosystem approach to the restoration and management of shallow lakes in the Netherlands, PhD Thesis, Wageningen University, The Netherlands.

Huisman, J. and Weissing, F.J. (1999) Biodiversity of plankton by species oscillations and chaos, *Nature* **402**, 407-410.

Huisman, J. and Weissing, F.J. (2001) Biological conditions for oscillations and chaos generated by multispecies competition, *Ecology* **82**, 2682-2695.

Jansson, M., Olsson, H. and Petterson, K. (1988) Phosphatases: origin, characteristics and function in lakes, *Hydrobiologia* **170**, 157-175.

Jürgens, U.J. and Weckesser, J. (1985) Carotenoid-containing outer membrane of *Synechocystis* sp. strain PCC 6714, *Journal of Bacteriology* **164**, 384-389.

Jürgens, U.J., Drews, G. and Weckesser, J. (1983) Primary structure of the peptidoglycan from the unicellular cyanobacterium *Synechocystis* sp. strain PCC 6714, *Journal of Bacteriology* **154**, 471-478.

Kaneko, T., Sato, S., Kotani, H., Tanaka, A., Asamizu, E., Nakamura, Y., Miyajima, N., Hirosawa, M., Sugiura, M., Sasamoto, S., Kimura, T., Hosouchi, T., Matsuno, A., Muraki, A., Nakazaki, N., Naruo, K., Okumura, S., Shimpo, S., Takeuchi, C., Wada, T., Watanabe, A., Yamada, M., Yasuda, M. and Tabata, S. (1996) Sequence analysis of the genome of the unicellular cyanobacterium *Synechocystis* sp. strain PCC6803. II. Sequence determination of the entire genome and assignment of potential protein-coding regions, *DNA Research* **3**, 109-136.

Klemme, J.H. (1976) Regulation of intracellular pyrophosphatase-activity and conservation of the phosphoanhydride-energy of inorganic pyrophosphatase in microbial metabolism, *Zeitung für Naturforschung* **31c**, 544-550.

Koch, A.L. (1998) The biophysics of the Gram-negative periplasmic space, *Critical Reviews in Microbiology* **24**, 23-59.

Kornberg, A. (1995) Inorganic polyphosphate: toward making a forgotten polymer unforgettable, *Journal of Bacteriology* **177**, 491-496.

La Roche, J., McKay, R.M.L. and Boyd, P. (1999) Immunological and molecular probes to detect phytoplankton responses to environmental stress in nature, *Hydrobiologia* **401**, 177-198.

Lahooti, M. and Harwood, C.R. (1999) Transcriptional analysis of the *Bacillus subtilis* teichuronic acid operon, *Microbiology* **145**, 3409-3417.

Lahti, R. (1983) Microbial inorganic pyrophosphatases, *Microbiological Reviews* **47**, 169-179.

Mann, N.H. and Scanlan, D.J. (1994) The SphX protein of *Synechococcus* species PCC 7942 belongs to a family of phosphate-binding proteins, *Molecular Microbiology* **14**, 595-596.

Mann, N.H., Scanlan, D.J., Chadd, H.E., Joint, I.R. and Carr, N.G. (1993) Nutrient acquisition by oceanic cyanobacterial picoplankters, *Trends in Microbial Ecology (Proceedings of the Sixth International Symposium on Microbial Ecology)*, pp. 51-54.

Marchesini, N., Ruiz, F.A., Vieira, M. and Docampo, R. (2002) Acidocalcisomes are functionally linked to the contractile vacuole of *Dictyostelium discoideum*, *Journal of Biological Chemistry 277*, 8146-8153.

Miller, S.R., Martin, M., Touchton, J. and Castenholz, R.W. (2002) Effects of nitrogen availability on pigmentation and carbon assimilation in the cyanobacterium *Synechococcus* sp. strain SH-94-5, *Archives of Microbiology* **177**, 392-400.

Mizuno, T., Kaneko, T. and Tabata, S. (1996) Compilation of all genes encoding bacterial two-component signal transducers in the genome of the cyanobacterium, *Synechocystis* sp. strain PCC 6803, *DNA Research* **3**, 407-414.

Nedoma, J., Štrojsová, A., Vrba, J., Komárková, J. and Šimek, K. (2003) Extracellular phosphatase activity of natural plankton studied with ELF97 phosphate: fluorescence quantification and labelling kinetics, *Environmental Microbiology* **5**, 462-472.

Olsson, H. and Jansson, M. (1984) Stability of dissolved ^{32}P-labelled phosphorus compounds in lake water and algal cultures – resistance to enzymatic treatment and algal uptake, *Verhandlungen der Internationalen Vereinigung für Theoretische und Angewandte Limnologie* **22**, 200-204.

Palenik, B. and Wood, A.M. (1998) Molecular markers of phytoplankton physiological status and their application at the level of individual cells, in K.E. Cooksey (ed.), *Molecular Approaches to the Study of the Ocean*, Chapman & Hall, London.

Palenik, B., Brahamsha, B., Larimer, F.W., Land, M., Hauser, L., Chain, P., Lamerdin, J., Regala, W., Allen, E.E., McCarren, J., Paulsen, I., Dufresne, A., Partensky, F., Webb, E.A. and Waterbury, J. (2003) The genome of a motile marine *Synechococcus*, *Nature* **424**, 1037-1042

Pant, H.K. and Reddy, K.R. (2001) Hydrologic influence of organic phosphorus in wetland detritus, *Journal of Environmental Quality* **30**, 668-674.

Parpais, J., Marie, D., Partensky, F., Morin, P. and Vaulot, D. (1996) Effect of phosphorus starvation on the cell cycle of the photosynthetic prokaryote *Prochlorococcus* spp., *Marine Ecology Progress Series* **132**, 265-274.

Parkinson, J.S. and Kofoid, E.C. (1992) Communication modules in bacterial signalling proteins, *Annual Review of Genetics* **26**, 71-112.

Pat, P.Y., Gillor, O., Post, A., Belkin, S., Schmid, R.D. and Bachmann, T.T. (2001) Monitoring of phosphorus bioavailability in water by an immobilized luminescent cyanobacterial reporter strain, *Biosensors and Bioelectronics* **16**, 811-818.

Paulsen, I.T., Sliwinski, M.K. and Saier Jr., M.H. (1998) Microbial genome analyses: global comparisons of transport capabilities based on phylogenies, bioenergetics and substrate specificities, *Journal of Molecular Biology* **277**, 573-592.

Pel, R., Hoogveld, H.L. and Floris, V. (2003) Using the hidden isotopic heterogeneity in phyto- and zooplankton to unmask disparity in trophic carbon transfer, *Limnology and Oceanography* **48**, 2200-2207.

Pick, U. and Weiss, M. (1991) Polyphosphate hydrolysis within acidic vacuoles in response to amine-induced alkaline stress in the halotolerant alga *Dunaliella salina*, *Plant Physiology* **97**, 1234-1240.

Raivio, T.L. and Silhavy, T.J. (2001) Periplasmic stress and ECF sigma factors, *Annual Review of Microbiology* **55**, 591-624.

Rao, N.N., Liu, S. and Kornberg, A. (1998) Inorganic polyphosphate in *Escherichia coli*: the phosphate regulon and the stringent response, *Journal of Bacteriology* **180**, 2186-2193.

Ray, J.M., Bhaya, D., Block, M.A. and Grossman, A.R. (1991) Isolation, transcription, and inactivation of the gene for an atypical alkaline phosphatase of *Synechococcus* sp. strain PCC 7942, *Journal of Bacteriology* **173**, 4297-4309.

Rengefors, K., Petterson, K., Blencker, T. and Anderson, D.M. (2001) Species-specific alkaline phosphatase activity in freshwater spring phytoplankton: application of a novel method, *Journal of Plankton Research* **23**, 435-443.

Rengefors, K., Ruttenberg, K.C., Haupert, C.L., Taylor, C., Howes, B.L. and Anderson, D.M. (2003) Experimental investigation of taxon-specific response of alkaline phosphatase activity in natural freshwater phytoplankton, *Limnology and Oceanography* **48**, 1167-1175.

Resch, C.M., and Gibson, J. (1983) Isolation of the carotenoid-containing cell wall of three unicellular cyanobacteria, *Journal of Bacteriology* **155**, 345-350.

Reynolds, C.S. (1998) What factors influence the species composition of phytoplankton in lakes of different trophic status? *Hydrobiologia* **369/370**, 11-26.

Rijkeboer, M., De Bles, F. and Gons, H.J. (1991) Role of sestonic detritus as a P-buffer, *Memorie Dell'Istituto Italiano di Idrobiologia* **48**, 251-260.

Ronson, C.W., Nixon, B.T. and Ausubel, F.M. (1987) Conserved domains in bacterial regulatory proteins that respond to environmental stimuli, *Cell* **49**, 579-581.

Rosenberg, H., Gerdes, R.G. and Chegwidden, K. (1977) Two systems for the uptake of phosphate in *Escherichia coli*, *Journal of Bacteriology* **131**, 505-511.

Ruiz, F.A., Marchesini, N., Seufferheld, M., Govindjee, and Docampo, R. (2001) The polyphosphate bodies of *Chlamydomonas reinhardtii* possess a proton-pumping pyrophosphatase and are similar to acidocalcisomes, *Journal of Biological Chemistry* **276**, 46196-46203.

Scanlan, D.J., Mann, N.H. and Carr, N.G. (1993) The response of the picoplanktonic marine cyanobacterium *Synechococcus* species WH7803 to phosphate starvation involves a protein homologous to the periplasmic phosphate-binding protein of *Escherichia coli*, *Molecular Microbiology* **10**, 181-191.

Scanlan, D.J., Silman, N.J., Donald, K.M., Wilson, W.H., Carr, N.G., Joint, I. and Mann, N.H. (1997) An immunological approach to detect phosphate stress in populations and single cells of photosynthetic picoplankton, *Applied and Environmental Microbiology* **63**, 2411-2420.

Schindler, D.W. (1977) Evolution of phosphorus limitation in lakes: natural mechanisms compensate for deficiencies of nitrogen and carbon in eutrophied lakes, *Science* **195**, 260-262.

Shapiro, J. (1988) Introductory lecture at the international symposium 'Phosphorus in freshwater ecosystems', Uppsala, Sweden in October 1985, *Hydrobiologia* **170**, 9-17.

Silver, S. and Walderhaug, M. (1992) Gene regulation of plasmid- and chromosome-determined inorganic ion transport in bacteria, *Microbiological Reviews* **56**, 195-228.

Simonis, W. and Urbach, W. (1973) Photophosphorylation in vivo, *Annual Review of Plant Physiology* **24**, 89-114.

Siuda, W. (1984) Phosphatases and their role in organic phosphorus transformation in natural waters, *Polskie Archiwum Hydrobiologii* **31**, 207-233.

Šmarda, J., Šmajs, D., Komrska, J. and Krzyžánek, V. (2002) S-layers on cell walls of cyanobacteria, *Micron* **33**, 257-277.

Sommer, U. (1989) The role of competition for resources in phytoplankton succession, in U. Sommer (ed.), *Phytoplankton Ecology: Succession in Plankton Communities*, Springer Verlag, Berlin, pp. 57-106.

Suzuki, S., Ferjani, A., Suzuki, I. and Murata, N. (2004) The SphS-SphR two component system is the exclusive sensor for the induction of gene expression in response to phosphate limitation in *Synechocystis*, *Journal of Biological Chemistry* **279**, 13234-13240.

Veldhuis, M.J.W., Kraay, G.W. and Timmermans, K.R (2001) Cell death in phytoplankton: correlation between changes in membrane permeability, photosynthetic activity, pigmentation and growth, *European Journal of Phycology* **36**, 167-177.

Vives-Rego, J., Lebaron, P. and Nebe-von Caron, G. (2000) Current and future applications of flow cytometry in aquatic microbiology, *FEMS Microbiology Reviews* **24**, 429-448.

Vollenweider, R.A. (1975) Input-output models, with special reference to the phosphorus loading concept in limnology, *Schweizerische Zeitschrift für Hydrobiologie* **37**, 53-84.

Vrieling, E.G. and Anderson, D.M. (1996) Immunofluorescence in phytoplankton research: applications and potential, *Journal of Phycology* **32**, 1-16.

Wagner, F., Falkner, R. and Falkner, G. (1995) Information about previous phosphate fluctuations is stored via an adaptive response of the high-affinity phosphate uptake system of the cyanobacterium *Anacystis nidulans*, *Planta* **197**, 147-155.

Wagner, F., Sahan, E. and Falkner, G. (2000) The establishment of coherent phosphate uptake behaviour by the cyanobacterium *Anacystis nidulans*, *European Journal of Phycology* **35**, 243-253.

Wagner, K.U., Masepohl, B. and Pistorius, E.K. (1995) The cyanobacterium *Synechococcus* sp. strain PCC 7942 contains a second alkaline phosphatase encoded by *pho*V, *Microbiology* **141**, 3049-3058.

Yeremenko, N. (2004) Functional flexibility of photosystem I in cyanobacteria, PhD thesis, University of Amsterdam, The Netherlands.

CHAPTER 5

NUTRIENT LIMITATION OF MARINE CYANOBACTERIA

Molecular ecology of nitrogen limitation in an oligotrophic sea

Anton F. POST

1 Introduction

Many seas in the (sub)tropical regions are characterized as warm, stratified water bodies, the surface layers of which are deplete in inorganic nutrients, low in chlorophyll *a* concentrations as well as in suspended particulate matter. The coastal areas of such seas are often decorated with diverse benthic communities, like seaweed meadows, coral reefs and sponge cover, many of them massive expanses of an astonishing beauty. These benthic communities with their dwelling fish populations form natural resources of great biodiversity and ecological value, which often harbor rare and endemic species. As these communities often thrive in overlying waters of high transparency, the single most important parameter influencing the viability and health of such communities is the quality of the pelagic waters that feed them. The main risks threatening (sub)tropical benthic communities are eutrophication and chemical pollution of the overlying waters due to the impact of ever increasing human activity. Whereas chemical pollution, like e.g. release of antifouling, minor oil spills, may have a localized effect in the direct vicinity of the point source, eutrophication effects extend into the marine food web as a whole. It thereby directly or indirectly affects phytoplankton productivity and community structure. Open waters of (sub)tropical seas carry phytoplankton communities that are adapted to thrive in nutrient-deplete waters. These photosynthetic communities are composed of a mixture of eukaryotic and prokaryotic species, most of which do not cause massive blooms, release significant amounts of toxin or are harmful in other ways. A single exception is formed by the infrequent occurrence of blooms of *Trichodesmium*, a filamentous, colony-forming, nitrogen-fixing cyanobacterium. Blooms of *Trichodesmium* appear at the sea surface and colony densities may be so high that they cause discoloration of surface waters (red tide). *Trichodesmium* blooms are potentially harmful due to their toxic potential and their limited trophic interactions with the marine food web. These blooms are short-lived and show a great extent of patchiness, but are assumed to have considerable effect on the marine

J. Huisman, H.C.P. Matthijs and P.M. Visser (eds.), Harmful Cyanobacteria, 87–107.

food web in their vicinity. Little is known about factors that determine the set of conditions for the development of *Trichodesmium* blooms. This chapter describes aspects of the molecular ecology of marine phytoplankton communities. Specifically, some of the interactions between the diazotrophic *Trichodesmium* and non nitrogen-fixing picophytoplankton communities in the Gulf of Aqaba, northern Red Sea, are examined here.

2 The Gulf of Aqaba, northern Red Sea

The northern Red Sea is a warm water body, stratified during both summer and winter, that carries oligotrophic, nutrient-depleted surface waters (Veldhuis and Kraay, 1993; Li et al., 1998). The Gulf of Aqaba is an extension of the northern Red Sea, a deep basin (600-1800 m), separated from the northern Red Sea by a shallow sill (240 m) at the Straits of Tiran. High evaporation rates drive a thermohaline circulation with a continuous advection of nutrient poor surface waters from the Red Sea into the Gulf, counterbalanced by an efflux of more dense deep waters (Klinker et al., 1976; Wolf-Vecht et al., 1992). The Gulf of Aqaba is subject to a distinct yearly cycle of stable stratification in summer and deep convective mixing in winter (Wolf-Vecht et al., 1992; Genin et al., 1995). Surface waters in summer are characterized by a shallow, but stable thermocline with surface temperatures of approximately 26 °C declining to 20.7 °C in the deep layers.

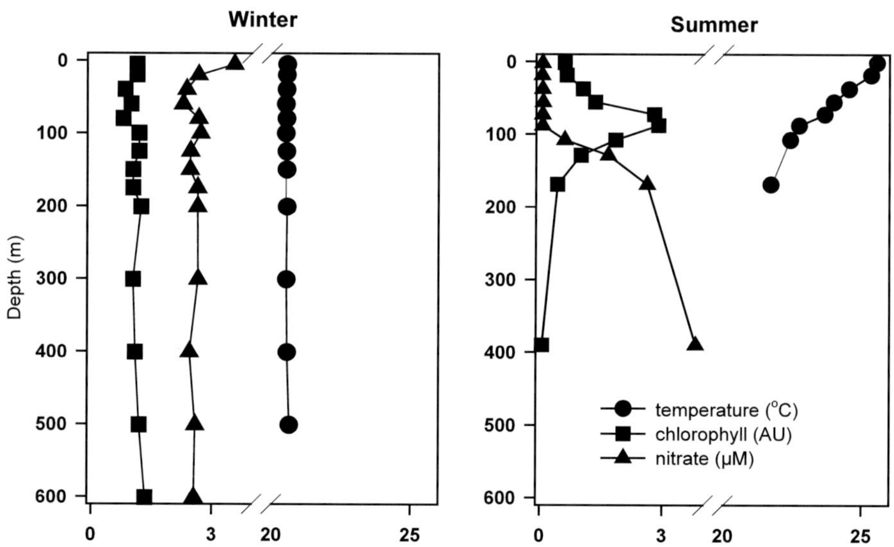

Figure 1. Typical depth profiles of temperature, nitrate and chlorophyll a *concentrations during winter mixing (left panel) and summer stratification (right panel) at sampling station A (29°28 N; 34°55 E) in the Gulf of Aqaba, northern Red Sea (Klinker et al., 1978; Genin et al., 1995; Lindell and Post, 1995).*

Lower air temperatures in fall cause a rapid erosion of the thermocline, leading to deep convective mixing during the winter months. Convective mixing may reach down to depths of 600 m or more in the northern part of the Gulf (Genin et al., 1995; Lindell and Post, 1995). This alternating pattern is reflected in a pronounced seasonal change in the trophic state of the Gulf, switching between mesotrophic to oligotrophic conditions as judged from inorganic nutrient and chlorophyll *a* concentrations in the surface layers (Fig. 1). Particularly interesting is the homogenous distribution of both nitrate (and other inorganic nutrients) and chlorophyll *a* during winter mixing. Phytoplankton and bacterial populations frequently show an even vertical distribution during this season, indicative of the extent and velocity of the deep mixing.

3 Phytoplankton communities

Phytoplankton chlorophyll *a* in surface layers of the Gulf of Aqaba is low in summer, with concentrations fluctuating between 0.02-0.04 μg per liter (Klinker et al., 1978; Genin et al., 1995; Yahel et al., 1998), considered characteristic for oligotrophic conditions. Chlorophyll *a* reaches maximal concentrations, in some years up to ~1.2 μg per liter, towards the end of the winter mixing period (Genin et al., 1995). Chlorophyll *a* concentrations in coastal surface waters fluctuate in a less dramatic fashion and averaged 0.45±0.12 μg per liter during winter 1997/1998 and declined to 0.22±0.06 μg per liter in the following summer.

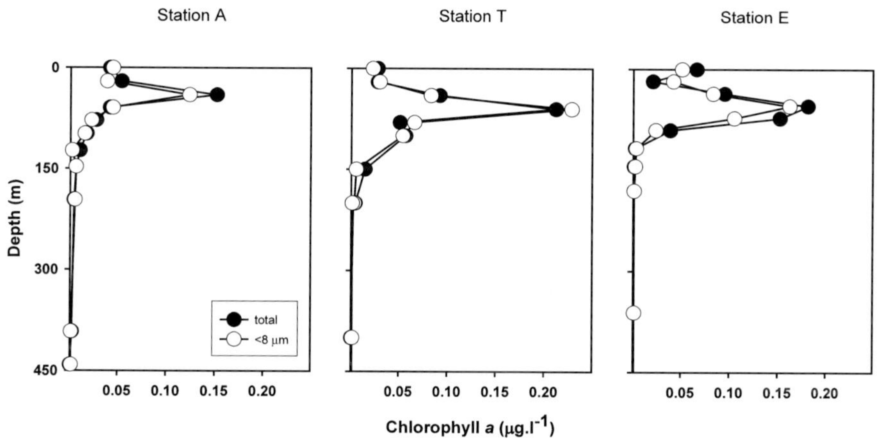

Figure 2. Depth distributions of chlorophyll a *concentrations at three distant locations in the Red Sea during June 1995. Station A is located in the northern tip of the Gulf of Aqaba, station T in the northern Red Sea and station E in the southern Red Sea. Note that the size fraction of < 8 μm (open circles) contains the bulk of chlorophyll* a *as determined on the non-filtered fraction (closed circles). 250 ml samples were filtered on 25 mm 0.2 μm pore size RC59 filters (Schleicher and Schuell) and extracted in 90% acetone.*

The phytoplankton community is in part made up of larger organisms (> 60 μm): various diatom and dinoflagellate species along with the filamentous, colony-forming, diazotrophic cyanobacterium *Trichodesmium* spp. Diatoms are more abundant during winter and early spring, but decline steadily during summer (Post et al., 2002). Dinoflagellates show only little variation in their cell numbers. *Trichodesmium* colonies are not observed during winter mixing, but they appear in surface waters during the stratification period and occasionally form blooms in both coastal and open waters (Post et al., 2002). However, more than 90% of phytoplankton chlorophyll *a* in coastal waters is contributed by phytoplankton cells of < 8 μm in diameter (Yahel et al., 1998); this percentage rises to > 95% in open waters (Fig. 2). Epifluorescence microscopy and flow cytometry later asserted that the bulk of this community was actually < 2 μm in cell size and thus belongs to the picophytoplankton size fraction. Picophytoplankton have a selective advantage over larger phytoplankton in (permanently) stratified, oligotrophic waters due to their high surface-to-volume ratios, which allows them to thrive at low nutrient concentrations (Chisholm, 1992), and their low sinking rates, which keep them suspended in the photic zone.

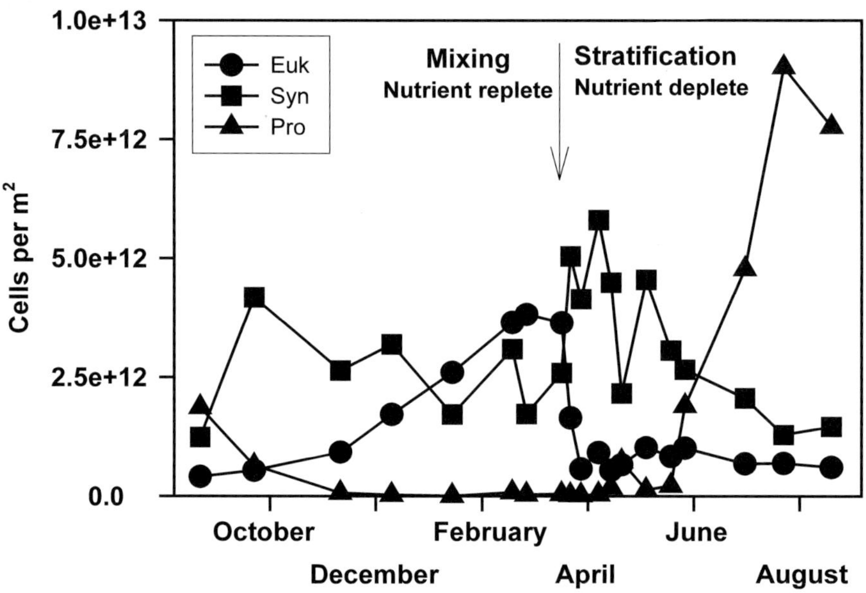

Figure 3. Seasonal change (1992-1993) in abundance of the major phytoplankton groups: eukaryotic algae (Euk), Synechococcus *(Syn) and* Prochlorococcus *(Pro) in the open water of the Gulf of Aqaba. The seasonal succession observed among these phytoplankton groups is typical for the Gulf of Aqaba after deep convective mixing (> 400 m) during winter. Abundance is presented as cell number integrated over the full depth of the water column. Arrow indicates the approximate time for cessation of mixing and onset of stratification.*

Thus the phytoplankton community in the Red Sea and Gulf of Aqaba is mostly composed of small eukaryotic algae (2-8 μm), and two groups of photosynthetic prokaryotes, the cyanobacteria *Synechococcus* (1-2 μm) and *Prochlorococcus* (0.6-1 μm). *Prochlorococcus* was found to be the dominant genus in the stably stratified Red Sea with high numbers (>100,000 cells per ml) during both summer and winter (Veldhuis and Kraay, 1993; Sommer et al., 2002). However, Figure 3 shows that in the Gulf of Aqaba the three phytoplankton groups are engaged in a distinct seasonal succession pattern as described by Lindell and Post (1995). Eukaryotic algae dominate phytoplankton in winter when the water column is deeply mixed and extensive amounts of nutrients are injected in the surface layers. At the onset of stratification eukaryotic algae decline strongly and a distinct *Synechococcus* bloom develops, fed by entrapped nutrients with nitrate as a prominent N-source. *Prochlorococcus* populations in the Gulf of Aqaba are most dynamic, they vary over 3 orders of magnitude, between > 250,000 cells per ml during summer stratification and virtual absence in winter, when waters of the Gulf are mixed down to 600 m depth (Lindell and Post, 1995). *Prochlorococcus* was rarely observed during the two months following the onset of stratification. After this period *Prochlorococcus* reappeared and rapidly increased their numbers to completely dominate the water column during the summer (Lindell and Post, 1995). The clear delineation between eukaryotic algal dominance under nutrient replete conditions and cyanobacterial dominance in stratified, nutrient deplete waters is especially striking. Thus the prokaryotic cyanobacterial groups among phytoplankton are most successful in nutrient-deplete surface waters in which unicellular *Synechococcus* and *Prochlorococcus* establish a stable abundance. In other (sub)tropical seas the two cyanobacterial groups also dominate in nutrient-depleted waters (Olson et al., 1990; Olson et al., 1990; Karl, 1999; Partensky et al., 1999). In addition, the nitrogen-fixing cyanobacterium *Trichodesmium* spp. is commonly found in such waters (Carpenter and Price, 1977; Carpenter and Romans, 1991; Karl et al., 1992; Capone et al., 1997; Karl, 1999) Lastly, diatom populations of oligotrophic waters often contain species which harbor the endosymbiotic, nitrogen-fixing cyanobacterium *Richellia* sp., like *Hemiaulus* spp. and *Rhizosolenia* spp. These diatom species were observed in the Gulf of Aqaba during summer stratification (Gordon et al., 1994). Overall, these findings suggest that cyanobacterial phytoplankton are better suited to cope with a limited supply of nutrients in marine environments. These observations raise questions with respect to how cyanobacteria cope with low nutrient supply and whether cyanobacterial populations are nutrient limited. Specifically, the study of nutrient effects on their biodiversity, adaptive potential and nutrient status will provide insights in phytoplankton population dynamics and conditions that lead up to blooms of the toxic cyanobacterium *Trichodesmium* spp. The semi-annual alternation in water column conditions of the Gulf of Aqaba with its changes in nutrient availability and the ensuing seasonal succession of phytoplankton are highly predictable and thus forms a unique model system for the study of phytoplankton - especially cyanobacterial - adaptation to nutrients in both short and sufficient supply.

4 Nutrient limitation

In general, marine phytoplankton productivity is thought to be limited by nutrients. The elements nitrogen (Kilham, 1988; Fanning, 1992; Tyrrell and Laws, 1997), phosphorus (Krom et al., 1991) and iron (Chavez et al., 1991; Kolber et al., 1994; Coale et al., 1996; McKay et al., 1997) are the macronutrients which are traditionally considered as potentially limiting phytoplankton productivity in oligotrophic seas. Low iron supply occurs in ocean provinces where desert dust deposition is low, specifically in High Nitrate Low Chlorophyll (HNLC) regions, e.g. provinces of the central Pacific (Chavez et al., 1991; Martin et al., 1994). Iron may also limit "new" primary production based on nitrogen fixation and specifically affect *Trichodesmium* productivity (Falkowski, 1997; Berman-Frank et al., 2001; Berman-Frank et al., 2001). In the Sargasso Sea such populations were found to experience iron stress (E. Webb, pers. comm.) and *Trichodesmium* in waters over the continental shelf off N-Australia had low iron quota (Kustka et al., 2003). The Gulf of Aqaba is considered to be iron-replete due to its proximity to land mass and high dust deposition (Post et al., 2002). Whether phytoplankton in the Gulf are controlled by nitrogen or phosphorus supply (or other factors) is still an open question. On the one hand, observations suggest prevailing N-limiting conditions. Inorganic nitrogen sources are at or below the detection limit in the photic layer (Lindell and Post, 1995), nitrogen contents of plankton biomass are low, short-lived blooms of the N_2-fixing cyanobacterium *Trichodesmium* develop (Post et al., 2002) and diatoms harboring diazotrophic endosymbionts appear in the stratified surface waters (Gordon et al., 1994). On the other hand, dominance of *Synechococcus* and *Prochlorococcus* with their low P contents (Bertilsson et al., 2003; Heldal et al., 2003), elevated alkaline phosphatase activities of picoplankton and of *Trichodesmium* colonies in surface waters of the Gulf (Li et al., 1998; Stihl et al., 2001) are indicative of low supply of inorganic phosphorus, possibly of phosphorus-limiting conditions. The establishment of N- versus P-limitation of phytoplankton is not a straightforward procedure as no available measurement provides an unequivocal answer. This becomes clear if one considers the following points:

1. More accurate and more sensitive methods for the chemical determination of dissolved nutrients are certainly important, but they do not directly address the question of nutrient limitation. Whereas e.g. ammonium concentrations in the sea can now be measured accurately (Holmes et al., 1999) at nanomolar levels (well below the Km values for known ammonium permeases), they do not provide direct information on the ammonium flux into cells and thus the nitrogen status of the phytoplankton community remains uncertain.
2. A wide variety of both P- and N-sources can be accessed by phytoplankton and not all of them are covered by standard determinations in water chemistry (e.g. phosphonate, urea, amino acids). The availability of other sources (dissolved organic N- and P-pools) is difficult to determine due to procedural problems and lack of information on the N-and P-species constituting these pools.
3. Bioassays based on nutrient additions (even low-level spikes of radioactively or stable-isotope labeled nutrients) to phytoplankton in enrichment experiments have

their own inherent problems. They may cause (co-)limitation by low trace metal availability, adverse effects due to sample enclosure (bottle effects, lack of turbulence, reduced gas exchange, exclusion of grazers, etc.).

There is thus a need for the development of an alternative method by which undisturbed natural samples can be interrogated regarding their nutrient status. Modern molecular techniques may provide the appropriate answer as they can directly assess this nutrient status. Firstly, one can identify the genes required for the assimilation of a given nutrient, e.g. the *narB* gene, which encodes nitrate reductase, is commonly found in marine *Synechococcus* isolates (Moore et al., 2002; Rocap et al., 2003). However, the gene is absent from all *Prochlorococcus* strains tested and apparently this genus is incapable of nitrate utilization altogether and thus distinct from *Synechococcus* (Moore et al., 2002; Rocap et al., 2003). Secondly, one can identify gene products (mRNA, protein) that are known to be induced and become strongly expressed under nutrient stress conditions. The principles of this approach have been reviewed in detail by Scanlan and Wilson (1999). At present, methods are available for detection of P-stress in marine cyanobacteria ranging from the expression of P-transport related periplasmic proteins in single cells (immunoassays using α-PstS antibodies) to the alkaline phosphatase activities of isolated phytoplankton aggregates and whole plankton community level (Scanlan et al., 1997; Li et al., 1998; Stihl et al., 2001; see Dignum et al., Chapter 4 in this volume). Below we summarize the stress responses of marine cyanobacteria to a low supply of combined nitrogen in the Gulf of Aqaba.

5 Cyanobacterial nitrogen stress responses

Marine cyanobacteria, like all phytoplankton groups, may either use "regenerated" N-sources like ammonium and urea or they may utilize "new" N-sources, e.g. by the assimilation of nitrate and fixation of molecular dinitrogen. Regenerated N-sources are those nitrogenous compounds that are recycled by heterotrophs (grazers, bacteria) within the photic layer. New N-sources are those N-compounds that originate from outside the photic layer (deep waters, atmosphere). Surprisingly little is known about cyanobacterial utilization of organic N-compounds, although their capability for urea and amino acid utilization has been recognized (Capone et al., 1994; Collier et al., 1999; Palinska et al., 2000; Lindell and Post, 2001; Valladares et al., 2002). Ammonium is the preferred source of inorganic nitrogen in cyanobacteria (Glibert and Ray, 1990; Flores and Herrero, 1994; Lindell et al., 1999; Lindell and Post, 2001). It may be imported from the environment into the cell by either passive diffusion of the uncharged ammonia or via active uptake of the ammonium ion. Ammonium is directly assimilated into organic matter via the concerted activities of glutamine synthetase (GS) and glutamate synthase (GOGAT) (Flores and Herrero, 1994). In the absence of sufficient ammonium, the cyanobacterial cell undergoes a series of adaptive processes in order to obtain the nitrogen required for growth and survival. The initial responses to ammonium deficiency include the induction of higher affinity ammonium uptake systems (Montesinos et al., 1998) and the synthesis of proteins required for the utilization of other nitrogenous compounds such as nitrate, nitrite, urea and amino acids (Flores and Herrero, 1994; Herrero et al., 2001). Most marine *Synechococcus* have the potential for utilization of nitrate, nitrite and urea (Moore et al., 2002), and the genome of the oceanic *Synechococcus* WH8102 (Palenik et al., 2003) contains additional genes and open reading frames with similarity to amino acid and oligopeptide transporters known from other organisms. However, the *nifHDK* genes, which encode the main constituents of the nitrogenase enzyme complex, are lacking in the genomes of both *Synechococcus* and *Prochlorococcus*. Thus, these extremely abundant cyanobacteria do not engage in nitrogen fixation. The utilization of alternative nitrogen sources is energetically more expensive than that of ammonium as, in most cases, they require both active transport over the cell membrane as well as the conversion to ammonium before assimilation into organic compounds (Guerrero and Lara, 1987; Flores and Herrero, 1994). Hence it is not surprising that elevated levels of ammonium prevent the utilization of alternative nitrogen sources, e.g. nitrate and nitrite, by directly inhibiting their assimilation and by repressing the protein synthesis required for their assimilation at the level of gene transcription (Guerrero and Lara, 1987; Suzuki et al., 1993; Flores and Herrero, 1994; Luque et al., 1994). Once all external nitrogen sources have been exploited, the cell enters a stage of nitrogen deprivation. During adaptation of the cell to nitrogen stress, growth may continue transiently as many physiological changes take place including the specific degradation of phycobiliproteins which results in chlorosis in both marine and freshwater *Synechococcus* (Wyman et al., 1985; Grossman et al., 1994). This process drives the reuse of cellular nitrogen for the synthesis of proteins required for survival under nitrogen-deprived conditions (Grossman et al., 1994; Gorl et al., 1998). Growth is halted once both external and internal nitrogen supplies have been exhausted.

Synthesis of the nitrogen regulatory protein, NtcA, is an essential step in cyanobacterial adaptation to ammonium-deplete conditions (Vega-Palas et al., 1990). This transcriptional activator is subject to negative control by ammonium at the level of gene expression (Luque et al., 1994; Lindell et al., 1998). *ntcA* expression is down-regulated to basal levels in the presence of ammonium. In the absence of ammonium, NtcA enhances the expression of its own gene as well as of those required for the uptake and assimilation of nitrogen sources like nitrate and nitrite (Luque et al., 1994). NtcA may further be involved in the expression of genes required for urea utilization (Collier et al., 1999; Valladares et al., 2002). A mutant strain of marine *Synechococcus* WH7803, carrying an inactivated *ntcA* gene, is incapable of growth on nitrate and nitrite (Fig. 4) and does not degrade phycobiliproteins in a timely manner under nitrogen-deplete conditions (Moyal and Post, unpublished results). A lack of phycobiliprotein degradation was also observed in a *ntcA* mutant strain of freshwater *Synechococcus* PCC7942 (Sauer et al., 1999). The mode of action of NtcA in the chlorosis process of marine *Synechococcus* under N-depleted conditions has yet to be elucidated. *ntcA* expression in marine *Synechococcus* WH7803 occurs at three distinctly different levels of transcript accumulation: a low basal level in ammonium-grown cells, an intermediate level in nitrate-grown cells and maximal expression in nitrogen-deprived cells (Lindell et al.,

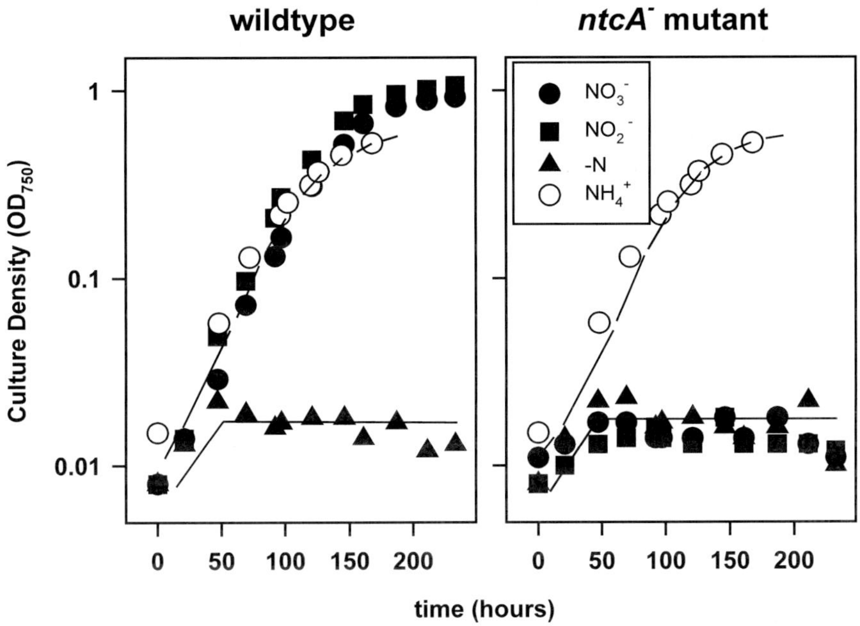

Figure 4. Autotrophic growth of marine Synechococcus *strain WH7803 wildtype and ntcA⁻ mutant cells in a mineral seawater medium with different inorganic N-sources (ammonium, nitrite and nitrate) or deplete of a combined N-source (-N). Data provided by A. Moyal and A.F. Post.*

1998; Lindell and Post, 2001). Such an expression pattern can form the basis for the development of molecular probes that assess the N-status of natural *Synechococcus* populations (see below).

Prochlorococcus, a genus closely related to *Synechococcus*, also carries the *ntcA* gene (Lindell et al., 2002). As it co-occurs with *Synechococcus* in overlapping niches in the surface ocean (Partensky et al., 1999), it experiences similar conditions with respect to N-nutrition. In contrast to *Synechococcus*, *Prochlorococcus* is much less versatile in its utilization of N-compounds. None of the many *Prochlorococcus* isolates studied is capable of nitrate utilization and only few isolates can utilize nitrite (Moore et al., 2002). Whole genome analyses indeed confirmed these physiological characteristics as *Prochlorococcus* spp. lack the genes needed for nitrate assimilation (Dufresne et al., 2003; Rocap et al., 2003). This would imply that, unlike *Synechococcus*, *Prochlorococcus* populations are virtually incapable of engaging in any form of so-called "new" primary production. In addition, the *Prochlorococcus* strain SS120 lacks the genes needed for urea utilization (Dufresne et al., 2003), thus pointing to ammonium as possibly being the major N-source for this abundant phytoplankter. At the molecular level, *Prochlorococcus* strain

PCC9511 (genetically nearly identical to strain MED4) expresses *ntcA* in response to N-stress (Lindell et al., 2002). However, replenishment of N-deprived cultures with fresh ammonium failed to induce an immediate reduction of *ntcA* transcript levels as was observed in *Synechococcus* WH7803 (Lindell et al., 1998; Lindell et al., 2002). The expression of *ntcA* showed some correlation with the N:C quota of the cells. The expression pattern observed for *ntcA* was not mirrored by that of the *amt* gene, which encodes the ammonium transporter. *amt* transcripts were high in N-replete and N-limited cells, but they declined sharply in N- and P-deprived cells (Lindell et al., 2002). Moreover, *amt* transcript levels did not correlate with N:C quota, but showed positive correlation with photosynthetic efficiencies of the cells.

The third marine cyanobacterium, globally important and abundantly present in the open waters of the northern Red Sea and the Gulf of Aqaba, is the filamentous, non-heterocystous *Trichodesmium* spp. Its presence is most pronounced in marine waters that are N-deplete in which it contributes significant input of "new" nitrogen due to its nitrogen fixation (Carpenter and Romans, 1991; Capone et al., 1997). Often *Trichodesmium* is considered as an obligate diazotroph, although it has the potential for utilizing combined N-sources. Cultures of *Trichodesmium* can be maintained with ammonium as the N-source (Kustka et al., 2003). *Trichodesmium* strains IMS101 and NIBB 11067 are capable of nitrate utilization (Ohki et al., 1991; Wang et al., 2000). The involvement of *ntcA* in N-stress responses of diazotrophic cyanobacteria is more complex. The *nirA* operon, encoding nitrate utilization, of strain IMS101 was found to have distinct promoter elements but lacks a clearly defined NtcA binding sequence (Wang et al., 2000). The promoter region of the *nifHDK* operon, which encodes nitrogenase, in *Trichodesmium* IMS101 shows sequence features with similarity to NtcA binding sites (Dominic et al., 1998). However, the fact that nitrogen fixation by *Trichodesmium* is accompanied by a release of ammonium from the filaments to the environment (Mulholland and Capone, 1999), predicts that *ntcA* expression is not simply controlled by ammonium levels. Indeed, it was found that *ntcA* transcript levels were not negatively affected by ammonium. *Trichodesmium* IMS101 cultures supplied with ammonium contained levels of *ntcA* mRNA just like those accumulated in N-deprived, nitrogen-fixing cultures (Wang and Post, unpublished results). Thus nitrogen fixation may require *ntcA* to be expressed, but *ntcA* and *nif* expression are not repressed by ammonium. On the other hand, transcript levels of the nitrate transport gene decreased with increasing external ammonium concentrations (Wang and Post, unpublished results). This decrease does not correlate with the *ntcA* transcription patterns, suggesting that nitrate utilization is ammonium repressed but not *ntcA* controlled.

In conclusion, the three abundant cyanobacterial taxa that often coexist in the same nutrient-deplete ocean waters, each have a distinctly different genetic potential for nitrogen acquisition and different nitrogen stress responses. One can thus propose a hypothesis regarding the coexistence of these taxa based on their location in the photic zone, available N-sources, the N-acquisition potential and the N-stress responses. Although *Trichodesmium* has a potential to acquire combined N-sources, the chemical composition of its natural environment, the very surface layer of stratified water bodies, suggests that it rarely employs this potential. Its nitrogen fixation produces intracellular ammonium, part of which is leaked to the environment (Mulholland and Capone, 1999), by which it may form a source of direct N-supply. *Trichodesmium* populations are mostly found to express their nitrogen fixation potential and thus require continued NtcA-mediated expression of the *nif* genes. The uncoupling between *ntcA* expression and ammonium repression is consistent with this strategy. *Trichodesmium* thus avoids N-limiting conditions by nitrogen fixation. Whole genome analyses of *Prochlorococus* spp. show that they are restricted in the N-sources they can acquire. They depend mostly on ammonium and urea, which are supplied through nutrient regeneration in the surface ocean. The *Prochlorococcus* types that utilize nitrite are indeed found at depths near the primary nitrite maximum (Rocap et al., 2002). *ntcA* expression in *Prochlorococcus* appears to respond both slowly and moderately to changes in N-nutrition. This is in good agreement with the stratified environments it thrives in, which are characterized by a predictably stable availability of N-sources at any depth. Moreover, *Prochlorococcus*, by means of its small cell size (< 1 μm), may successfully compete for nutrients at the low end of the concentration range of any nutrient (Chisholm, 1992). Possibly, *Prochlorococcus* encounters limited N-supply only rarely, if at all. Findings so far indicate that *Synechococcus* spp. are the most versatile of the three as they have the potential to utilize a wide range of N-sources. Of the three cyanobacteria, *Synechococcus* is the only one to thrive and bloom in water bodies that have recently been subjected to upwelling or deep convective mixing. Such water bodies may show thermal stratification, but they are still enriched with nutrients. The nature of such environments, reflected in dynamic changes in the distribution and concentration of inorganic and organic N-species, requires a genetic capability to meet these changes. *Synechococcus* can express its adaptive capability by means of a rapid modulation of *ntcA* transcription (Lindell et al., 1998). Since *Synechococcus* cells (> 1 μm) are larger than those of *Prochlorococcus*, it is less suited in the competition for nutrients at extremely low concentrations. *Synechococcus* may thus actually meet N-limiting conditions and need to express its adaptive potential. Therefore, this non-nitrogen-fixing, unicellular cyanobacterium forms the candidate group for N-stress studies among natural phytoplankton communities.

6 Probing the N-status of natural populations

Once sufficient information has been gathered on the ability of marine cyanobacteria to adapt to changes in their N-nutrition in culture, the true challenge lies in determining their N-status in their natural environment. Given the multitude of N-compounds and the little known order of preference in their utilization, one ideally would choose a gene that is both informative for the N-status irrespective of the N-source(s) being utilized and specific to cyanobacteria.

The global nitrogen regulator gene *ntcA* is such a gene. The responsiveness of *ntcA* to nitrogen availability and the pivotal role the gene plays in the adaptation of *Synechococcus* cells to ammonium- and nitrogen-depleted conditions suggests that different levels of *ntcA* expression may be useful indicators of the N-status of

Synechococcus populations among marine phytoplankton. In order to develop molecular tools based on *ntcA* expression, a few more characteristics of *ntcA* need to be established:

1. Ideally, *ntcA* is present on the cyanobacterial genome as a single copy gene. Multiple copies of the gene would likely be expressed differently as they may have different promoter regions, and respond to different input signals, including different environmental stresses. Southern analysis of *Synechococcus* WH7803 genomic DNA indicated that only a single copy of the *ntcA* gene was present (Lindell et al., 1998). Whole genome sequences for oceanic *Synechococcus* and *Prochlorococcus* strains (Dufresne et al., 2003; Palenik et al., 2003; Rocap et al., 2003) also contained a single *ntcA* copy suggesting that nitrogen stress responses, conveyed by the global nitrogen regulator NtcA, are expressed from a single gene copy.
2. Equally important, *ntcA* expression is required to respond to nitrogen stress and not be activated under other stresses. Since virtually all studies of *ntcA* were performed as part of laboratory experiments of nitrogen acquisition and metabolism by cyanobacterial cultures, little information is available on this subject. However, the only reported promoter element of cyanobacterial genes involved in nitrogen stress response, including *ntcA* itself, is the binding site for NtcA (Flores and Herrero, 1994). This binding site was also found upstream of nitrogen stress genes in marine cyanobacteria (Lindell et al., 1998; Wang et al., 2000; Lindell et al., 2002). In addition, *ntcA* expression in *Synechococcus* WH7803 remained at basal level when cells were subjected to either iron or phosphate stress (Lindell and Post, 2001).
3. Furthermore, *ntcA* expression needs to reflect rapid changes in the availability of nitrogenous compounds, especially the down regulation of *ntcA* expression by increased ammonium levels. As discussed above, this condition is not met by either *Prochlorococcus* PCC9511 nor by *Trichodesmium* IMS101 (Lindell et al., 2002; Wang and Post, unpublished results) and may extend to related genotypes of these marine cyanobacteria. However, marine *Synechococcus* WH7803 was shown to respond rapidly to both upward and downward changes in ammonium availability (Lindell and Post, 2001).
4. Natural levels of ammonium in the marine environment should permit induction and repression of *ntcA* transcription. *Synechococcus* WH7803 cultures express *ntcA* at < 1 μM, a threshold concentration that falls in the range of ammonium concentrations in the sea. This range spans nanomolar concentrations in the oligotrophic surface ocean and concentrations of > 1 μM in more productive coastal regions. Marine phytoplankton reportedly utilize ammonium > 1 μM preferentially over other N-compounds irrespective of their concentrations. The alternative compounds are utilized when ammonium concentrations drop below 0.5-1 μM (Glibert and Ray, 1990; McCarthy et al., 1999). Therefore, the ammonium concentration range of *ntcA* expression, and thus utilization of alternative N-sources by *Synechococcus*, matches that of the natural environment. Moreover, the immediate down regulation of *ntcA* transcript levels upon ammonium addition has potential for the identification of point sources of

ammonium in the marine environment: exudation by grazer communities, fish, benthic systems including coral reef communities, but also release resulting from anthropogenic activity like wastewater discharge and mariculture activities.

5. Lastly, probing of the N-status of natural phytoplankton communities based on *ntcA* expression should be informative and specific of the targeted group, namely the marine cyanobacteria. *ntcA* presence in other marine planktonic organisms or unrelated RNA sequences with a close similarity to that of *ntcA* would obviously yield inaccurate information on the cyanobacterial N-status. A PCR protocol based on degenerate primers targeting conserved regions of the *ntcA* gene identified *ntcA* in cyanobacterial templates, but not in DNA templates provided by other organisms, be it from heterotrophic and photosynthetic bacteria or eukaryotic algae (Lindell and Post, unpublished results). PCR amplification of environmental templates yielded specific products, the sequences of which - without exception – showed high similarity to *ntcA* sequences of *Prochlorococcus* and *Synechococcus* culture isolates (Fig. 5). The differences in *ntcA* sequence between these two closely related groups were very distinct and allow for specific probing of the *Synechococcus* genotypes.

The general fulfillment of the five criteria listed above is a requirement for any molecular probe designed to study a given environmental stress response. The next steps towards probing of natural communities are the design of a specific probe and the establishment of a protocol for probing. Foremost, genes of which the transcription does not produce high cellular levels of mRNA cannot be studied by standard northern blots or RNAse protection assays. Expression of such genes in lowly populated oligotrophic waters is quantified by reverse transcriptase PCR (RT-PCR) protocols, The expression of *ntcA* among natural *Synechococcus* populations was thus studied by RT-PCR and primers were designed which amplify *Synechococcus ntcA* mRNA but not that of *Prochlorococcus* or other cyanobacteria (Lindell and Post, 2001). Since *Synechococcus* population size, RNA extraction efficiency, sample quality, etc., may fluctuate with depth and time, one should judge actual *ntcA* expression of each sample relative to a minimal and maximal standard to be determined. Minimal or basal expression can be rapidly induced by addition of 1 mM NH_4^+, whereas maximal expression is obtained by the addition of methionine-sulfoximine, a glutamine synthetase inhibitor (Lindell and Post, 2001). This protocol has been applied to surface samples from the Gulf of Aqaba that contained ammonium (0.6 μM), nitrite (0.2 μM) and nitrate (0.6 μM) and *Synechococcus* expressed ntcA at basal levels (Lindell and Post, 2001). Subsequent analysis of surface samples over a yearly cycle revealed that *Synechococcus* populations expressed *ntcA* at levels that were significantly lower than those for induced N-stress (Fig. 6). These populations were thus nitrogen sufficient, both during periods of deep winter mixing and of stable summer stratification. Whether these populations utilize ammonium or alternative N-sources was determined in a separate study.

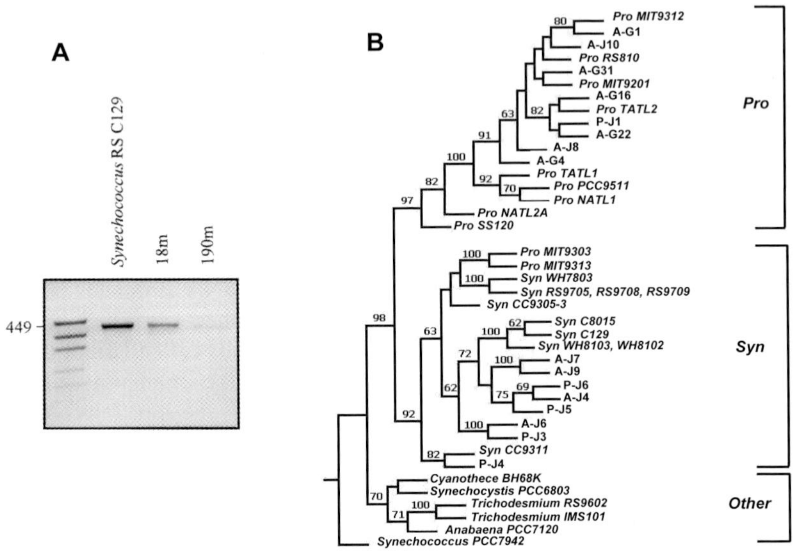

Figure 5. (A) PCR amplification of a 449 bp ntcA *product from stratified, nutrient depleted waters of the Gulf of Aqaba. DNA was obtained from 18 and 190 m depth, within and just below the photic zone. Genomic DNA of the* Synechococcus *isolate C129 from the Red Sea was used for the positive control. (B) PCR products were cloned into the pGEM-T vector (Promega) and sequenced. All products (A- and P- codes) yielded* ntcA *sequence and fell into the* Synechococcus *(Syn) and* Prochlorococcus *(Pro) clusters of the* ntcA *gene tree (Lindell, 2000).*

Throughout the year, with exception of the final stages of winter mixing, *Synechococcus* was shown to be ammonium sufficient (Lindell, 2000) in surface waters with ambient concentrations fluctuating between 20-300 nM (David and Post, unpublished results). However, *Synechococcus* expressed *ntcA* at intermediate levels in nitrate enriched waters, towards the end of the deep mixing event (Lindell, 2000). The conclusion of these findings is that *Synechococcus* and – based on the arguments discussed above – *Prochlorococcus* populations in the Gulf of Aqaba are nitrogen sufficient, even in surface waters which have been nutrient deplete for an extended period. This conclusion does not necessarily extend to eukaryotic algae and thus not to the phytoplankton community as a whole. The question thus arises whether the marine cyanobacteria in the Gulf of Aqaba are nutrient limited at all or whether their populations are controlled by other factors, e.g. mortality due to viral infection and grazing. Based on a number of additional observations a hypothesis has been developed which couples the ecology of the unicellular picoplanktonic cyanobacteria to that of filamentous microplanktonic *Trichodesmium* populations. Firstly, an underlying aspect of cyanobacterial N-stress studies, largely ignored as mentioned above, is the change in the genotypic make-up of the *Synechococcus* and

Prochlorococcus population in response to changes in nitrogen availability. Clear evidence was found that *Prochlorococcus* genotypes occupied different niches. Genotypic diversity of *Prochlorococcus* populations has been studied from *in situ* hybridization using 16S rDNA targeted oligonucleotides as well as from functional genes like *ntcA*, *cpeB*, *fstZ* and *psbA* (Lindell, 2000; West et al., 2001; Holtzendorff et al., 2002; Steglich et al., 2003; Zeidner et al., 2003). These studies showed that one distinct genotypic group occupies the upper layers of the photic zone (0-70 m) and may extend below that but at much lower abundances. The second genotypic group is found in the bottom half of the photic layer (60-150 m). Representatives of these two groups were shown to belong to distinct ecotypes (Moore and Chisholm, 1999; West and Scanlan, 1999) that differ in their utilization capability of nitrogen sources (Moore et al., 2002). Surface populations of the high-light adapted *Prochlorococcus* ecotype maintain their genotypic make-up over seasons despite strong changes in nitrogen availability (Zandbank and Post, unpublished results). Globally, the genotypic diversity of *Synechococcus* is extensive: 6-10 distinct clusters belonging to the abundant group A of marine *Synechococcus* have been identified (Rocap et al., 2002; Fuller et al., 2003). These clusters are coherent in their characteristics like pigmentation, motility, utilization of N-compounds, etc. Cultured isolates of *Synechococcus* from the Gulf of Aqaba belong to clusters II, III, VII, VIII and IX, but none of the other clusters (Fuller et al., 2003). Cluster II was found to be by far the most abundant among the *Synechococcus* populations of the Gulf of Aqaba year-round (Fuller et al., 2003). In conclusion, the genotypic composition - and by consequence the physiology - of the *Prochlorococcus* and *Synechococcus* are maintained over the year despite the extreme changes in water column properties in the Gulf of Aqaba.

The last point to consider in this chapter is whether another nutrient might be limiting cyanobacterial, and possibly phytoplankton, productivity in the Gulf of Aqaba. Of the inorganic nutrients phytoplankton have the highest demand for phosphate in waters where carbon and nitrogen supplies are replete. Phosphate is found as various species of inorganic and organic phosphate, but only orthophosphate is being taken up by phytoplankton (Björkman and Karl, 1994). Organic phosphate can be utilized by phytoplankton following hydrolysis of these compounds, a process mediated by extracellular enzyme activity, e.g. that of alkaline phosphatase, a mono-esterase with a broad range of organic phosphate substrates (see Dignum et al., Chapter 4 in this volume). Both transcription and activity of alkaline phosphatase are rapidly induced upon phosphate depletion of *Synechococcus* cells (Gillor et al., 2002). The plankton community of the Gulf of Aqaba showed elevated levels of alkaline phosphatase activity in early summer, approximately one month after the cessation of the deep mixing event (Li et al., 1998). The bulk of this activity was associated with picoplankton (Li et al., 1998). The spatial distribution of alkaline phosphatase activity showed the highest correlation with that of *Synechococcus* populations (Li et al., 1998). *Trichodesmium* populations, which appear during the same period, were also characterized by strong alkaline phosphatase activities (Stihl et al., 2001). Moreover, populations of the picoplanktonic cyanobacteria and *Trichodesmium* were shown to have elevated levels of the phosphate transport protein PstS (Fuller et al., 2005; Scanlan, pers.

comm.). Thus, the invasion of stratified waters, recently depleted of combined nitrogen, by populations of nitrogen-fixing cyanobacteria like *Trichodesmium* spp. injects new nitrogen into the surface layers. Hence, an imbalance is created between N and P-supply to the phytoplankton community leading to a draw down of the inorganic phosphate pool. *Trichodesmium* has been implied in P-depletion in open oceanic systems as well (Wu et al., 2000; Sanudo-Wilhelmy et al., 2001; Dyhrman et al., 2002). This imbalance then causes a P-stress response in the cyanobacterial species which is detected by immunoassays (Scanlan et al., 1997; Fuller et al., 2005) and enzyme activities (Li et al., 1998; Stihl et al., 2001; Dyhrman et al., 2002). Possibly, this scenario has application for the ocean as a whole. Recent findings that both *Prochlorococcus* and *Synechococcus* have lower P quota than expected from the Redfield ratio (Bertilsson et al., 2003; Heldal et al., 2003), suggest that these phytoplankters are adapted to environments with low phosphate availability.

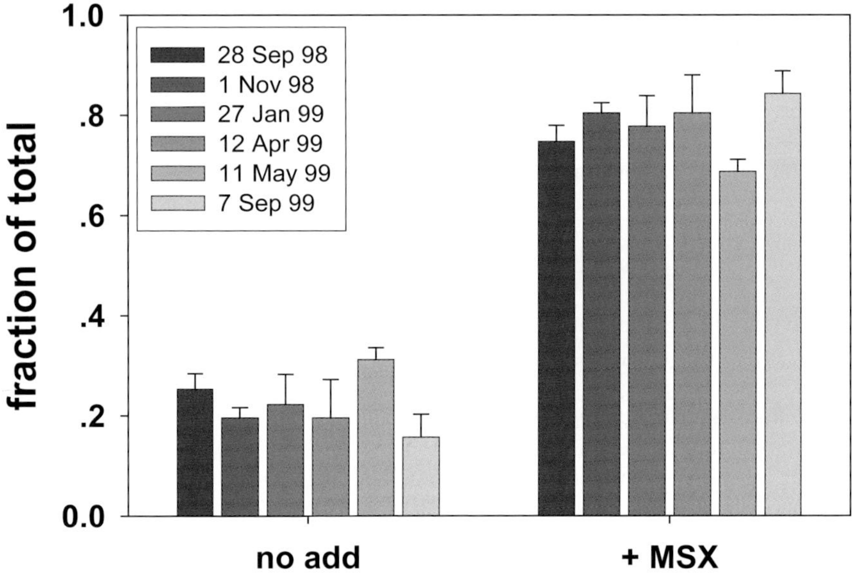

Figure 6. Actual (no add) and maximal (+ Methionine-Sulfoximine) levels of ntcA *expression for* Synechococcus *genotypes in surface samples (5 m) at sampling station A in the northern Gulf of Aqaba over a yearly cycle (1998-1999).*

7 Concluding remarks

In contrast to their freshwater counterparts, marine cyanobacteria are indicators of pristine oligotrophic conditions. Hence, their biomass is likely controlled by a limited supply of nutrients. The recent development and use of molecular tools allows the probing of nutrient stress responses in marine cyanobacterial species within plankton communities. This approach has provided an extra dimension to techniques that are common to biological oceanography, e.g. determination of nutrient levels and tracer studies of stable isotopes. For the Gulf of Aqaba it was shown from molecular studies that cyanobacterial phytoplankton are N-replete, but show P-stress responses, especially in periods when *Trichodesmium* is abundant. A potential risk is formed by the fact that the Gulf of Aqaba receives additional phosphate through anthropogenic activities, specifically due to fertilizer shipping via the ports of Aqaba and Eilat, but also by mariculture, tourist activities and wastewater disposal. This phosphate loading could potentially stimulate and enhance *Trichodesmium* blooms. Since *Trichodesmium* has a distinct toxic potential and forms an ecological nuisance due to its limited interaction in the marine food web, such blooms are highly undesired. Conversely, a combined phosphorus and nitrogen loading may suppress cyanobacterial abundance near pollution sources (as is observed during deep winter mixing events in the Gulf) and provide favorable conditions for the development of harmful algal populations (e.g., dinoflagellates). The questions pertaining to the latter scenario require the application of molecular techniques to eukaryotic phytoplankton (studies of nutrient stress, toxin production, allelopathy). This field provides exciting challenges for the development of novel molecular tools in marine ecology.

8 Acknowledgements

Amotz Osri, Jake Asher, and Dorit Golan provided technical assistance on various cruises. Debbie Lindell, Aliza Moyal, Qingfeng Wang, Sigrid Penno, Efrat David, Li Hong, Ruti Gotlieb, and Andrea Stihl provided data on nutrient concentrations, phytoplankton abundance and diversity as well as characterization of N- and P-stress responses of marine cyanobacteria in culture and natural environments. Financial support was provided by US Binational Science Foundation grants 94-146 and 99-194, Israel Science Foundation grant 525/98, the "Enrico Berman" Foundation for Solar Energy, the EU projects PROMOLEC (MAS3-CT97-0128) and MARGENES (QLRT 2001-01226), and the "Moshe Shilo" Minerva Center for Marine Biogeochemistry.

9 References

Berman-Frank, I., Cullen, J.T., Shaked, Y., Sherrel, R.M. and Falkowski, P.G. (2001) Iron availability, cellular iron quotas and nitrogen fixation in *Trichodesmium*, *Limnology and Oceanography* **46**, 1249-1260.

Berman-Frank, I., Lundgren, P., Chen, Y.B., Kuepper, H., Kolber, Z., Bergman, B. and Falkowski, P.G. (2001) Segregation of nitrogen fixation and oxygenic photosynthesis in the marine cyanobacterium *Trichodesmium*, *Science* **294**, 1534-1537.

Bertilsson, S., Berglund, O., Karl, D.M. and Chisholm, S.W. (2003) Elemental composition of marine *Prochlorococcus* and *Synechococcus*: implications for the ecological stochiometry of the sea, *Limnology and Oceanography* **48**, 1721-1731.

Björkman, K. and Karl, D.M. (1994) Bioavailability of inorganic and organic phosphorus compounds to natural assemblages of microorganisms in Hawaiian coastal waters, *Marine Ecology Progress Series* **111**, 265-273.

Capone, D.G., Ferrier, M.D. and Carpenter, E.J. (1994) Amino-acid cycling in colonies of the planktonic marine cyanobacterium *Trichodesmium thiebautii*, *Applied and Environmental Microbiology* **60**, 3989-3995.

Capone, D.G., Zehr, J.P., Paerl, H.W., Bergman, B. and Carpenter, E.J. (1997) *Trichodesmium*, a globally significant marine cyanobacterium, *Science* **276**, 1221-1229.

Carpenter, E.J. and Price, C.C. (1977) Nitrogen fixation, distribution and production of *Oscillatoria (Trichodesmium) thiebautii* in the eastern Sargasso Sea, *Limnology and Oceanography* **20**, 381-401.

Carpenter, E.J. and Romans, K. (1991) Major role of the cyanobacterium *Trichodesmium* in nutrient cycling in the North Atlantic Ocean, *Science* **254**, 1356-1358.

Chavez, F.P., Buck, K.R., Coale, K.H., Martin, J.H., DiTullio, G.R., Welschmeyer, N.A., Jacobson, A.C. and Barber, R.T. (1991) Growth rates, sinking, and iron limitation of equatorial Pacific phytoplankton, *Limnology and Oceanography* **36**, 1816-1833.

Chisholm, S.W. (1992) Phytoplankton size, in P.G. Falkowski and A.D. Woodhead (eds.), *Primary Productivity and Biogeochemical Cycles in the Sea*, Plenum Press, New York, USA, pp. 213-237.

Coale, K.H., Johnson, K.S., Fitzwater, S.E., Gordon, R.M., Tanner, S., Chavez, F.P., Ferioli, L., Sakamoto, C., Rogers, P., Millero, F., Steinberg, P., Nightingale, P., Cooper, D., Cochlan, W.P., Landry, J., Constantinou, J., Rollwagen, G., Trasvina, A. and Kudela, R. (1996) A massive phytoplankton bloom induced by an ecosystem-scale iron fertilization experiment in the equatorial Pacific Ocean, *Nature* **383**, 495-501.

Collier, J., Brahamsha, B. and Palenik, B. (1999) The marine cyanobacterium *Synechococcus* sp. WH7805 requires urease (urea amido hydrolase, E.C. 3.5.1.5) to utilize urea as a nitrogen source: molecular-genetic and biochemical analysis of the enzyme, *Microbiology* **145**, 447-459.

Dominic, B., Chen, Y.B. and Zehr, J.P. (1998) Cloning and transcriptional analysis of the *nifUHDK* genes of *Trichodesmium* sp. IMS101 reveals stable *nifD, nifDK* and *nifK* transcripts, *Microbiology* **144**, 3359-3368.

Dufresne, A., Salanoubat, M., Partensky, F., Artiguenave, F., Axmann, I.M., Barbe, V., Duprat, S., Galperin, M.Y., Koonin, E.V., Le Gall, F., Makarova, K.S., Ostrowskii, M., Oztas, S., Robert, C., Rogozin, I.B., Scanlan, D.J., Tandeau de Marsac, N., Weissenbach, J., Wincker, P., Wolf, Y.I. and Hess, W.R. (2003) Genome sequence of the cyanobacterium *Prochlorococcus marinus* SS120, a nearly minimal oxyphototrophic genome, *Proceedings of the National Academy of Sciences USA* **100**, 10020-10025.

Dyhrman, S.T., Webb, E.A., Anderson, D.M., Moffett, J.W. and Waterbury, J.B. (2002) Cell specific detection of phosphate stress in *Trichodesmium* from the Western North Atlantic, *Limnology and Oceanography* **47**, 1832-1836.

Falkowski, P.G. (1997) Evolution of the nitrogen cycle and its influence on the biological sequestration of CO_2 in the ocean, *Nature* **387**, 272-275.

Fanning, K.A. (1992) Nutrient provinces in the sea: Concentration ratios, reaction rate ratios, and ideal covariation, *Journal of Geophysical Research* **97**, 5693-5712.

Flores, E. and Herrero, A. (1994) Assimilatory nitrogen metabolism and its regulation, in D.A. Bryant (ed.), *The Molecular Biology of Cyanobacteria*, Kluwer Academic Publishers, Dordrecht, the Netherlands, pp. 487-517.

Fuller, N.J., Marie, D., Partensky, F., Vaulot, D., Post, A.F. and Scanlan, D.J. (2003) Clade-specific 16S ribosomal DNA oligonucleotides reveal the predominance of a single marine *Synechococcus* clade throughout a stratified water column in the Red Sea, *Applied and Environmental Microbiology* **69**, 2430–2443.

Fuller, N.J., Marie, D., Yallop, M., Rivlin, T., West, N.J., Post, A.F. and Scanlan, D.J. (2005) Dynamics of community structure and P status of picocyanobacterial populations in the Gulf of Aqaba, Red Sea, during 1999-2000, *Limnology and Oceanography* (in press).

Genin, A., Lazar, B. and Brenner, S. (1995) Vertical mixing and coral death in the Red Sea following the eruption of mount Pinatubo, *Nature* **377**, 507-510.

Gillor, O., Hadas, O., Post, A.F. and Belkin, S. (2002) Phosphorus bioavailability monitoring by a bioluminescent cyanobacterial sensor strain, *Journal of Phycology* **38**, 107-105.

Glibert, P.M. and Ray, R.T. (1990) Different patterns of growth and nitrogen uptake in two clones of marine *Synechococcus* spp., *Marine Biology* **107**, 273-280.

Gordon, N., Angel, D.L., Neori, A., Kress, N. and Kimor, B. (1994) Heterotrophic dinoflagellates with symbiotic cyanobacteria and nitrogen limitation in the Gulf of Aqaba, *Marine Ecology Progress Series* **107**, 83-88.

Gorl, M., Sauer, J., Baier, K. and Forchhammer, K. (1998) Nitrogen-starvation induced chlorosis in *Synechococcus* PCC 7942: adaptation to long-term survival, *Microbiology* **144**, 2449-2458.

Grossman, A.R., Schaefer, M.R., Chiang, G.G. and Collier, J.L. (1994) The responses of cyanobacteria to environmental conditions: light and nutrients, in D.A. Bryant (ed.), *The Molecular Biology of Cyanobacteria*, Kluwer Academic Publishers, Dordrecht, the Netherlands, pp. 641-675.

Guerrero, M.G. and Lara, C. (1987) Assimilation of inorganic nitrogen, in P. Fay and C. Van Baalen (eds.), *The Cyanobacteria,* Elsevier Science Publishers, Amsterdam, the Netherlands, pp. 165-186.

Heldal, M., Scanlan, D.J., Norland, S., Thingstad, F. and Mann, N.H. (2003) Elemental composition of single cells of various strains of marine *Prochlorococcus* and *Synechococcus* using X-ray microanalysis, *Limnology and Oceanography* **48**, 1732-1743.

Herrero, A., Muro-Pastor, A.M. and Flores, E. (2001) Nitrogen control in cyanobacteria, *Journal of Bacteriology* **183**, 411-425.

Holmes, R.M., Aminot, A., Kerouel, R., Hooker, B.A. and Peterson, B.J. (1999) A simple and precise method for measuring ammonium in marine and freshwater ecosystems, *Canadian Journal of Fisheries and Aquatic Sciences* **56**, 1801-1808.

Holtzendorff, J., Marie, D., Post, A.F., Partensky, F., Rivlin, A. and Hess,W.R. (2002) Synchronized expression of *ftsZ* in natural *Prochlorococcus* populations of the Red Sea, *Environmental Microbiology* **4**, 644–653.

Karl, D.M. (1999) A sea of change: biogeochemical variability in the North Pacific subtropical gyre, *Ecosystems* **2**, 181-274.

Karl, D.M., Letelier, R., Hebel, D.V., Bird, D.F. and Winn, C.D. (1992) *Trichodesmium* blooms and new nitrogen in the north Pacific gyre, in E.J. Carpenter, D.G. Capone and J.G. Reuter (eds.), *Marine Pelagic Cyanobacteria: Trichodesmium and other Diazotrophs*, Kluwer Academic Publishers, Dordrecht, the Netherlands, pp. 219-237.

Kilham, P. (1988) Comparative ecology of marine and freshwater phytoplankton, *Limnology and Oceanography* **33**, 776-795.

Klinker, J., Reiss, Z., Kropach, C., Levanon, I., Harpaz, H., Halicz, E. and Assaf, G. (1976) Observations on the circulation pattern in the Gulf of Aqaba (Elat), Red Sea, *Israel Journal Earth Science* **25**, 85-103.

Klinker, J., Reiss, Z., Levanon, I., Harpaz, H. and Shapiro, Y. (1978) Nutrients and biomass distribution in the Gulf of Aqaba (Elat), Red Sea, *Marine Biology* **45**, 53-64.

Kolber, Z.S., Barber, R.T., Coale, K.H., Fitzwater, S.E., Green, R.M., Johnson, K.S., Lindley, S. and Falkowski, P.G. (1994) Iron limitation of phytoplankton photosynthesis in the equatorial Pacific Ocean, *Nature* **371**, 145-149.

Krom, M.D., Kress, N., Brenner, S. and Gordon, L.I. (1991) Phosphorus limitation of primary productivity in the eastern Mediterranean Sea, *Limnology and Oceanography* **36**, 424-432.

Kustka, A.B., Sanudo-Wilhelmy, S.A., Carpenter, E.J., Capone, D., Burns, J. and Sunda, W.J. (2003) Iron requirements for dinitrogen and ammonium supported growth in cultures of *Trichodesmium* (IMS101): comparison with nitrogen fixation rates and iron:carbon ratios of field populations, *Limnology and Oceanography* **48**, 1869-1884.

Li, H., Veldhuis, M.J.W. and Post, A.F. (1998) Alkaline phosphatase activities among planktonic communities in the northern Red Sea, *Marine Ecology Progress Series* **173**, 107-115.

Lindell, D. (2000) Assessing the nitrogen status of marine prokaryotic phytoplankton using molecular methods, *Department of Microbial and Molecular Ecology Report,* Hebrew University, Jerusalem, Israel.

Lindell, D., Erdner, D., Marie, D., Claustre, H., Prasil, O., Le Gall, F., Rippka, R., Partensky, F., Scanlan, D.J. and Post, A.F. (2002) Contrasting expression of *ntcA* and *amt* in *Prochlorococcus* sp. strain PCC 9511, *Journal of Phycology* **38**, 1113-1124.

Lindell, D., Padan, E. and Post, A.F. (1998) Regulation of *ntcA* expression and nitrite uptake in the marine *Synechococcus* sp. strain WH 7803, *Journal of Bacteriology* **180**, 1878-1886.

Lindell, D., Padan, E. and Post, A.F. (1999) The effect of ammonium on *ntcA* expression, nitrate and nitrite uptake in marine *Synechococcus* sp. strain WH 7803, in L. Charpy and A.W.D. Larkum (eds.), *Marine Cyanobacteria, Bulletin de l'Institut Oceanographique, Monaco*, special issue N° 19, pp. 273-278.

Lindell, D. and Post, A.F. (1995) Ultraphytoplankton succession is triggered by deep winter mixing in the Gulf of Aqaba (Eilat), Red Sea, *Limnology and Oceanography* **40**, 1130-1141.

Lindell, D. and Post, A.F. (2001) Ecological aspects of *ntcA* gene expression and its use as an indicator of the nitrogen status of marine *Synechococcus* spp., *Applied and Environmental Microbiology* **67**, 3340-3349.

Luque, I., Flores, E. and Herrero, A. (1994) Molecular mechanism for the operation of nitrogen control in cyanobacteria, *EMBO Journal* **13**, 2862-2869.

McCarthy, J.J., Garside, C. and Nevins, J.L. (1999) Nitrogen dynamics during the Arabian Sea northeast monsoon, *Deep-Sea Research* **46**, 1623–1664.

McKay, R.M.L., Geider, R.J. and LaRoche, J. (1997) Physiological and biochemical responses of the photosynthetic apparatus of two marine diatoms to Fe stress, *Plant Physiology* **114**, 615-622.

Montesinos, M.L., Muro-Pastor, A.M., Herrero, A. and Flores, E. (1998) Ammonium/methylammonium permeases of a cyanobacterium: identification and analysis of three nitrogen regulated *amt* genes in *Synechocystis* sp. PCC 6803, *Journal of Biological Chemistry* **273**, 31463-31470.

Moore, L.R. and Chisholm, S.W.(1999) Photophysiology of the marine cyanobacterium *Prochlorococcus*: ecotypic differences among cultured isolates, *Limnology and Oceanography* **44**, 628-638.

Moore, L.R., Post, A.F., Rocap, G. and Chisholm, S.W. (2002) Differential nitrogen utilization of the marine cyanobacteria *Prochlorococcus* and *Synechococcus*, *Limnology and Oceanography* **47**, 989-996.

Mulholland, M.R. and Capone, D.G. (1999) Nitrogen fixation, uptake and metabolism in natural and cultured populations of *Trichodesmium* spp., *Marine Ecology Progress Series* **188**, 33-49.

Ohki, K., Zehr, J.P., Falkowski, P.G. and Fujita, Y. (1991) Regulation of nitrogen-fixation by different nitrogen sources in the marine non-heterocystous cyanobacterium *Trichodesmium* sp. NIBB1067, *Archives of Microbiology* **156**, 335-337.

Olson, R.J., Chisholm, S.W., Zettler, E.R., Altabet, M.A. and Dusenberry, J.A. (1990) Spatial and temporal distribution of prochlorophyte picoplankton in the North Atlantic Ocean, *Deep-Sea Research* **37**, 1033-1051.

Olson, R.J., Chisholm, S.W., Zettler, E.R. and Armbrust, E.V. (1990) Pigments, size, and distribution of *Synechococcus* in the North Atlantic and Pacific Oceans, *Limnology and Oceanography* **35**, 45-58.

Palenik, B., Brahamsha, B., Larimer, F.W., Land, M., Hauser, L., Chain, P., Lamerdin, J., Regala, W., Allen, E.E., McCarren, J., Paulsen, I., Dufresne, A., Partensky, F., Webb, E.A. and Waterbury, J. (2003) The genome of a motile marine *Synechococcus*, *Nature* **424**, 1037-1042.

Palinska, K.A., Jahns, T., Rippka, R. and Tandeau de Marsac, N. (2000) *Prochlorococcus marinus* strain PCC 9511, a picoplanktonic cyanobacterium, synthesizes the smallest urease, *Microbiology* **146**, 3099-3107.

Partensky, F., Hess, W.R. and Vaulot, D. (1999) *Prochlorococcus*, a marine photosynthetic prokaryote of global significance, *Microbiology and Molecular Biology Reviews* **63**, 106-127.

Post, A.F., Dedej, Z., Gottlieb, R., Li, H., Thomas, D., El-Absawy, M., El-Naggar, A., El-Gharabawi, M. and Sommer, U. (2002) Spatial and temporal distribution of *Trichodesmium* spp. in the stratified waters of the Gulf of Aqaba (northern Red Sea), *Marine Ecology Progress Series* **239**, 241-250.

Rocap, G., Distel, D.L., Waterbury, J.B and Chisholm, S.W. (2002) Resolution of *Prochlorococcus* and *Synechococcus* ecotypes by using 16S-23S ribosomal DNA internal transcribed space sequences, *Applied and Environmental Microbiology* **68**, 1180-1191.

Rocap, G., Larimer, F.W., Lamerdin, J., Malfatti, S., Chain, P., Ahlgren, N.A., Arellano, A., Coleman, M., Hauser, L., Hess, W.R., Johnson, Z.I., Land, M., Lindell, D., Post, A.F., Regala, W., Shah, M., Shaw, S.L., Steglich, C., Sullivan, M.B., Ting, C.S., Tolonen, A., Webb, E.A., Zinser, E.R. and Chisholm, S.W. (2003) Genome divergence in two *Prochlorococcus* ecotypes reflects oceanic niche differentiation, *Nature* **424,** 1042-1047.

Sanudo-Wilhelmy, S.A., Kustka, A.B., Gobler, C.J., Hutchins, D.A., Yang, M., Lwiza, K., Burns, J., Capone, D.G., Raven, J.A. and Carpenter, E.J. (2001) Phosphorus limitation of nitrogen fixation by *Trichodesmium* in the central Atlantic Ocean, *Nature* **411**, 66-69.

Sauer, J., Margit,G. and Forchhammer, K. (1999) Nitrogen starvation in *Synechococcus* PCC 7942: involvement of glutamine synthetase and NtcA in phycobiliprotein degradation and survival, *Archives of Microbiology* **172**, 247-255.

Scanlan, D.J., Silman, N.J., Donald, K.M., Wilson, W.H., Carr, N.G., Joint, I. and Mann, N.H. (1997) An immunological approach to detect phosphate stress in populations and single cells of photosynthetic picoplankton, *Applied and Environmental Microbiology* **63**, 2411-2420.

Scanlan, D.J. and Wilson, W.H. (1999) Application of molecular techniques to addressing the role of P as a key effector in marine ecosystems, *Hydrobiologia* **401**, 149-175.

Sommer, U., Berninger, U.G., Bottger-Schnack, R., Cornils, A., Hagen,W., Hansen, T., Al-Najjar, T., Post, A.F., Schnack-Schiel, S.B., Stibor, H., Stubing, D. and Wickham, S. (2002) Grazing during the spring bloom in the Gulf of Aqaba and the Northern Red Sea, *Marine Ecology Progress Series* **239**, 251-261.

Steglich, C., Post, A.F. and Hess, W.R. (2003) Analysis of natural populations of *Prochlorococcus* spp. in the northern Red Sea using phycoerythrin gene sequences, *Environmental Microbiology* **5**, 681–690.

Stihl, A., Sommer, U. and Post, A.F. (2001) Alkaline phosphatase activities among populations of the colony-forming, diazotrophic cyanobacterium *Trichodesmium* (Cyanobacteria) in the Red Sea, *Journal of Phycology* **62**, 310-317.

Suzuki, I., Sugiyama, T. and Omata, T. (1993) Primary structure and transcriptional regulation of the gene for nitrite reductase from the cyanobacterium *Synechococcus* PCC 7942, *Plant and Cell Physiology* **34**, 1311-1320.

Tyrrell, T. and Laws, C.S. (1997) Low nitrate:phosphate ratios in the global ocean, *Nature* **387**, 793-796.

Valladares, A., Montesinos, M.L., Herrero, A. and Flores, E. (2002) An ABC-type, high affinity urea permease identified in cyanobacteria, *Molecular Microbiology* **43**, 703-715.

Vega-Palas, M.A., Madueno, F., Herrero, A. and Flores, E. (1990) Identification and cloning of a regulatory gene for nitrogen assimilation in the cyanobacterium *Synechococcus* sp. strain PCC 7942, *Journal of Bacteriology* **172**, 643-647.

Veldhuis, M.J.W. and Kraay, G.W. (1993) Cell abundance and fluorescence of picoplankton in relation to growth irradiance and nitrogen availability in the Red Sea, *Netherlands Journal of Sea Research* **31**, 135-145.

Wang, Q., Li, H. and Post, A.F. (2000) The nitrate assimilation genes of the marine diazotrophic cyanobacterium *Trichodesmium* sp. strain WH9601, *Journal of Bacteriology* **182**, 1764-1767.

West, N.J. and Scanlan, D.J. (1999) Niche-partitioning of *Prochlorococcus* populations in a stratified water column in the eastern North Atlantic Ocean, *Applied and Environmental Microbiology* **65**, 2585-2591.

West, N.J., Schoenhuber, W.A., Fuller, N.J., Walsby, A.E., Amann, R.I., Rippka, R., Post, A.F. and Scanlan, D.J. (2001) Identification and enumeration of *Prochlorococcus* sp. in natural communities by in situ hybridization using 16S rRNA-targeted oligonucleotides and tyramide signal amplification, *Microbiology* **147**, 1731-1744.

Wolf-Vecht, A., Paldor, N. and Brenner, S. (1992) Hydrographic indications of advection/convection in the Gulf of Eilat, *Deep-Sea Research* **39**, 1393-1401.

Wu, J., Sunda, W., Boyle, E.A. and Karl, D.M. (2000) Phosphate depletion in the western North Atlantic Ocean, *Science* **289**, 759-762.

Wyman, M., Gregory, R.P.F. and Carr, N.G. (1985) Novel role for phycoerythrin in a marine cyanobacterium, *Synechococcus* strain DC2, *Science* **230**, 818-230.

Yahel, G., Post, A.F., Fabricius, K., Marie, D., Vaulot, D. and Genin, A. (1998) Phytoplankton distribution and grazing near coral reefs, *Limnology and Oceanography* **43**, 551-563.

Zeidner, G., Preston, C.M., DeLong, E.F., Massana, R., Post, A.F., Scanlan, D.J. and Béjà, O. (2003) Molecular diversity among marine picophytoplankton as revealed by *psbA* analyses, *Environmental Microbiology* **5**, 212–216.

CHAPTER 6

THE ECOPHYSIOLOGY OF THE HARMFUL CYANOBACTERIUM *MICROCYSTIS*

Features explaining its success and measures for its control

Petra M. VISSER, Bas W. IBELINGS, Luuc R. MUR & Anthony E. WALSBY

1 Introduction

1.1 HARMFUL CYANOBACTERIA

Cyanobacteria (blue-green algae) are the scourge of water management. Several species are toxic and on forming scums generate dense cell accumulations in which the toxins concentrate many fold. Reynolds (1991) has calculated that as little as 5 µL of a dense scum could contain a lethal dose of cyanotoxin. Many types of harmful cyanobacteria have been described, mostly from the freshwater phytoplankton community. We focus here on *Microcystis*, which is cosmopolitan, usually harmful (it produces toxins and surface blooms), and one of the best studied genera of the cyanobacteria. In selecting *Microcystis* for a case study, the intention is to find explanations that apply to toxic cyanobacteria in general. We first compare the main forms of cyanobacteria that inhabit lakes and reservoirs, in order to determine the common denominators.

The principal genera of planktonic cyanobacteria in lakes are listed in Table 1. There are three basic morphological forms: picoplanktonic unicells (e.g., *Synechococcus* sp.); separate filamentous forms (e.g., *Planktothrix rubescens*); and colonies, either of unicells (e.g., *Microcystis aeruginosa*) or of filaments (*Aphanizomenon flos-aquae*). Against this morphological diversity, what do these organisms have in common? All of these cyanobacteria grow by photoautotrophy, which is possible only within the euphotic zone of lakes. Most of the filamentous forms and all of the colonial forms produce gas vesicles, which confer buoyancy. In many other respects, however, they differ. Even the toxins they produce vary: they include cyclic peptides, heterocyclic compounds and lipids.

J. Huisman, H.C.P. Matthijs and P.M. Visser (eds.), Harmful Cyanobacteria, 109–142.

1.2 VARIATION IN *MICROCYSTIS* SPP.

The taxonomy of the genus *Microcystis* was originally based on the morphology of cyanobacteria present in natural waters to include coccoid, unicellular cyanobacteria that form colonies. Bourrelly (1970) restricted *Microcystis* to spp. with spherical cells, suggesting that those with ellipsoidal cells (like *M. elabens*) are put in *Aphanothece*. The genus *Microcystis* (first described by Kützing in 1833) was made synonymous with *Aphanocapsa* on the grounds that the distinguishing character that separated them (cells scarcely or densely packed) was difficult to define, but Komàrek and Anagnostidis (1986) kept the genera separate on the grounds that *Aphanocapsa* cells divide in only two successive planes whereas *Microcystis* cells appear to divide in three. *Microcystis* is distinguished from the colonial *Gomphosphaeria*, in which the cells are localised radially in a surface layer and joined by stalks radiating from the centre (Komárek and Hindák, 1988), and from various solitary unicellular forms, such as *Synechocystis*. Examples isolated into culture, however, usually lose their ability to form colonies. It is possible that the strains that grow in culture are not the principal components of the natural colonies but other minor components of the mixed population. The majority of *Microcystis* spp. listed by Geitler (1932) contains gas vacuoles (though some, e.g., *M. pallida*, do not). It is now often assumed that gas vacuoles are diagnostic; in the Pasteur Culture Collection of cyanobacteria (PCC) a putative *Microcystis* isolate that did not retain its gas vesicles in culture was re-classified as *Synechocystis* sp. (Rippka et al., 1979).

Table 1. Examples of planktonic cyanobacterial genera and species, grouped by morphology. The presence of gas vacuoles is noted with +.

Morphology *Genus or species*	***Gas vacuoles***	***Colony form***
Picoplanktic unicells		
Synechococcus sp.	–	–
Single filaments		
Aphanizomenon ovalisporum	+	–
Planktothrix (Oscillatoria)	+	–
Tychonema (O. borrellyi)	–	–
Limnothrix	+	
Colonies of unicells		
Microcystis spp.	+	pleomorphic
Coelosphaerium	+	oval
Gomphosphaerium	+	oval
Colonies of filaments		
Gloeotrichia echinulata	+	radial colony
Nodularia spumigena	+	tangle; some single
Aphanizomenon flos-aquae	+	trichomes in rafts

Microcystis belongs to the order Chroococcales and to the family *Microcystaceae* Elenkin 1933, together with *Eucapsis* spp., *Gloeocapsa* spp. and *Chondrocystis* spp. *Microcystis* is the only genus in this family without gelatinous envelopes around individual cells or small groups of cells. The coccal *Microcystis* cells always grow in colonies in natural populations. Cells are spherical and cell division occurs regularly in three planes perpendicular one to another in successive generations. Consequently, daughter cells are hemispherical but they grow into the original spherical form and size before the next division.

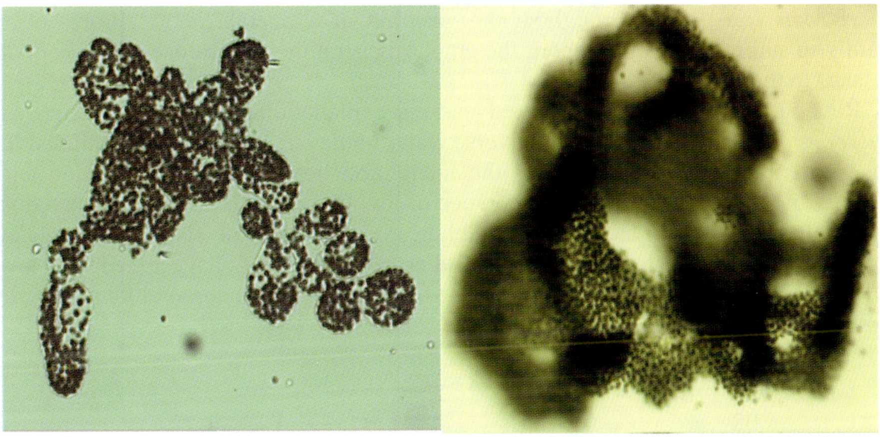

Figure 1. Two Microcystis *species:* M. wesenbergii *(left) and* M. aeruginosa *(right). Photographs by P.M. Visser.*

Natural populations of *Microcystis* spp. are often referred to as *Microcystis aeruginosa* though taxonomists distinguish many more species separated by characters such as the size and shape of the cells and the colonies (Fig. 1; see Komárek and Anagnostidis, 1986, 1999; Komárek and Hindák, 1988). The non-taxonomist is daunted by the plasticity of the salient characters and by reports of seasonal changes in cell size and colony type (Reynolds et al., 1981). The observations made in this article may refer to a rather diverse group of unicellular colonial cyanobacteria.

1.3 THE OCCURRENCE OF *MICROCYSTIS*

What is the niche that *Microcystis* occupies and what are the factors that explain its dominance in the phytoplankton community of so many lakes? Figure 2 shows an example of a surface bloom of *Microcystis*. Surface blooms of *Microcystis* have been described in water bodies ranging in size from small ponds to the largest lakes, distributed geographically from cold-temperate regions to the tropics. Examples that have been studied include the following: Lake Limmaren in Sweden (Brunberg and Blomqvist, 2003), Crose Mere in England (Reynolds, 1976); Lake Vinkeveen (Ibelings et al., 1991a, 1991b), Lake Nieuwe Meer (Visser et al., 1996a) and Lake IJsselmeer (Ibelings et al., 2003) in the Netherlands; fish ponds in Israel (Van Rijn and Shilo, 1985); a hypertrophic shallow lake in Cameroon (Kemka et al., 2003); Lake George, on the equator in Uganda (Ganf, 1974); Hartbeespoort Dam in South Africa (Robarts and Zohary, 1984); lakes in China (Xie et al., 2003; Chen et al., 2003); Lake Biwa in Japan (Tsujimura et al., 2000); the lower Great Lakes in USA (Murphy et al., 2003); Mount Bold Reservoir in South Australia (Ganf et al., 1989), and Lake Okaro in New Zealand (Walsby and Macallister, 1987). In tropical lakes, *Microcystis* can persist throughout the year but in temperate lakes blooms occur mainly in summer. In this chapter, the emphasis is on the occurrence of *Microcystis* in lakes in the temperate zone.

Reynolds et al. (1981) suggested various explanations for the occurrence of *Microcystis* water blooms in such diverse water bodies: i) high nutrient loadings; ii) raised pH and decreased CO_2; iii) the interaction between light attenuation and stability of the water column; iv) tolerance of low O_2-concentrations and low redox potential affecting the availability of sulphur, iron and other metals; v) resistance of colonies to grazing, either through size and or toxicity. Others have added vi) a low N:P ratio as a step in the pathway leading to noxious cyanobacterial blooms (Smith, 1983; Elser, 1999).

Figure 2. A surface bloom of Microcystis *(Lake Nieuwe Meer, Amsterdam, The Netherlands). Photograph by P.M. Visser.*

1.4 SCOPE OF THIS CHAPTER

In this chapter the basic features of *Microcystis* are related to the conditions under which mass developments of *Microcystis* occur. We attempt to explain the adaptations that make *Microcystis* such a conspicuous member of the phytoplankton of many lakes. We conclude by reviewing the ways in which knowledge of its ecology has been applied to the development of measures for preventing *Microcystis* blooms.

2 Features of *Microcystis*

2.1 BUOYANCY OF *MICROCYSTIS* COLONIES

During the summer, *Microcystis* often develops in the epilimnion of lakes that are subject to frequent mixing. When a lake is strongly mixed by wind, all suspended phytoplankton are stirred through the epilimnion. In eutrophic lakes, the light will be steeply attenuated by the circulating phytoplankton and their growth will become light-limited. When calm conditions return, phytoplankton left in the surface layers will receive enough light for growth but those becalmed at depth may be below the photosynthetic compensation point. Small-celled phytoplankton will be incapable of moving up the water column but buoyant *Microcystis* colonies will float up into the higher irradiance near the water surface. The buoyancy of these cyanobacteria is provided by gas vesicles. To understand the mechanisms (and costs) of buoyancy regulation it is useful first to summarise some of the properties of the gas vesicle and the way in which it is formed.

2.1.1 Gas vesicles

Gas vesicles have the form of minute hollow cylinders closed with conical end caps. The structure is rigid and waterproof. Gases dissolved in the surrounding solution diffuse freely through the wall, so that the gas inside is usually air at atmospheric pressure (Walsby, 1969). Surface tension at the hydrophobic inner surface prevents liquid water from entering. The gas vesicle wall is made entirely of protein that self-assembles to form first a biconical structure, which, when it has reached a certain diameter, then continues to grow by the extension of the cylindrical middle section. The wall of both the cylinder and the cones resembles a coil pot. The coil is formed by a small hydrophobic protein, GvpA, the main constituent. On the outside surface is another protein, GvpC, with a repeating structure that binds to GvpA and stabilises the structure (Walsby, 1994).

The rigid wall of the gas vesicle will withstand moderate pressures but at a certain critical pressure (p_c) it collapses flat: the wall is broken on collapse and cannot be reinflated. For mechanical reasons, the mean critical pressure depends mainly on the cylinder diameter, which varies in different cyanobacteria. Strains of *Planktothrix rubescens* from deep lakes have narrow gas vesicles (d = 50 nm) with a p_c of about 1.1 MPa (11 bar, equivalent to a water column about 110 m deep); in shallower lakes are found cyanobacteria with progressively wider and weaker gas vesicles; in the shallowest freshwater lakes strains of *Anabaena, Aphanizomenon* and *Gloeotrichia* are found with gas vesicles of width 85 nm and of p_c 0.55 MPa. There has evidently been natural selection for gas vesicles of increasing p_c in lakes of increasing depth. This production of narrower gas vesicles carries a penalty, however, because the wider gas vesicles have a lower buoyant density (only 120 kg m^{-3} in *Anabaena* compared with 190 kg m^{-3} in *Planktothrix*): in shallower lakes the efficient provision of buoyancy provides counter selection for the widest gas vesicles allowed by the depth requirement. The widths of the gas vesicles in different strains of cyanobacteria are determined at least in part by differences in the length and sequence of GvpC (Beard et al., 1999, 2000). These esoteric considerations are central to the understanding of buoyancy regulation and the design of devices for gas vesicle collapse used in the control of cyanobacteria.

Isolates of *Microcystis* from different lakes have gas vesicles whose mean p_c ranges from 0.6 to 0.9 MPa. One strain investigated in detail, *Microcystis* BC 8501, has gas vesicles with a diameter of 65 nm, a p_c of 0.85 MPa and a buoyant density of 158 kg m^{-3}.

2.1.2 Provision of buoyancy

Most of the principal constituents of cyanobacteria have densities (protein 1330 kg m^{-3}, carbohydrate 1600 kg m^{-3}, glycolipid 1050 kg m^{-3}) that are greater than that of water (998 kg m^{-3} at 20 °C). The overall density of the cells depends on the relative proportions of these constituents and the cell water, 70 – 80% of wet mass. The buoyant density of cyanobacteria, without gas vesicles, is usually in the region of 1060 kg m^{-3}. To reduce this density to that of water (neutral buoyancy) the cells must accumulate sufficient gas vesicles: the amount required depends on the width-dependent density of the gas vesicles: for this neutral buoyancy, the *Microcystis* cell needs to accumulate a volume of gas equivalent to 6.2% of the cell volume, which requires gas vesicle protein equivalent to 6.6% of the cell protein (4.9% of the dry mass). These are the irreducible costs of buoyancy. With the stronger, narrower gas vesicles in *Planktothrix* the costs rise to 8.2% of the cell protein, whereas with the weaker, wider *Anabaena* gas vesicles they fall to 6.2%. When *Microcystis* cells become over-buoyant in response to low irradiance they may devote as much as 10% of cell protein to gas vesicle production (Walsby, 1994).

2.1.3 Buoyancy changes in response to irradiance and nutrients

All gas-vacuolate cyanobacteria vary their buoyancy in response to irradiance: at low irradiance they become buoyant and on transfer to high irradiance they lose their buoyancy (Dinsdale and Walsby, 1972; Oliver, 1994). The proportion of buoyant cyanobacteria in natural populations has been shown to increase with depth and at night: both are explicable by the irradiance response (Walsby et al., 1983; Visser et al., 1996b; Porat et al., 2001). In laboratory cultures kept on light-dark (L:D) cycles there is a similar rise and fall in buoyancy state through the dark and light phases (Thomas and Walsby, 1986; Konopka et al., 1987b). Cultures of *P. rubescens* are on average neutrally buoyant over the L:D cycle when they receive a photon irradiance of between 7 and 11 μmol m^{-2} s^{-1} during the 12-h light phase, equivalent to a *neutral buoyancy insolation* (Q_n) of 0.3 – 0.5 mol m^{-2} over the 24-h cycle (Walsby et al., 1983; Walsby et al., 2004). In *Anabaena flos-aquae* Q_n is higher, 1.1 mol m^{-2} (Walsby and Booker, 1980), a consequence of the lower photosynthetic affinity coefficient for light (α) for this cyanobacterium, which lacks phycoerythrin. There has been no measurement of Q_n in *Microcystis* spp. but it is likely to be closer to the *Anabaena* value.

Irradiance is not the only factor that affects buoyancy; it is also affected by nutrient availability. The addition of combined nitrogen to a transparent plastic column lowered in a stratified lake caused a population of *Planktothrix* (*Oscillatoria*) *agardhii*, neutrally buoyant in the metalimnion, to become more buoyant and rise to the water surface (Klemer, 1978; Klemer et al., 1982). Cyanobacteria in culture also become more buoyant at higher nutrient concentrations. Cultures of *Microcystis aeruginosa*, which increased in buoyancy at low irradiance were less buoyant when phosphate-limited and increased in buoyancy when the limitation was relieved (Konopka et al., 1987a). These interactions of irradiance and nutrient availability can be understood with knowledge of the mechanisms of buoyancy change (see next section).

There are two main consequences of light-driven buoyancy regulation: stratification of cyanobacteria and vertical migration. They are the outcome of the same physiological response performed by cyanobacteria of different size (Kromkamp and Walsby, 1990). According to Stokes's Law, large colonies move faster through the water column than small colonies or single filaments. Filamentous cyanobacteria capable of buoyancy regulation may therefore stratify at the metalimnion of oligotrophic lakes. The water of the epilimnion must be sufficiently transparent to allow enough light for buoyancy regulation at the depths of the stable thermocline region. During the daytime the cyanobacteria lose buoyancy and slowly sink; at night they regain buoyancy and slowly float up. Only single filaments can do this: large colonies move too quickly and are lost from the stable density gradient. In eutrophic lakes insufficient light reaches the metalimnion for buoyancy regulation and filaments rising into the epilimnion become entrained in the surface mixed layer.

Vertical migration is performed by colonial cyanobacteria in the epilimnion of lakes. They also sink down during the day and float up at night, but, with their inherently greater velocity, they move over much greater depths. They may perform daily excursions to and from the lake surface, as observed with a population of *Microcystis* in Lake Okaro, New Zealand (Walsby and McAllister, 1987). It has been suggested that by performing buoyancy regulation, the colonies may benefit by avoiding prolonged exposure to potentially damaging high irradiance at the surface and perhaps also by gaining access to nutrient-rich waters below the thermocline at night (Fogg and Walsby, 1971), but a more recent study (Bormans et al., 1999) addressing this last question could not demonstrate this.

Models provide important tools for understanding the distribution of buoyant cyanobacteria. Vertical migration by *Planktothrix agardhii* has been modelled by Kromkamp and Walsby (1990). Visser et al. (1997) adapted this model to simulate migration of *Microcystis* in a quiescent water column and validated the model with the use of cultures. This *Microcystis*-model has been used in other publications with further modifications (Wallace and Hamilton, 1999; Rabouille et al., 2003).

2.1.4 Mechanisms of buoyancy regulation

Changes in the buoyant density of a cell must be brought about by changes in the relative proportions of gas vesicles, which are less dense, and the other cell components, which are denser than water. The loss of buoyancy in high irradiance may have three causes.

Gas vacuole collapse. In *Anabaena flos-aquae* the loss of buoyancy on exposure to high irradiance is correlated with an increase in cell turgor pressure sufficient to cause loss of the weaker gas vesicles present (Dinsdale and Walsby, 1972). Oliver and Walsby (1984) measured the volume of gas vesicles lost and showed that it could explain the buoyancy loss (but see below). The maximum turgor pressure is about 0.5 MPa, however, and in other cyanobacteria with stronger gas vesicles, turgor pressure alone is insufficient to collapse gas vesicles.

Regulation of gas vesicle production. In a red-coloured *Planktothrix* whose gas vesicles are too strong to be collapsed by turgor pressure, Utkilen et al. (1985) found that the relative gas vesicle content of the cells decreased at high irradiance. There was no loss of gas vesicles but the cells stopped making new gas vesicles: gas vesicles were diluted out by growth.

Carbohydrate production. The high rates of photosynthesis at (moderately) high irradiances produce an excess of carbohydrate, which is stored as polyglucose granules. This accumulates at high irradiances and counteracts the buoyancy provided by gas vesicles. In the dark, the store decreases as carbohydrate is respired and converted into less dense protein. These changes are universal and in gas-vacuolate strains this is the main cause of buoyancy loss. In *Anabaena flos-aquae* the buoyancy loss due to carbohydrate synthesis at high irradiance occurs even faster than that due to collapse of gas vesicles by turgor rise (Kinsman et al., 1991). In *Microcystis* sp., there is no down-regulation of gas vesicle production or collapse by turgor pressure at high irradiance and carbohydrate accumulation is the principal cause of buoyancy loss (Thomas and Walsby, 1985).

2.1.5 Benefits of buoyancy in lakes

The obvious benefit of buoyancy is that it lifts phototrophic cyanobacteria closer to the water surface where the higher irradiance supports a higher rate of photosynthesis. Humphries and Lyne (1988) used a mathematical model to investigate the effect of positive buoyancy on the vertical distribution of *Microcystis* colonies circulating in a sporadically mixed epilimnion, and analysed its effect on growth. They concluded that colonies that floated would receive more light and would always be more productive than those that sank but that the magnitude of the advantage depended on the ratio of the euphotic depth, z_{eu}, to the mixed depth, z_m. Humphries and Lyne (1988) described the behaviour of *Microcystis* in water columns of varying stability as "tracking" the near-surface mixed layer, though this can be misunderstood as suggesting a behavioural response rather than the outcome of passive buoyancy.

Walsby (1997) used a computer spreadsheet to calculate, from measured values of the varying surface irradiance and the vertical profiles of light attenuation and temperature, the potential photosynthesis of a cyanobacterial population at each

1-m depth through a 30-m water column and 5-min time interval over 24 h. In a population of *Aphanizomenon* followed for 9 days in the Baltic Sea, the potential photosynthesis of colonies that were mixed down during windy periods but floated up during calm periods, was twice that of a population uniformly distributed through the water column (Walsby et al., 1997). This doubling of production was contrasted with the cost of providing buoyancy, only 10% of protein production. A similar analysis has been made on a population of *Anabaena circinalis* in the turbid, slowly-flowing Darling River in Australia, in which the potential photosynthesis of the colonies that floated up exceeded five-fold that of a population evenly dispersed through the depths (Mitrovic et al., 2001).

For cyanobacteria that float up in calm periods following episodes of mixing, the advantage they obtain depends on their floating velocity. For example, in the Baltic Sea the *Aphanizomenon* population was uniformly mixed down to a depth of 16 m after a short storm, giving a mean depth for the population of 8 m: over the next few days the mean depth decreased to 4 m as the population was telescoped into a shallower surface layer. The accumulation depended on a high floating velocity of the colonies, about 20 m day^{-1} (Walsby et al., 1997). This is important not only in relation to the distance that the colonies must move but also in relation to upward movement within the surface mixed layer (see Oliver and Ganf, 2000).

The high floating velocity of cyanobacterial colonies is strongly dependent on size. According to the Stokes Equation, the floating velocity (U_s) of a sphere is proportional to the square of its radius (r): a tenfold increase in r results in a 100-fold increase in U_s. Except under very calm conditions, only cyanobacteria that form large colonies, such as *Anabaena*, *Aphanizomenon* and *Microcystis,* are able to produce surface blooms (Reynolds and Walsby, 1975; Reynolds, 1987). An example of surface bloom formation by *Planktothrix* was explained by the unusual aggregation of the separate filaments into urchin-like colonies (Walsby et al., 1983).

2.1.6 Buoyancy and surface scums

Consequences of being colonial. Microcystis species grow as colonies in their natural habitats. Colony formation obviously increases the size of these cyanobacteria. This has some important consequences for their ecology. First, the larger size (through a reduction in the ratio of surface area to volume) apparently has negative effects on nutrient uptake, on light harvesting, on photosynthesis, and on overall growth rate (Reynolds, 1997). In fact, Reynolds (1997) ranks *Microcystis* amongst the poorest resource competitors and slowest-growing species in the phytoplankton. This is probably due not only to the colony size, since they are also bad competitors in single cell cultures (Huisman et al., 1999). Second, larger size may offer protection against grazing, one of the main loss factors in phytoplankton populations. And third, large size enables *Microcystis* to translate buoyancy into rapid upward movement in the absence of mixing. This rapid flotation may even result in the formation of surface blooms with potentially enhanced access to light and CO_2 (Paerl and Ustach, 1982), but also with the risk of photodamage by the exposure to full sunlight (Ibelings, 1996). Is surface bloom formation a strategy of buoyant species aimed at intercepting light and CO_2 at the lake surface, as suggested by Paerl and Ustach (1982), or should scum formation be considered as a fortuitous consequence of being buoyant, the price *Microcystis* has to pay for the benefits of floating into the surface mixed layer?

The phenomenon of noxious surface blooms of decaying cyanobacteria has suggested there are disadvantages in buoyancy and has prompted explanations that they represented failure of the buoyancy regulating mechanism that might prevent their occurrence (Reynolds and Walsby, 1975). There are, however, mitigating circumstances near the water surface: the high irradiance will support a higher photosynthetic rate and the supply of CO_2 for the process is ultimately dependent for its replenishment on diffusion from the overlying atmosphere (Walsby, 1970; Ibelings and Maberly, 1998).

Scum thickness: flotation and skimming. The total population of cyanobacteria integrated through the water column of a lake does not usually much exceed an areal cell biovolume concentration of 200 cm^3 m^{-3}. If all of the population floats to the water surface it will form a continuous layer of cells only 0.2 mm thick (equivalent to 40 cells deep): even at this areal concentration there is severe self-shading. In practice, the surface layer may appear thicker because the cells are embedded in mucilage whose volume may be 5-fold greater (giving a 1-mm layer) and interstitial spaces must occur between even closely packed colonies, but by growth and flotation alone, the layer thickness should not much exceed 2 mm. See Figure 2 for an example of such a surface scum.

What explains the formation of scums several centimetres thick is an additional lateral concentration by wind-induced movements in the surface layer. Paradoxically, the highest surface current velocities (U_s) velocities occur at moderate wind speeds (U_a). Within the wind speed range of 0.5 to 5.0 m s^{-1}, the surface current velocity is given by the empirically determined relationship: $U_s = 0.03U_a - 0.005U_a^2$ (George and Edwards, 1976); the maximum U_s of 0.045 m s^{-1} (equivalent to 3.9 km day^{-1}) is given by a wind speed of 3 m s^{-1} (Oliver and Ganf, 2000). A skimming of colonies at the surface can cause the substantial build up of thick cyanobacterial scums along leeward shores. Scums found at the lee shores of lakes can be considered as an important part of the production in a lake and in very large lakes with a high state of eutrophication they may constitute a huge biomass. Extreme examples are the hyperscums formed in the South African Hartbeespoort Dam Reservoir (Zohary and Robarts, 1989) where scums of *Microcystis* persist over a period of ten months of the year. In moderate climate zones, blooms and scums are found only in summer time (but see Howard, 2001).

2.1.7 Conditions in scums

Cyanobacteria like *Microcystis* can take up HCO_3^- to provide CO_2 for photosynthesis and they are thought to possess particularly efficient carbon concentrating mechanisms (CCM) that elevate the concentration of CO_2 around the photosynthetic carboxylating enzyme Rubisco (Talling, 1976; Kaplan and Reinhold, 1999). In the absence of a CCM, Rubisco would mainly work as an oxygenase resulting in photorespiration – one of the processes that result in the uptake of oxygen in the presence of light. The buoyancy of *Microcystis* may help in assuring its access to carbon. A surface bloom positions cells close to re-supply of CO_2 from the air. Yet, because of the high biomass in a scum, ensuring strong local demand for CO_2, carbon may actually become depleted in scums (Ibelings and Maberly, 1998). Inorganic carbon is present in water as dissolved CO_2, HCO_3^-, and CO_3^{2-}, in proportions controlled by the master variable, pH. When the demand for carbon outstrips the re-supply, pH will rise shifting the equilibrium away from CO_2 towards HCO_3^-, and even CO_3^{2-}. Ibelings and Maberly (1998) studied scums of floating cyanobacteria in water with high and low alkalinity and a varying CO_2 concentration (between 0 and 3500 ppm) in the headspace above the bloom. Recovery of cells from exposure to high irradiance at noon was impeded when CO_2 was absent in the headspace, indicating that the combinations of limiting inorganic carbon and high irradiance overwhelmed the photoprotective mechanisms of the cyanobacteria in the bloom. Earlier, Whitelam and Codd (1983) also demonstrated experimentally that photoinhibition of photosynthesis in *Microcystis* is augmented under CO_2 depletion.

The main consequence for *Microcystis* colonies of transition from mixed conditions to static conditions at the lake surface is the exposure to full sunlight. In addition, the temperature may increase by several degrees inside a scum. Remote sensing from NOAA satellites shows that temperatures of the surface water in areas of scums may be up to 5°C higher than areas outside the scum (Ibelings et al., 2003). Ibelings (1996) showed that this combination of high irradiance and high temperature is more damaging to the cells in a scum than that of either factor on its own. The higher temperature explained why *Microcystis* cells in a scum do not recover from photoinhibition when irradiance decreased later in the afternoon (Ibelings, 1996). In the absence of such stresses, cells should recover from photoinhibition in a matter of minutes.

The combined effects of extreme irradiance and temperature suggest scum formation constitutes a potential source of loss for buoyant species. Loss factors are especially costly for slow growing species like *Microcystis*. The fate of *Microcystis* cells in a scum will be highly dependent on its recent (light) history. If the cells prior to bloom formation were already acclimated to high irradiance, they may survive. Cells acclimated to low irradiance, however, lack the necessary level of photoprotection (see next section). Zohary and Robarts (1990) found long-term survival of *Microcystis* in the extreme hyperscums they studied. In these systems the upper crust of the scum protected cells in deeper layers against photooxidation and desiccation.

2.2 PHOTOACCLIMATION AND PHOTOINHIBITION

Photoacclimation is the phenotypic adjustment to changes in the availability of light, most notably the up- or down-regulation of cellular pigment content, but also in the components of the electron transport or enzymes of the Calvin cycle (Falkowski and Laroche, 1991; MacIntyre et al., 2002). The role of photoacclimation is not just to maximize the rate of photosynthetic carbon assimilation, but also to protect the cells against damage by an excess of energy. Photoacclimation is not complete until conditions of balanced growth have been established, i.e., the specific rate of change of all measures of cell mass are equal

(MacIntyre et al., 2002). Some cyanobacteria for instance may be grown at full sunlight, but only when acclimated gradually to these extreme light conditions.

For buoyant, bloom-forming phytoplankton species like *Microcystis* the time spend at or near the lake surface determines the risk of photoinhibition. Cells will float to the surface only if they become disentrained from turbulent mixing, i.e. only at low wind speeds (generally below 3 m s^{-1}). Photoinhibition of photosynthesis will occur if cells are exposed to an irradiance that is (much) higher than the level to which they were acclimated (MacIntyre et al., 2002). Light stress results from irradiance in excess of that which can be used directly in photosynthesis (Powles, 1984). Under these conditions safe dissipation of the excess of excitation energy is required to protect the photosystems from long-term damage (Niyogi, 2000). Surface bloom formation by buoyant cyanobacteria is perhaps one of the best natural examples of a process bringing about the abrupt increase in irradiance that may cause severe photoinhibition, even in nutrient replete or otherwise unstressed cells. The effect is all the more damaging since surface bloom formation is more likely after a period of deep mixing / low average irradiance. Cyanobacteria synthesise new gas vesicles at low irradiances. Once mixing subsides the resulting over-buoyant colonies can no longer loose buoyancy and they become trapped at the surface (Reynolds and Walsby, 1975). Photoinhibition sets in quickly and there is then a real risk of photoinhibition causing long lasting damage.

Photosynthesis in cyanobacteria and higher plants is carried out by two sequentially placed photosystems, PS2 and PS1, linked by a chain of redox carriers. Light energy is captured by the antennae of both photosystems and transferred to the reaction centres. The central target of photoinhibition is the D1 protein of PS2. Safety valves include nonphotochemical mechanisms for quenching of excited chlorophyll, as well as alternative electron acceptors like oxygen (Niyogi, 2000), and changes in pigment composition. With light saturation of growth rate, the proportion of photosynthetic pigments decreases and the proportion of photoprotective pigments increases; the capacity for energy dissipation is thereby greatly enhanced (MacIntyre et al., 2002). In eukaryotic algae excess irradiance induces the conversion of the xanthophyll violaxanthin (a carotenoid), ultimately to zeaxanthin (Demmig-Adams and Adams, 1996). This reaction is reversed under low irradiance. The conversion to zeaxanthin is accompanied by a lower pH within the lumen of the photosynthetic membranes and an increase in non-photochemical quenching of fluorescence, indicative of the harmless dissipation of excess energy as heat. Low pH is required not only for the conversion of violaxanthin to zeaxanthin but also for the xanthophyll-cycle dependent energy dissipation itself. Cyanobacteria lack a xanthophyll cycle; they can produce zeaxanthin but they do so much more slowly. Other carotenoids typically found in cyanobacteria are myxoxanthophyll, echinenone and beta-carotene. Like zeaxanthin, these carotenoids are mainly involved in photoprotection: in cyanobacteria the carotene is present in the reaction centres and zeaxanthin is located in the cytoplasmic membranes (MacIntyre et al., 2002).

Ibelings et al. (1994) directly compared acclimation of *Microcystis* and the (non-buoyant) green alga *Scenedesmus protuberans* to high and fluctuating irradiances. The light regimes mimicked those received in lakes with various speeds and depths of mixing. *Microcystis* was more sensitive to photoinhibition than its green algal competitor *Scenedesmus*. At first sight this may seem surprising. Because of its buoyant nature and its migration behaviour, colonies of *Microcystis* will have a greater chance of being exposed to high irradiance at or near the lake surface. If *Microcystis* has evolved mechanisms to encounter high irradiances, then why the greater sensitivity? It has been argued that photoinhibition is actually part of the normal physiology of cells that are exposed to high irradiance. Photoinhibition that occurs without delay would protect cells from more damaging effects; it provides a mechanism for the long-term protection of photosystem 2 (Oquist et al., 1992). If protection is too slow or incomplete, cells that are caught in a surface bloom for longer periods risk photooxydative death (Zohary and Pais-Madeira, 1990; Abeliovich and Shilo, 1972).

Excess excitation energy is dissipated harmlessly as heat, thereby protecting the photosystem. The fact that photoinhibition occurred more readily in *Microcystis* may in fact be an adaptation to life as a buoyant species. Ibelings et al. (1994) argued that the constitutive presence of zeaxanthin was instrumental in the prompt quenching. Likewise Demmig-Adams et al. (1990) noted that cyanobacterial lichens containing zeaxanthin readily increased non-photochemical quenching of fluorescence. In contrast, Campbell et al. (1998) put forward the suggestion that the redistribution of excitation energy between PS2 and PS1 – so called state transitions - are the more important regulatory mechanism in cyanobacteria exposed to high irradiance.

Photoinhibition that is readily provoked may protect PS2. Nevertheless, Ibelings et al. (1994) found that *Microcystis* apparently depressed its rate of photosynthesis at times when the green alga (*Scenedesmus*) maintained uninhibited rates of photosynthesis. Campbell et al. (1998) argued that non-photochemical quenching of fluorescence in cyanobacteria does not necessarily cause a lower overall photochemical efficiency, since energy may be redirected towards PS1. It appears, however, that *Microcystis* is not as well adapted to fluctuating light as its eukaryotic competitors and is unable to benefit optimally from the saturating irradiance levels that are temporarily available when mixing takes cells to the upper layers of the watercolumn. Others have also found that mixing not only prevents surface bloom formation but also arrests growth of the bloom forming species (e.g. Reynolds et al., 1983). This re-emphasises the dependence of *Microcystis* on its buoyancy in lakes with a partially stable water column. Buoyancy enables the colonies to maintain themselves in a shallow near-surface mixed layer where light fluctuations are reduced and where irradiance levels are constantly (relatively) high, so that the continuous presence of zeaxanthin would be beneficial to the cells.

2.3 TOXINS

The hepatotoxic microcystins (MC) were first characterised in *Microcystis* (Botes et al., 1984) and named after this genus, but they were later also found in other cyanobacteria. *Microcystis* produces not only microcystins but also a variety of other cyclic or linear bioactive oligopeptides (Namikoshi and Rinehart, 1996). Some are known peptides (Erhard et al., 1999; Clemens et al., 1995), but others remain unidentified. The oligopeptide composition of different species was investigated in single colonies isolated from a natural bloom and analysed directly with matrix-assisted laser desorption ionization-time of flight mass spectrometry (MALDI-TOF MS) by Fastner et al. (2001). Strains of *M. wesenbergii, M. ichthyoblabe* and also *M. novacekii* seldomly produce microcystins (Watanabe et al., 1988; Neilan et al., 1997; Otsuka et al., 2000).

Detailed information on microcystins and other aspects of toxicity in cyanobacteria are reviewed in other chapters of this book (see Chapters 1-3).

2.4 RESISTANCE TO GRAZING

2.4.1 Grazing by zooplankton

Is *Microcystis* too large to be eaten by zooplankton? Recently, Ghadouani et al. (2003) concluded that zooplankton communities are negatively affected by cyanobacterial blooms and therefore the potential to reduce such blooms through herbivorous grazing is only limited. Colonial cyanobacteria are deemed unsuitable food for zooplankton for three reasons: i) mechanical interference; ii) inadequate chemical composition that reduces growth and reproduction of zooplankton; iii) the production of cyanobacterial toxins. There are insufficient field-based, experimental studies to allow firm conclusions to be drawn (Ghadouani et al., 2003). The uptake of filamentous and colonial cyanobacteria by filter-feeding zooplankton is still a matter of discussion (Gulati et al., 2001). Evidence has been presented for positive, negative and no relationships between cyanobacterial abundance and zooplankton density (Paterson et al., 2002); the negative effects of cyanobacteria on zooplankton growth have been discussed by Paerl et al. (2001). The fact that microcystin is found in zooplankton (e.g. Thostrup and Christoffersen, 1999; Ferrao et al., 2002) during *Microcystis* blooms clearly indicates that cyanobacteria are taken up from the seston: without ingestion and assimilation of toxic cells, microcystins would not be found in *Daphnia* and other zooplankters. Grazing by zooplankton must therefore be a loss factor for *Microcystis*. Lampert (1987) concluded that large *Microcystis* colonies cannot be handled by zooplankton, but neither do these colonies seriously interfere with the filtration. Smaller colonies, however, (analogous to filamentous cyanobacteria) may severely reduce filtration rates through mechanical interference. The correlation between the clearance rate of *Daphnia galeata* and total phytoplankton biomass in Bautzen Reservoir was negative only if biomass of *Microcystis aeruginosa* was excluded. This suggested that *Microcystis aeruginosa* was the main grazing-resistant phytoplankton species in the reservoir (Boing et al., 1998).

Cyanobacterial toxins have been shown to have adverse effects on filtration, growth, reproduction and survival of zooplankton (e.g. Thostrup and Christoffersen, 1999; Rohrlack et al., 1999a, 1999b). *Daphnia* populations present during a bloom of toxic *Microcystis* may decrease in numbers and in individual body growth. Through these effects, blooms of *Microcystis* may affect the overall functioning of a lake ecosystem, since *Daphnia* is a key organism in the foodweb. The microcystin-producer *Microcystis* PCC 7806 was found to be poisonous to *Daphnia galeata*, whereas the non-producing mutant did not have lethal effects. Both variants of PCC7806, however, were able to reduce the *Daphnia* ingestion rate. Rohrlack et al. (1999c) conclude from this that while microystins may poison *Daphnia* (cf. Jungmann and Benndorf, 1994), these toxins are not responsible for inhibiting the ingestion process. It is important to distinguish in these experiments the three separate ways that *Microcystis* may affect zooplankton growth: through colony size interfering with filtration, through unidentified substances inhibiting feeding, and through direct toxic effects of microcystins. The possibility of correlations between the separate factors should be investigated, e.g., the occurrence of microcystin genes might be correlated with colony size: the *Microcystis* colonies of largest size (>100 μm) can have the highest proportion of microcystin-producing genotypes, the lowest proportion cells that lack microcystin, and the highest microcystin cell quotas (Kurmayer et al., 2003).

Larger cladocerans are often negatively correlated with blooms of toxic cyanobacteria, and hence these blooms may indirectly promote the smaller cladocerans, rotifers and copepods. Filter feeding and a capacity to ingest relatively large phytoplankton particles may increase the exposure of large *Daphnia* species to blooms of toxic *Microcystis,* which in addition to potential toxic effects provide little nutritional value. Yet it is not the colonial form per se that makes *Daphnia* more vulnerable than smaller species of zooplankton, which tend to leave large colonial *Microcystis* untouched. Fulton and Paerl (1987) found that unicellular strains of *Microcystis* inhibited the clearance rate of *Daphnia* on other algae more than colonial *Microcystis*. Rohrlack et al. (1999b), on the other hand, demonstrated the role of the mucilaginous envelope of colonial *Microcystis* strains in the inhibition of ingestion by daphnids. Results by DeMott (1999) further complicate the picture. Toxic strains of *Microcystis* inhibited feeding by *Daphnia* species, but *D. magna*, in which the inhibition was strongest after one hour incubation, subsequently recovered after 24 h incubation. These results seem consistent with a foraging strategy that needs to balance the benefits of a reduced microcystin intake with the disadvantage of reduced energy intake in a non-selective filter feeder. Epp (1996) does not agree with the view that *Daphnia* is non-selective. In his experiments *Daphnia pulicaria* was able selectively to avoid (filamentous) cyanobacteria while successfully feeding on other species. The sensitivity of zooplankton to cyanobacterial toxins varies widely. Selection of zooplankton strains with lower susceptibility to cyanobacterial toxins may increase in populations over several decades of eutrophication (Hairston et al., 1999). Hence, over time, *Microcystis* would have to rely more and more on morphological, rather than chemical, deterrents of zooplankton grazing. Rohrlack et al. (2001), however, conclude that differences in survival among *Daphnia* clones were due to variations in microcystin intake rather than due to differences in susceptibility to the toxins. It is not only microcystins that harm zooplankton: cyanobacteria contain many other bioactive peptides that might be toxic in some way. One example is the peptide microviridin J, which has been found in *Microcystis* and appeared to cause a lethal molting disruption in *Daphnia* spp., upon ingestion of living cyanobacterial cells (Rohrlack et al., 2003).

Overall, we can paraphrase Lampert (1987) in writing that there is no general answer to the question whether zooplankton is affected by cyanobacteria (or vice versa). Although in most cases cyanobacteria have an adverse effect on growth and reproduction of zooplankton, there are no universal patterns. The effects seem to depend on the strains of zooplankton and cyanobacteria present. Nevertheless, it seems that grazing by zooplankton affects *Microcystis* less than it affects most other phytoplankton species (since *Microcystis* is protected by size and/or toxin production) although grazing losses for *Microcystis* by herbivorous zooplankton cannot be neglected.

2.4.2 Grazing by zebra mussels

Filter-feeders, like the zebra mussel *Dreissena polymorpha*, may have a similar or even larger effect on *Microcystis* populations than zooplankton. In Lake IJsselmeer (The Netherlands), at places where their densities are high enough, zebra mussels keep the water clear of cyanobacteria. As a consequence, cyanobacteria and *Dreissena* occupy different areas of the lake (see Vos et al., 2003; Pires and van Donk, 2002; Ibelings et al., 2003). Grazing experiments with seston from the IJsselmeer showed that *Dreissena* was capable of removing particles in a size range that included colony-forming cyanobacteria – different size classes were cleared at equal rates (Pires et al., unpublished manuscript) – and cyanobacteria were not discriminated against. The dominant *Daphnia* species in the IJsselmeer (*D. galeata*), on the other hand, showed an optimum for clearing particles with a diameter between 1 and 20 μm, which includes only some of the smaller *Microcystis* colonies. Interestingly, the amount of microcystin in the mussels in the IJsselmeer is an order of magnitude lower than that in the zooplankton (Ibelings et al., 2004).

Vanderploeg et al. (2001) state that *Dreissena* has reversed progress made by nutrient control measures. Whereas in Lake IJsselmeer *Dreissena* is believed to moderate the negative effects of eutrophication (and may even stimulate a shift from *Microcystis* to green algae), in Lake Erie and Lake Huron (USA), zebra mussels are believed to promote succession in the opposite direction, resulting in *Microcystis* blooms. Even if *Dreissena* was not instrumental in the return of *Microcystis* to Lake Erie, at least it can be said that zebra mussels were incapable of preventing the blooms of *Microcystis* (Vanderploeg et al., 2001). Of course, during periods of water column stability mussels refilter water from a benthic boundary layer (Ackerman et al., 2001), whilst floating cyanobacteria like *Microcystis* concentrate in a near-surface mixed layer, escaping grazing by the mussels.

Vanderploeg et al. (2001) put forward the idea that selective filtration by the mussels has promoted development of *Microcystis* in Lake Erie and Lake Huron. *Dreissena* is capable of sorting particles on the pallial organs (Baker et al., 1998).

Viable *Microcystis* was rejected via the pseudofaeces of the mussel, i.e. the (toxic) colonies did not enter the digestive tract (Vanderploeg et al., 2001). A similar result was obtained for a unicellular strain isolated from Lake Erie.

Size alone seems to be an unlikely criterion for rejection since Vanderploeg et al. (2001) found that colonies up to a diameter of 153 μm did not present a great problem. Ten Winkel and Davids (1982) observed that cyanobacteria even as large as 750 μm could be ingested. Pires and van Donk (2002) showed that the presence of toxic *Microcystis* cells inhibited clearance rates of the green alga *Chlamydomonas*. Non-toxic *Microcystis* cells had no such effect. Toxic *Microcystis* enhanced the production of pseudofaeces, containing viable cells. Perhaps contrary to what may be expected, the pseudofaeces mainly contained *Chlamydomonas* cells, whilst toxic *Microcystis* cells were assimilated. *Chlamydomonas* has a thick cell wall, which makes it hard to digest. Baker et al. (1998) also found that *Microcystis* was ingested preferentially over green algae and diatoms.

In most experiments in which toxic strains are assimilated, the mussels did not seem to suffer from exposure to microcystin, with the exception of a reduction in clearance rate (Pires et al., 2003). Yet Vanderploeg et al. (2001) suggested that a reduction in filtration activity of natural *Dreissena* populations (54% of the time instead of more than 90%) may be related to chronic exposure to toxic *Microcystis*. The larvae of *Dreissena* seem highly sensitive to microcystin, even during short-term exposure (Pires et al., 2003). Survival of larvae fed on toxic *Microcystis* was lower than for larvae feeding on non-toxic *Microcystis* (but higher than that of starved larvae).

2.5 THE LIFE CYCLE OF *MICROCYSTIS*

2.5.1 Summer and autumn

Microcystis may minimise its population losses in two ways: gas vesicles prevent sedimentation losses and colony formation may reduce grazing losses. And yet throughout the growing season cells are lost, by photooxidation in scums, by stranding on lee shores or in reed beds, and by flushing from lakes. Nevertheless in many temperate lakes large populations persist throughout the summer, only to disappear in autumn. Small numbers of colonies may survive the winter and form the inoculum for the next season's growth but there is also evidence that colonies overwinter on the sediments (Reynolds et al., 1981; Takamura et al., 1984; Oliver et al., 1985; Thomas and Walsby, 1986; Tsujimura et al., 2000). There are several explanations for what causes *Microcystis* colonies to sink out of suspension in the autumn.

The loss of buoyancy in autumn has been related to events accompanying autumnal overturn, such as an increase in the ballast of storage products in cyanobacteria (Reynolds and Rogers, 1976; Reynolds et al., 1981). Experiments with a laboratory strain of *Microcystis* showed that whilst cells kept at 20 °C regained buoyancy at night by metabolising the dense stores of carbohydrate accumulated during the day, those kept at 8 °C failed to metabolise the carbohydrate and remained negatively buoyant (Thomas and Walsby, 1986). Such changes in carbohydrate levels suggest differences in relative rates of respiration and photosynthesis at lower temperatures. From a study in laboratory cultures at lower temperatures, Visser et al. (1995) concluded that the accumulation of carbohydrate at reduced temperature was the result of a lowered rate of protein synthesis during the light period. Although the photosynthetic rate itself decreased at reduced temperature, the decreased incorporation of carbon into protein resulted in more of the fixed CO_2 being stored as carbohydrate, thus increasing the ballast of the cells.

Another explanation for the autumnal sedimentation is co-precipitation with suspended particles. In Blelham Tarn, English Lake District, *Microcystis* colonies were confined to the aerobic epilimnion throughout the summer and were absent from the hypolimnion, which became anoxic from late May to early October. At holomixis in mid-October, reduced iron compounds, which were freely soluble in the anaerobic water of the hypolimnion, became oxidised when mixed with the aerated water from the epilimnion, and formed a yellow precipitate, which adhered to the *Microcystis* colonies causing them to become denser than water and to sink (Oliver et al., 1985). Recent findings (Verspagen et al., 2004) in the rather shallow Lake Volkerak, The Netherlands, showed a similar co-precipitation of buoyant *Microcystis* colonies with clay particles that caused sedimentation of the colonies. In this case, sedimentation took place at all times of the year.

2.5.2 Winter and spring

Some of the *Microcystis* colonies from the summer blooms overwinter on the lake sediment, not in special akinetes but as vegetative cells (Brunberg and Böstrom, 1992; Fallon and Brock, 1981; Reynolds et al., 1981; Takamura et al., 1984). In spring, these benthic colonies may reinvade the water column (Preston et al., 1980). Various conditions responsible for the recruitment have been suggested. Low oxygen concentrations, resulting from the onset of stratification, may enhance the recruitment of cyanobacteria (Càceres and Reynolds, 1984; Reynolds et al., 1981; Trimbee and Harris, 1984; Trimbee and Prepas, 1988). Oliver et al. (1985) suggested that some colonies regain buoyancy under anaerobic conditions, when the ferric iron compounds that weigh them down are reduced to soluble ferrous compounds. Càceres and Reynolds (1984) concluded from their experiments that an increase in water temperature may be critical for the renewed activity of *Microcystis,* but anoxic conditions and light may also play a role. Penetration of relatively high irradiance levels to the sediment seems necessary for this process, since gas vesicle formation in the dark occurs only after earlier accumulation of energy reserves (Deacon and Walsby, 1990). Reynolds and Bellinger (1992) showed on the basis of an 18-year data set that a population of *Microcystis* in the water column could be established more quickly when the water was clear after the onset of stratification, allowing light to penetrate to the lake bottom.

An alternative is that recruitment may not be triggered by specific conditions, but may be a more or less continuous process during the winter and spring as result of the ongoing decrease of the carbohydrate content in the colonies. In autumn, colonies that sink down may initially have a high carbohydrate content but will metabolise the carbohydrate in the very low irradiance at or near the sediment. At a certain time in winter or spring, dependent on the initial carbohydrate concentration, oxygen conditions and temperature, buoyancy will be regained if the gas vesicles have remained intact. When the temperature is high enough to sustain regular protein synthesis, *Microcystis* colonies will be able to stay in the epilimnion, grow and re-establish a planktonic population.

Buoyant colonies buried in the sediment, however, may find it hard to free themselves from the lake bottom: numerous buoyant colonies can sometimes be found in the sediment of lakes. In Lake Volkerak, recruitment of *Microcystis* from the sediment was probably not the result of an active buoyancy change in the benthic *Microcystis* population, but rather resulted from resuspension after wind-induced mixing and/or bioturbation (Verspagen et al., 2004). This happens to a larger extent in shallower areas of the lake, which consequently will contribute more of the *Microcystis* recruited (Brunberg and Blomqvist, 2003).

3 Competition

3.1 *MICROCYSTIS* VERSUS ALGAE

The characteristics of cyanobacteria that give them their competitive strength over algae are summarized and discussed in Chapter 3 (Kardinaal and Visser) and Chapter 7 (Huisman and Hulot) of this book. In this paragraph we will try to sort out under which conditions *Microcystis* has the greatest competitive strength.

Cyanobacteria are known to be strong competitors for light. However, competition experiments in light-limited continuous cultures revealed that *Microcystis* is not a particularly strong competitor for light (Huisman et al., 1999). It loses the competition when grown in mixed cultures with the green alga *Chlorella,* and also when grown in mixed cultures with the cyanobacterium *Aphanizomenon.*

Microcystis has been considered to be a K-strategist (Reynolds, 1984) because it has a low maximum growth rate but succeeds when the losses by grazing and sedimentation are low. K-strategists, however, often win the competition by their high affinity for some resource. This does not seem to be the case for *Microcystis*. Rather, its relatively low affinity for light is compensated by its buoyancy, which enables it to increase its daily light dose. In a stable water column, buoyancy thereby provides *Microcystis* with a direct competitive advantage.

3.2 *MICROCYSTIS* VERSUS *PLANKTOTHRIX*

In the temporate climate zones *Microcystis* and *Planktothrix* are arguably among the most abundant harmful cyanobacteria in freshwater systems. *Planktothrix* can be found in two distinctly different niches: (1) stratifying on the thermocline (*P. rubescens*) and (2) homogenously suspended in the epilimnion (mainly *P. agardhii*). We discuss here the principal niche differences between the epilimnetic species of *Planktothrix* and *Microcystis*. Although *Microcystis* and *Planktothrix* are usually found in different lakes, they sometimes coexist or alternate in the same waterbodies, e.g. in Lake IJsselmeer, *Microcystis* sp. and *Planktothrix agardhii* dominate in different years.

By floating up, *Microcystis* colonies obtain a higher daily light dose (provided the water column is sufficiently stable). Moreover, if they subsequently accumulate sufficient carbohydrate ballast, *Microcystis* will migrate rapidly down the water column and avoid photooxidation early in the day. *Planktothrix agardhii* on the other hand, being filamentous, can only regulate its vertical migration in the stable metalimnion (Section 2.1.3). Once entrained in the epilimnion the light it receives is determined by circulation through the surface layers. It is more sensitive to light than *Microcystis* and suffers photoinhibition and photodamage at lower irradiances (Eloff et al., 1976; Van Liere and Mur, 1980; Paerl et al., 1983). If the light is steeply attenuated, however, *Planktothrix* will avoid prolonged exposure to high irradiance near the surface. The steepest light gradients are commonly found in shallow waters with dense populations of phytoplankton.

Planktothrix populations occur most commonly in waters less than 4 m deep. In more northern climate zones with their lower maximum irradiance, *Planktothrix* is also found in deeper lakes. Very shallow lakes usually do not have the right conditions for *Microcystis*, which prefers somewhat deeper, diurnally or seasonally stratifying lakes where it can profit most from its buoyancy. In lakes with a depth of 4-6 m, like Lake IJsselmeer, the two niches overlap and interesting patterns of competition can be found. Differences in the affinity for light between the competitors become more important. *Planktothrix* has a higher affinity for light than *Microcystis* and if the phytoplankton biomass is sufficiently concentrated, it creates the 'shade' conditions in which *Planktothrix agardhii* flourishes (Scheffer et al., 1997); *Microcystis* then loses the competition. At lower biomass concentrations, however, the growth of *Planktothrix* is adversely affected by photoinhibition and under such conditions *Microcystis* may win the competition.

4 Control of *Microcystis* in lakes and reservoirs

We summarise here some of the methods developed to control cyanobacterial blooms. Many options are available to the lake manager: "bottom up" control refers to limiting the availability of nutrients or light; "top down" refers to foodweb control, encouraging predation by animals. The latter option, a form of biomanipulation, is often used as an adjunct to phosphorus reduction and may help to shift eutrophic lakes from the turbid to the clear state (Perrow et al., 1997; Meijer et al., 1999). Biomanipulation has had some success with *Microcystis* (e.g., Anadotter et al., 1999), although small blooms of *Microcystis* may persist in the clear state (Korner, 2001) or once the more-or-less permanent blooms of *Planktothrix agardhii* have disappeared (Meijer et al., 1999). Ghadouani et al. (2003) have questioned the usefulness of herbivory to control phytoplankton in lakes dominated by *Microcystis*.

Lake flushing has been successfully used to reduce the amount of cyanobacteria, and especially filamentous species, in lakes. Davis et al. (2003) compared the observed changes in the *Planktothrix* population in Blelham Tarn (English Lake District) with the changes calculated from growth rate and found good agreement if allowance was made for the proportion lost from the surface mixed layer by flushing, which was calculated from the rainfall and catchment area around the lake. Walsby et al. (1989) suggested that one explanation for the absence of faster-floating colonial cyanobacteria in Lake Rotongaio (New Zealand) was that a proportion of the cyanobacteria in the surface layer was lost in the outflow into Lake Taupo each day and that the losses would have been greater for colonies concentrated in the surface film than for the small *Anabaena minutissima* filaments that dominated the phytoplankton. Because of their relatively low growth rates *Microcystis* blooms may often be correlated with reduced flushing rates (e.g. Jacoby et al., 2000). In the remainder of this section on control of *Microcystis* we focus on those measures that are specific for control of bloom-forming cyanobacteria like *Microcystis* by targeting their special adaptations.

4.1 REMOVAL BY PRESSURE DEVICES

4.1.1 Loss and recovery of gas vesicles in darkness

Without gas vesicles, colonial cyanobacteria like *Microcystis* not only lack an advantage over other phytoplankton, they are at a considerable disadvantage. Their large colony size, which confers the necessary high floating velocity when gas vesicles provide buoyancy, confers a high sinking velocity when the gas vesicles are collapsed. The colonies are then rapidly lost to the hypolimnion or the sediment. This potentially offers a method of controlling these organisms. We consider briefly the consequences of sedimentation, the pressures that must be applied, and some methods of destroying gas vesicles.

An organism that sediments to a depth exceeding the compensation depth will not easily re-enter the euphotic zone, unless it is resuspended by strong turbulent mixing, or regains its buoyancy by making new gas vesicles. Kromkamp et al. (1989) found that *Microcystis* cells required light to support *de novo* gas vesicle synthesis (e.g., when phosphate inhibition was relieved); Deacon and Walsby (1990) found that after all existing gas vesicles had been collapsed by pressure, there was some gas vesicle production in the dark but that this occurred only if the cells had been pre-incubated at high irradiance and thereby increased their energy reserves. Even then, insufficient new gas vesicles were made to render the cells buoyant. The general conclusion, therefore, is that if, after gas vesicle collapse, cells sink into the aphotic zone they will not recover buoyancy and will not easily rejoin the population in the epilimnion.

In engineering a device for collapsing gas vesicles, the cost will rise with the pressure that must be applied and it is therefore necessary to determine the minimum pressure required. The mean critical pressure of gas vesicles in *Microcystis* varies from 0.6 to 0.9 Mpa, but in each cell the critical collapse pressure (p_c) of individual gas vesicles varies and that of the strongest may be as much as 0.2 MPa above the mean value. The maximum p_c may therefore be in the range of 0.8 to 1.1 MPa. From this may be subtracted the cell turgor pressure, which usually exceeds 0.2 MPa. To collapse all of the gas vesicles in turgid cells will therefore require application of 0.6 to 0.9 MPa, depending on the strain.

To minimise the pressure costs, it might be argued that it is not necessary to collapse all of the gas vesicles: usually, not more than 50% need to be collapsed to make cells sink, requiring application of the median critical pressure, which is close to the mean p_c (Walsby and Bleything, 1988). The danger, however, is that some cells will be left close to neutral buoyancy and with the production of few more gas vesicles will become positively buoyant again.

The recommendation, therefore, is to design for the maximum required, 0.9 MPa for *Microcystis* spp.

4.1.2 Devices for collapsing gas vesicles

A number of different devices have been trialled for the lake-scale collapse of gas vesicles in cyanobacteria. There are considerations of efficacy and cost in the construction and operation of the devices, which are made in more detail by Walsby (1992).

Ultrasonic transducer. D.A. Hill and R.F. Packman (Lowestoft Water Company, U.K., 1957: see Walsby, 1992) described an attempt to use gas vesicle collapse in the removal of floating cyanobacteria. The cyanobacteria passing through a pipe were subjected to ultrasonic radiation from a transmitter in the pipe wall. Gas vesicle collapse and buoyancy loss occurred but was incomplete in the device used. There are problems in treating large volumes of water because the intensity of the radiation decreases as the square of distance. In small waterbodies this method has been used in combination with mixing and flushing to combat *Microcystis* blooms (Nakano et al., 2001).

Explosives. The detonation of a firework over a water bath containing floating *Microcystis* colonies was shown to generate sufficient pressure to collapse the gas vesicles and cause the colonies to sink (Walsby, 1968). The method was scaled up by Menday and Buck (1972) who detonated a charge of the explosive Cordex in a flooded quarry; it caused gas vesicle collapse and buoyancy loss of cyanobacteria up to 30 m away. They analysed the costs of treating a reservoir with a grid of explosive charges and assessed the collateral damage, e.g. to fish, some of which were killed.

Deep concentric pipes. The excess hydrostatic pressure (above the overlying atmospheric pressure) at a depth h in a water column is equal to $p = h\rho g$, where ρ is the density of water (998 kg m^{-3}) and g is gravitational acceleration (9.81 m s^{-2}). With each additional meter in depth, the pressure therefore rises by 9790 Pa. The pressure of 0.9 MPa suggested for treating *Microcystis* is hence obtained at a depth of 88 m. Clarke and Walsby (1988) described a deep concentric pipe through which water could be circulated to expose it to hydrostatic pressure sufficient to collapse gas vesicles in cyanobacteria. Only a small amount of energy is required to drive the water through the pipe. A concentric pipe 86-m deep sunk at the Lound Reservoir near Lowestoft removed all of the gas vesicles from *Microcystis* colonies entering the treatment works and prevented cyanobacteria floating above the primary filtration system. Pressurised cyanobacteria could be removed by sedimentation in shallow ponds as a pretreatment to reduce the cyanobacterial loading before entering the treatment plant. Colonial cyanobacteria could also be removed from a lake by cycling the water through such a deep pipe.

An unintended deep-pipe treatment: the Lake Kinneret pipe. The water from Lake Kinneret (the Sea of Galilee), which is 210 m below mean sea level, is lifted through a height of 256 m as it enters the distribution system of the Israeli National Water Carrier. In 1994, a bloom of the toxic gas-vacuolate cyanobacterium *Aphanizomenon ovalisporum* appeared in the lake. As it was pumped into the system it was subjected to a pressure of 2.5 MPa, greatly exceeding the critical pressure of the *Aphanizomenon* gas vesicles, which were collapsed. This caused a loss of turbidity in the water at the top of the pipe, giving a spurious indication that the water quality had improved. In the time it took for water to flow from the lake to the storage reservoirs the *Aphanizomenon* was subjected twice more to pressures exceeding 0.8 MPa as it passed through siphons carrying the water across deep gorges. The *Aphanizomenon* entering the reservoir was not buoyant and what few cyanobacteria remained at the treatment works were easily removed (Porat et al., 1999, 2000).

4.2 ARTIFICIAL MIXING

4.2.1 Effects of artificial mixing on Microcystis *and non-buoyant algae*

Buoyancy is crucial for *Microcystis* but the advantages of buoyancy are largely removed when colonies are entrained in intensely circulating water. Artificial mixing of lakes and reservoirs aims to reduce *Microcystis*. Decreased *Microcystis* biomass due to artificial mixing was found by Toetz (1981), Visser et al. (1996a) and Lindenschmidt (1999); cyanobacterial dominance shifted to negatively buoyant green algae and diatoms. Diatoms and green algae may in fact profit from artificial mixing, as artificial mixing reduces their sedimentation losses (Visser et al., 1996c) and may lower the pH thus shifting the inorganic carbon complex to CO_2.

In a recent study, turbulence was recorded in Lake Nieuwe Meer, The Netherlands, using a Self-Contained Autonomous Micro-Profiler (SCAMP) in a situation with and without artificial mixing (Huisman et al., 2004; see also Chapter 7). The data were compared with a competition model that included turbulent mixing of the water column, buoyancy of *Microcystis,* as well as sedimentation losses of diatoms and green algae. The model predicts that changes in turbulent diffusivities may shift the competitive balance between buoyant and non-buoyant phytoplankton. For an average water-column depth of 18 m, as in Lake Nieuwe Meer, a critical turbulent diffusivity of about 3.4 $cm^2\ s^{-1}$ was predicted. If vertical turbulent diffusivities remain below this critical value, *Microcystis* profits from its buoyancy and is predicted to become dominant. Conversely, if vertical turbulent diffusivities exceed this critical value, surface blooms of *Microcystis* would be prevented, and diatoms and green algae are predicted to become dominant. The critical vertical diffusivities predicted by the model matched well with the diffusivities achieved by artificial mixing of Lake Nieuwe Meer. This novel quantitative approach allows prediction of the mixing intensities required for the prevention of surface blooms of buoyant cyanobacteria.

4.2.2 Unsuccessful artificial mixing

Artificial mixing is not always successful in reducing the cyanobacterial biomass (e.g. Knoppert et al., 1970; Lackey, 1973; Osgood and Stiegler, 1990). Pastorok et al. (1980) summarised the results of a large number of destratification experiments: out of twenty-four cases examined, the abundance of cyanobacteria decreased in twelve, increased in eight, and showed no change in four cases. The following explanations may be offered for the failure of artificial mixing: i) mixing relieved nutrient limitation; (ii) insufficient account was taken of lake bathymetry; iii) mixing was insufficient; iv) changes in the mixing regime did not produce the required downshift in pH. These reasons will be further discussed below.

Artificial mixing is most successful in reducing the biomass of light-limited cyanobacterial populations, because deep and permanent mixing will increase light limitation. When cyanobacteria are not light-limited but nutrient-limited prior to mixing, the cyanobacterial biomass per surface area of lake may even increase as a result of artificial mixing, due to a higher nutrient availability when water from the nutrient-rich hypolimnion gets mixed into the epilimnion.

A study in a reservoir where *Microcystis* remained dominant despite artificial mixing (Visser et al., 1996b) showed that growth of *Microcystis* occurred mostly in a large, relatively shallow (6-m depth) area of the reservoir. No aerators were present to mix the water in this shallow area. *Microcystis* colonies demonstrated a higher buoyancy loss during the day in the shallow area, indicating that they received a higher light dose than those at the deep site. Apparently, the large shallow, unmixed area provided a source of *Microcystis* that interfered with mixing of the cyanobacteria in the deeper parts of the reservoir.

Another reason for the failure to decrease the amount of *Microcystis* is that the mixing velocity of the water is insufficient to prevent the (larger) colonies from escaping entrainment as occurs when the installed equipment is too weak for the lake area. The work in Lake Nieuwe Meer revealed that there is a critical threshold value for the turbulent diffusion coefficient that must be exceeded to shift the competitive dominance from buoyant cyanobacteria to sinking phytoplankton (Huisman et al., 2004) The distribution of the aeration tubes over the lake is important as well: in Lake Nieuwe Meer, it was found that the mixing efficiency diminished rapidly with distance from the aeration tubes.

Forsberg and Shapiro (1980) showed that in most cases with rapid mixing in enclosures, a shift from cyanobacteria to eukaryotic algae occurred. The shift from a system dominated by cyanobacteria to a system dominated by green algae may be augmented if mixing lowers the pH and increases the relative availability of free CO_2 (Shapiro, 1997). Deppe et al. (1999) showed that a strategy that combined phosphorus reduction with the transport of hypolimnetic water rich in free CO_2 to the epilimnion completely suppressed *Microcystis*.

4.2.3 Mixing devices
There are various devices for artificial mixing of lakes: usually, air is injected close to the lake bottom by a perforated pipe, or by a special diffuser producing very small bubbles. Mechanical destratification devices make use of centrifugal pumps or propellers. Mechanical mixing is less efficient than aeration (Symons et al., 1967, 1970). The depth of the air inlet is important: the greater the depth the more efficient the aeration and mixing (Cooke et al., 1993). Lorenzen and Fast (1977) concluded that an air flow rate of 9.2 m^3 min^{-1} km^{-2} lake area should be sufficient to accomplish adequate mixing in most lakes.

4.3. REDUCTION OF NUTRIENT LOADING

In many lake restoration programmes, decreasing the phosphorus (P) loading in a lake has resulted in a lower biomass of all phytoplankton, including *Microcystis*. There may, however, be hysteresis effects such that reduced P-loading does not result in reduced phytoplankton biomass, for instance because of internal loading of P stored in the sediment or the presence of a large benthic *Microcystis* population. A nice example is Lake Trummen (USA): only a modest reduction in *Microcystis* blooms occurred after reducing the external P-load, but the blooms were greatly diminished after a second step, dredging the sediment (Cronberg et al., 1975).

In lakes where eutrophication has been reduced, *Microcystis* populations may lose their dominance and be partly replaced by other organisms like diatoms (*Asterionella*, *Fragillaria*), chrysophytes (*Mallomonas*, *Dinobryon*, *Synura*) and dinoflagellates (*Ceratium hirundinella*) (see for examples Sas, 1989). With further P-reduction *Microcystis* blooms may disappear altogether, or remain at a reduced biomass after lake restoration.

In several lakes, however, *Microcystis* blooms are a natural phenomenon and they occurred long before the peak in anthropogenic eutrophication in the second half of the 20th century. Although nutrient reduction measures may eventually reduce the problem of noxious blooms of *Microcystis*, it is perhaps wishful thinking that they will disappear altogether. Surface blooms have a tendency to appear during periods of warm, stable weather, i.e., at those moments when lakes are used most intensively by the public.

5 Outlook: effects of climate change on *Microcystis* blooms

There is evidence that the increase in atmospheric CO_2 has caused global warming by approximately 0.6 °C over the last century and that this trend is likely to continue (Houghton et al., 2001). Consequences of global warming are changes in ice-cover, wind, solar insolation and precipitation. Water column stability, nutrient-loading and lake residence time may all be affected, with implications for phytoplankton. Higher temperatures will affect physiological processes such as photosynthesis, respiration and growth of the phytoplankton (Reynolds, 1997). In combination with i) the anticipated increase in the availability of phosphorus, ii) a reduction in some loss processes like sedimentation, and iii) a lengthened growing season, global warming is expected to result in an increase in phytoplankton biomass (e.g. Kilham et al., 1996)

Elevated water temperature may promote *Microcystis* blooms via different mechanisms, including an enhanced growth rate. *Microcystis* seems to have an exceptionally large Q_{10} for growth: 9-10 compared to 1-3 for other phytoplankton species (Reynolds, 1997). Indeed, growth rate of *Microcystis* increased 9-fold when temperature was increased from 10 to 20 ^{o}C, whereas growth rates of *Aphanizomenon* and *Planktothrix* increased 3-4 fold (Laboratory of Aquatic Microbiology, University of Amsterdam, unpublished results).

Higher temperatures from global warming may enhance the stability of the water column and thus suppress turbulent mixing, which would be highly advantageous for buoyant cyanobacteria like *Microcystis*. Conversely, the predicted increase in cloud cover and wind speed would weaken water-column stability and this would reduce the competitive strength of cyanobacteria. Howard and Easthope (2002) produced a model of cyanobacterial growth that incorporated climate change, and concluded that cyanobacterial production will (slightly) diminish in the next 90 years, mainly as a consequence of light limitation under an increased cloud cover; more intense short blooms were predicted during periods of higher insolation and high water temperature, although scum formation should decrease in a windier future (Ibelings et al., 2003). In reality, however, there are so many uncertainties involved in the prediction of climate scenarios that it is hard to foretell how water-column stability in different areas of the world will be affected by global warming.

6 Conclusions

We have discussed several adaptations that support the widespread distribution of *Microcystis* species and the attendant problems in lakes of different types. Arguably, the most important characteristics of *Microcystis* are its buoyancy and large colony size. Without buoyancy the large colonies would sink and would thus be unable to form surface blooms. Without large size the organism would not float fast enough to regain the euphotic zone after a mixing event. Moreover, without large size, grazing losses would be greater. Smaller algae must sustain larger grazing losses whilst these losses seem to be much reduced for colony-forming cyanobacteria like *Microcystis*. There are still many questions on the functional role of the toxins produced by *Microcystis*, but it seems these might deter potential predators. Other adaptations of *Microcystis* (and other cyanobacteria) to life in eutrophic lakes with a dense biomass of phytoplankton include its ability to use inorganic carbon efficiently and to grow at high pH. However, compared to other phytoplankton species, *Microcystis* species have a low specific growth rate and a relatively low affinity for phosphorus, which make them less successful in waters with a short residence time and in oligotrophic waters. All these features help to explain the potential dominance of *Microcystis* over other phytoplankton species in a wide range of eutrophic lakes, and they indicate measures that can be taken by water management to combat these harmful cyanobacteria.

7 Acknowledgements

We like to thank Geoff Codd, Jef Huisman and an anonymous reviewer for critical reading of the manuscript and their valuable suggestions. Part of this work was supported by a grant from the UK Natural Environment Research Council to AEW. PMV was supported by a MEERVOUD grant of the Earth and Life Sciences Foundation (ALW), which is subsidized by the Netherlands Organization for Scientific Research (NWO).

8 References

Abeliovich, A. and Shilo, M. (1972) Photooxidative death in blue-green algae, *Journal of Bacteriology* **111**, 682-689.

Annadotter, H., Cronberg, G., Aagren, R., Lundstedt, B., Nilsson, P.A. and Strobeck, S. (1999) Multiple techniques for lake restoration, *Hydrobiologia* **396**, 77-85.

Anagnostidis, K. and Komárek, J. (1988) Modern approach to the classification system of cyanophytes. 3 – Oscillatoriales, *Archiv für Hydrobiologie (Algological Studies)* **80**, 327-472.

Ackerman, J.D., Loewen, M.R. and Hamblin, P.F. (2001) Benthic-pelagic coupling over a zebra mussel reef in western Lake Erie, *Limnology and Oceanography* **46**, 892-904.

Agustí, S. and Phlips, E.J. (1992) Light absorption by cyanobacteria: implications of the colonial growth form, *Limnology and Oceanography* **37**, 434-441.

Baker, S.M., Levinton, J.S., Kurdziel, J.P. and Shumway, S.E. (1998) Selective feeding and biodeposition by zebra mussels and their relation to changes in phytoplankton composition and seston load, *Journal of Shellfish Research* **17**, 1207-1213.

Beard, S.J., Handley, B.A., Hayes, P.K. and Walsby, A.E. (1999) The diversity of gas vesicle genes in *Planktothrix rubescens* from Lake Zürich, *Microbiology* **145**, 2757-2768.

Beard, S.J., Davis, P.A., Iglesias-Rodríguez, D., Skulberg, O.M. and Walsby, A.E. (2000) Gas vesicle genes in *Planktothrix* spp. from Nordic lakes: strains with weak gas vesicles possess a longer variant of *gvpC*, *Microbiology* **146**, 2009-2018.

Boing, W.J., Wagner, A., Voigt, H., Deppe, T. and Benndorf, J. (1998) Phytoplankton responses to grazing by *Daphnia galeata* in the biomanipulated Bautzen reservoir, *Hydrobiologia* **389**, 101-114.

Bourelly, P. (1970) Les algues d'eau douce **3**, N. Boubée and Cie, Paris (in French)

Bormans, M., Sherman, B.S. and Webster, I.T. (1999) Is buoyancy regulation in cyanobacteria an adaptation to exploit separation of light and nutrients?, *Marine and Freshwater Research* **50**, 897-906.

Botes, D.P., Tuiman, A.A., Wessels, P.L, Viljoen, C.C., Kruger, H., Williams, D.H., Santikarn, S., Smith, R.J. and Hammond, S.J. (1984) The structure of cyanoginosin-LA, a cyclic peptide from the cyanobacterium *Microcystis aeruginosa, Journal of the Chemical Society - Perkin Transactions* **I**, 2311-2319.

Brunberg, A.K. and Blomqvist, P. (2003) Recruitment of *Microcystis* (Cyanophyceae) from lake sediments: the importance of littoral inocula, *Journal of Phycology* **39**, 58-63.

Brunberg, A.K. and Böstrom, B. (1992) Coupling between benthic biomass of *Microcystis* and phosphorus release from the sediments of a highly eutrophic lake, *Hydrobiologia* **235/236**, 375-385.

Cáceres, O. and Reynolds, C.S. (1984) Some effects of artificially-enhanced anoxia on the growth of *Microcystis aeruginosa* Kutz. emend. Elenkin, with special reference to the initiation of its annual growth cycle in lakes, *Archiv für Hydrobiologie* **99**, 379-397.

Campbell, D., Hurry, V., Clarke, A.K., Gustafsson, P. and Oquist, G. (1998) Chlorophyll fluorescence analysis of cyanobacterial photosynthesis and acclimation, *Microbiology and Molecular Biology Reviews* **62**, 667-671.

Chen, Y.W., Qin, B.Q., Teubner, K. and Dokulil, M.T. (2003) Long-term dynamics of phytoplankton assemblages: *Microcystis*- domination in Lake Taihu, a large shallow lake in China, *Journal of Plankton Research* **25**, 445-453.

Clarke, K.B. and Walsby, A.E. (1988) Improvements in the treatment of water to remove gas vacuolate cyanobacteria, U.K. Patent Application 2196330A.

Clemens, J., Oberer, L., Quiquerez, C., Konig, W.A. and Weckesser, J. (1995) Cyanopeptolin-S, a sulfate-containing depsipeptide from a water bloom of *Microcystis* sp., *FEMS Microbiology Letters* **129**, 129-133.

Cooke, G.D., Welch, E.B., Peterson, S.A. and Newroth, P.R. (1993) Artificial circulation, in G.D. Cooke, E.B. Welch, S.A. Peterson and P.R. Newroth (eds.), *Restoration and Management of Lakes and Reservoirs,* Lewis Publishers, pp. 419-449.

Cronberg, G., Gelin, C. and Larsson, K. (1975) Lake Trummen restoration project. II. Bacteria, phytoplankton and phytoplankton productivity, *Verhandlungen der Internationalen Vereinigung für Theoretische und Angewandte Limnologie* **19**, 1088-1096.

Davis, P.A., Dent, M., Parker, J., Reynolds, C.S. and Walsby, A.E. (2003) The annual cycle of growth rate and biomass change in *Planktothrix* spp. in Blelham Tarn, English Lake District, *Freshwater Biology* **48**, 852-867.

Deacon, C. and Walsby, A.E. (1990) Gas vesicle formation in the dark, and in light of different irradiances, by the cyanobacterium *Microcystis* sp., *British Phycological Journal* **25**, 133-139.

Demmig-Adams, B. and Adams, W.W. (1996) The role of the xanthophyll cycle carotenoids in the protection of photosynthesis, *Trends in Plant Science* **1**, 21-26.

Demmig-Adams, B., Maguas, C., Adams, W.W., Meyer, A., Kilian, E. and Lange O.L. (1990) Effect of high light on the efficiency of photochemical energy-conversion in a variety of lichen species with green and blue-green phycobionts, *Planta* **180**, 400-409.

DeMott, W.R. (1999) Foraging strategies and growth inhibition in five daphnids feeding on mixtures of a toxic cyanobacterium and a green alga, *Freshwater Biology* **42**, 263-274.

Deppe, T., Ockenfeld, K., Meybohm, A., Opitz, M. and Benndorf, J. (1999) Reduction of *Microcystis* blooms in a hypertrophic reservoir by a combined ecotechnological strategy, *Hydrobiologia* **409**, 31-38.

Dinsdale, M.T. and Walsby, A.E. (1972) The interrelations of cell turgor pressure, gas-vacuolation, and buoyancy in a blue-green alga, *Journal of Experimental Botany* **23**, 561-570.

Elser, J.J. (1999) The pathway to noxious cyanobacteria blooms in lakes: the food web as the final turn, *Freshwater Biology* **42**, 537-543.

Eloff, J.N., Steinitz, Y. and Shilo, M. (1976) Photooxydation of cyanobacteria in natural conditions, *Applied and Environmental Microbiology* **31**, 119-126.

Epp, G.T. (1996) Grazing on filamentous cyanobacteria by *Daphnia pulicaria*, *Limnology and Oceanography* **41**, 560-567.

Erhard, M., von Döhren, H. and Jungblut, P.R. (1999) Rapid identification of the new anabaenopeptin G from *Planktothrix agardhii* HUB 011 using matrix-assisted laser desorption/ionization time-of-flight mass spectrometry, *Rapid Communications in Mass Spectrometry* **13**, 337-343.

Fallon, R.D. and Brock, T.D. (1981) Overwintering of *Microcystis* in Lake Mendota, *Freshwater Biology* **11**, 217-226.

Falkowski, P.G. and LaRoche, J. (1991) Acclimation to spectral irradiance in algae, *Journal of Phycology* **27**, 8-14.

Fastner, J., Erhard, M. and von Döhren, H. (2001) Determination of oligopeptide diversity within a natural population of *Microcystis* spp. (Cyanobacteria) by typing single colonies by matrix-assisted laser desorption ionization-time of flight mass spectrometry, *Applied and Environmental Microbiology* **67**, 5069-5076.

Ferrão, A.D., Kozlowsky-Suzuki, B. and Azevedo, S.M.F.O. (2002) Accumulation of microcystins by a tropical zooplankton community, *Aquatic Toxicology* **59**, 201-208.

Fogg, G.E. and Walsby, A.E. (1971) Buoyancy regulation and the growth of planktonic blue-green algae, *Mitteilungen der Internationale Vereinigung für Theoretisch und Angewandte Limnologie* **19**, 182-188.

Forsberg, B.R. and Shapiro, J. (1980) Predicting the algal response to destratification, in *Restoration of Lakes and Inland Waters, EPA 440/5-81-010, United States Environmental Protection Agency, Washington, D.C.*, pp. 134-145.

Fulton, R.S. and Paerl, H.W. (1987) Toxic and inhibitory effects of the blue-green alga *Microcystis aeruginosa* on herbivorous zooplankton, *Journal of Plankton Research* **9**, 837-855.

Ganf, G.G. (1974) Diurnal mixing and the vertical distribution of phytoplankton in a shallow equatorial lake (Lake George, Uganda), *Journal of Ecology* **62**, 611-629.

Ganf, G.G., Oliver, R.L. and Walsby, A.E. (1989) Optical properties of gas-vacuolate cells of *Microcystis* in relation to light attenuation in a turbid, stratified reservoir (Mount Bold Reservoir, South Australia), *Australian Journal of Marine and Freshwater Research* **40**, 595-611.

Ghadouani, A., Pinel-Alloul, B. and Prepas, E.E. (2003) Effects of experimentally induced cyanobacterial blooms on crustacean zooplankton communities, *Freshwater Biology* **48**, 363-381.

Geitler, L. (1932) *Cyanophyceae*, in Rabenhorst's *Kryptogamen-Flora* Vol. **14**, Akademische Verlagsgesellschaft, Leipzig.

George, D.G. and Edwards, R.W. (1976) The effect of wind on the distribution of chlorophyll *a* and crustacean plankton in a shallow eutrophic reservoir, *Journal of Applied Ecology* **13**, 607-690.

Gulati, R.D., Bronkhorst, M. and Van Donk, E. (2001) Feeding in *Daphnia galeata* on *Oscillatoria limnetica* and on detritus derived from it, *Journal of Plankton Research* **23**, 705-718.

Hairston, N.G., Lampert, W., Cáceres, C.E., Holtmeier, C.L., Weider, L.J., Gaedke U., Fischer, J.M., Fox, J.A. and Post, D.M. (1999) Lake ecosystems: rapid evolution revealed by dormant eggs, *Nature* **401**, 446-446.

Houghton, J.T., Ding, Y., Griggs, D.J., Noguer, M., Van der Linden, P.J. and Xiaosu, D (eds). (2001) *Climate Change 2001: The Scientific Basis. Contribution of Working Group I to the Third Assessment Report of the IPCC*, Cambridge University Press, Cambridge, U.K., 944 pp.

Howard, A. (2001) Modeling movement patterns of the cyanobacterium *Microcystis*, *Ecological Applications* **11**, 304-310.

Howard, A. and Easthope, M.P. (2002) Application of a model to predict cyanobacterial growth patterns in response to climatic change at Farmoor Reservoir, Oxfordshire, UK, *Science of the Total Environment* **282**, 459-469.

Huisman, J., Jonker, R.R., Zonneveld, C. and Weissing, F.J. (1999) Competition for light between phytoplankton species: experimental tests of mechanistic theory, *Ecology* **80**, 211-222.

Huisman, J., Sharples, J., Stroom, J.M., Visser, P.M., Kardinaal, W.E.A., Verspagen, J.M.H. and Sommeijer, B. (2004) Changes in turbulent mixing shift competition for light between phytoplankton species, *Ecology* **85**, 2960-2970.

Humphries, S.E. and Lyne, V.D. (1988) Cyanophyte blooms: the role of buoyancy, *Limnology and Oceanography* **33**, 79-91.

Ibelings, B.W. (1996) Changes in photosynthesis in response to combined photon irradiance and temperature stress in cyanobacterial surface waterblooms, *Journal of Phycology* **32**, 549-557.

Ibelings, B.W. and Maberly, S.C. (1998) Photoinhibition and the availability of inorganic carbon restrict photosynthesis by surface blooms of cyanobacteria, *Limnology and Oceanography* **43**, 408-419.

Ibelings, B.W., Mur, L.R. and Walsby, A.E. (1991a) Diurnal changes in buoyancy and vertical distribution in populations of *Microcystis* in two shallow lakes, *Journal of Plankton Research* **13**, 419-436.

Ibelings, B.W., Mur, L.R., Kinsman, R. and Walsby, A.E. (1991b) *Microcystis* changes its buoyancy in response to the average irradiance in the surface mixed layer, *Archiv für Hydrobiologie* **120**, 385-401.

Ibelings, B.W., Kroon, B.M.A. and Mur, L.R. (1994) Acclimation of photosystem II in a cyanobacterium and a eukaryotic green alga to high and fluctuating photosynthetic photon flux densities, simulating light regimes induced by mixing in lakes, *New Phytologist* **128**, 407-424.

Ibelings, B.W., Vonk, M., Los, H.F.J., van der Molen, D.T. and Mooij, W.M. (2003) Fuzzy modeling of cyanobacterial surface waterblooms: validation with NOAA-AVHRR satellite images, *Ecological Applications* **13**, 1456–1472.

Ibelings, B.W., Bruning, K., de Jonge, J., Wolfstein, K., Pires, L.M.D., Postma, J. and Burger, T. (2004) Distribution of microcystins in the foodweb of a large shallow lake: no evidence for bioaccumulation or biomagnification, *Microbial Ecology*, in press.

Jacoby, J.M., Collier, D.C., Welch, E.B., Hardy, F.J. and Crayton, M. (2000) Environmental factors associated with a toxic bloom of *Microcystis aeruginosa*, *Canadian Journal of Fisheries and Aquatic Sciences* **57**, 231-240.

Jungmann, D. and Benndorf, J. (1994) Toxicity to *Daphnia* of a compound extracted from laboratory and natural *Microcystis* spp., and the role of microcystins, *Freshwater Biology* **32**, 13-20.

Kaplan, A. and Reinhold, L. (1999) CO_2 concentrating mechanisms in photosynthetic microorganisms, *Annual Review of Plant Physiology and Plant Molecular Biology* **50**, 539-570.

Kemka, N., Njine, T., Togouet, S.H.Z., Niyitegeka, D., Monkiedje, A. and Menbohan, S.F. (2003) Quantitative importance of cyanobacteria populations in a hypertrophic shallow lake in the subequatorial African region (Yaounde Municipal Lake, Cameroon), *Archiv für Hydrobiologie* **156**, 495-510.

Kilham, S.S., Theriot, E.C. and Fritz, S.C. (1996) Linking planktonic diatoms and climate change in the large lakes of the Yellowstone ecosystem using resource theory, *Limnology and Oceanography* **41**, 1052-1062.

Kinsman, R., Ibelings, B.W. and Walsby, A.E. (1991) Gas vesicle collapse by turgor pressure and its role in buoyancy regulation by *Anabaena flos-aquae*, *Journal of General Microbiology* **137**, 1171-1178.

Klemer, A.R. (1978) Nitrogen limitation of growth and gas vacuolation in *Oscillatoria rubescens*, *Verhandlungen der Internationalen Vereinigung für Theoretische und Angewandte Limnologie* **20**, 2293-2297.

Klemer, A.R., Feuillade, J. and Feuillade, M. (1982) Cyanobacterial blooms: carbon and nitrogen limitation have opposite effects on the buoyancy of *Oscillatoria*, *Science* **215**, 1629-1631.

Knoppert, P.L., Rook, J.J., Hofker, T. and Oskam, G. (1970) Destratification experiments in Rotterdam, *Journal of American Water Works Association* **62**, 448-454.

Komárek, J. and Anagnostidis, K. (1986) Modern approach to the classification system of cyanophytes. 2 – Chroococcales, *Archiv für Hydrobiologie (Algological Studies)* **43**, 157-226.

Komárek, J. and Hindák, F. (1988) Taxonomic review of natural populations of the cyanophytes from *Gomphosphaeria*-complex, *Archiv für Hydrobiologie (Algological Studies)* **80**, 203-225.

Komárek, J. and Anagnostides, K. (1999) *Cyanoprokaryota 1. Teil Chroococcales,* Gustav Fischer Verlag Jena, Jena, Germany.

Konopka, A., Kromkamp, J. and Mur, L.R. (1987a) Buoyancy regulation in phosphate-limited cultures of *Microcystis aeruginosa*, *FEMS Microbiology Ecology* **45**, 135-142.

Konopka, A., Kromkamp, J. and Mur, L.R. (1987b) Regulation of gas vesicle content and buoyancy in light- or phosphate-limited cultures of *Aphanizomenon flos-aquae* (Cyanophyta), *Journal of Phycology* **23**, 70-78.

Korner, S. (2001) Development of submerged macrophytes in shallow Lake Muggelsee (Berlin, Germany) before and after its switch to the phytoplankton-dominated state, *Archiv für Hydrobiologie* **152**, 395-409.

Kromkamp, J. and Walsby, A.E. (1990) A computer model of buoyancy and vertical migration in cyanobacteria, *Journal of Plankton Research* **12**, 161-183.

Kromkamp, J., Van den Heuvel, A. and Mur, L.R. (1989) Formation of gas vesicles in phosphorus-limited cultures of *Microcystis aeruginosa*, *Journal of General Microbiology* **135**, 1933-1939.

Kurmayer, R., Christiansen, G. and Chorus, I. (2003) The abundance of microcystin-producing genotypes correlates positively with colony size in *Microcystis* sp. and determines its microcystin net production in Lake Wannsee, *Applied and Environmental Microbiology* **69**, 787-795.

Lackey, R.T. (1973) Artificial reservoir destratification effects on phytoplankton, *Journal of the Water Pollution Control Federation* **45**, 668-673.

Lampert, W. (1987) Laboratory studies on zooplankton-cyanobacteria interactions, *New Zealand Journal of Marine and Freshwater Research* **21**, 483-491.

Lindenschmidt, K.E. (1999) Controlling the growth of *Microcystis* using surged artificial aeration, *International Review for Hydrobiology* **84**, 243-254.

Lorenzen, M.W. and Fast, R. (1977) A guide to aeration/circulation techniques for lake management, *Research Series EPA-600/3-77-004, United States Environmental Protection Agency, Washington D.C.*

MacIntyre, H.L., Kana, T.M., Anning, T. and Geider, R.J. (2002) Photoacclimation of photosynthesis irradiance response curves and photosynthetic pigments in microalgae and cyanobacteria, *Journal of Phycology* **38**, 17-38.

Meijer, M.L., de Boois, I., Scheffer, M., Portielje, R. and Hosper, H. (1999) Biomanipulation in shallow lakes in The Netherlands: an evaluation of 18 case studies, *Hydrobiologia* **409**, 13-30.

Menday, D.C. and Buck A.A. (1972) A shock-wave technique to collapse vacuoles of blue-green algae, *Water Research* **6**, 279-284.

Mitrovic, S.M., Bowling, L.C. and Buckney, R.T. (2001) Vertical disentrainment of *Anabaena circinalis* in the turbid, freshwater Darling River, Australia: quantifying potential benefits from buoyancy, *Journal of Plankton Research* **23**, 47-55.

Murphy, T.P., Irvine, K., Guo, J., Davies, J., Murkin, H., Charlton, M. and Watson, S.B. (2003) New microcystin concerns in the lower Great Lakes, *Water Quality Research Journal of Canada* **38**, 127-140.

Nakano, S., Hayakawa, K., Frenette, J.J., Nakajima, T., Jiao, C.M., Tsujimura, S. and Kumagai, M. (2001) Cyanobacterial blooms in a shallow lake: a large-scale enclosure assay to test the importance of diurnal stratification, *Archiv für Hydrobiologie* **150**, 491-509.

Namikoshi, M. and Rinehart, K.L. (1996) Bioactive compounds produced by cyanobacteria, *Journal of Industrial Microbiology and Biotechnology* **17**, 373-384.

Neilan, B.A., Jacobs, D., DelDot, T., Blackall, L.L., Hawkins, P.R., Cox, P.T. and Goodman, A.E. (1997) rRNA sequences and evolutionary relationships among toxic and nontoxic cyanobacteria of the genus *Microcystis*, *International Journal of Systematic Bacteriology* **47**, 693-697.

Niyogi, K.K. (2000) Safety valves for photosynthesis, *Current Opinions in Plant Biology* **3**, 455-460.

Oliver, R.L. (1994) Floating and sinking in gas-vacuolate cyanobacteria, *Journal of Phycology* **30**, 161-173.

Oliver, R.L. and Ganf, G.G. (2000) Freshwater blooms, in B.A. Whitton and M. Potts (eds.), *The Ecology of Cyanobacteria*, Kluwer Academic Publishers, Dordrecht, The Netherlands, pp. 149-194.

Oliver, R.L. and Walsby, A.E. (1984) Direct evidence for the role of light-mediated gas vesicle collapse in the buoyancy regulation of *Anabaena flos-aquae* (cyanobacteria), *Limnology and Oceanography* **29**, 879-886.

Oliver, R.L., Thomas, R.H., Reynolds, C.S. and Walsby, A.E. (1985) The sedimentation of buoyant *Microcystis* colonies caused by precipitation with an iron-containing colloid, *Proceedings of the Royal Society of London* B **223**, 511-528.

Oquist, G., Chow, W.S. and Anderson, J.M. (1992) Photoinhibition of photosynthesis represents a mechanism for the long-term protection of photosystem II, *Planta* **186**, 450-460.

Osgood, R.A. and Stiegler, J.E. (1990) The effects of artificial circulation on a hypereutrophic lake, *Water Research Bulletin* **26**, 209-217.

Otsuka, S., Suda, S., Li, R.H., Matsumoto, S. and Watanabe, M.M. (2000) Morphological variability of colonies of *Microcystis* morphospecies in culture, *Journal of General and Applied Microbiology* **46**, 39-50.

Paerl, H.W. and Ustach J.F. (1982) Blue-green algal scums: an explanation for their occurrence during freshwater blooms, *Limnology and Oceanography* **27**, 212-217.

Paerl, H.W., Tucker, J. and Bland, P.T. (1983) Carotenoid enhancement and its role in maintaining blue-algal (*Microcystis aeruginosa*) surface blooms, *Limnology and Oceanography* **28**, 847-857.

Paerl, H.W., Fulton, R.S., Mosiander, P. and Dyble, J. (2001) Harmful freshwater algal blooms, with an emphasis on cyanobacteria, *The Scientific World* **1**, 76-113.

Pastorok, R.A., Ginn, T.C. and Lorenzen, M.W. (1980) Review of aeration/circulation for lake management, in *Restoration of Lakes and Inland Waters, EPA 440/5-81-010, United States Environmental Protection Agency, Washington D.C.*, pp. 124-133.

Paterson, M.J., Findlay, D.L., Salki, A.G., Hendzel, L.L. and Hesslein, R.H. (2002) The effects of *Daphnia* on nutrient stoichiometry and filamentous cyanobacteria: a mesocosm experiment in a eutrophic lake, *Freshwater Biology* **47**, 1217-1233.

Perrow, M.R., Meijer, M.L., Dawidowicz, P. and Coops, H. (1997) Biomanipulation in the shallow lakes: state of the art, *Hydrobiologia* **342**, 355-365.

Pires, L.M.D. and Van Donk, E. (2002) Comparing grazing by *Dreissena polymorpha* on phytoplankton in the presence of toxic and non-toxic cyanobacteria, *Freshwater Biology* **47**, 1855-1865.

Pires, L.M.D., Kusserow, R. and Van Donk, E. (2003) Influence of toxic and non-toxic phytoplankton on feeding and survival of *Dreissena polymorpha* (Pallas) larvae, *Hydrobiologia* **491**, 193-200.

Porat, R., Teltsch, B., Mosse, R., Dubinsky, Z. and Walsby, A.E. (1999) Turbidity changes caused by collapse of cyanobacterial gas vesicles in water pumped from Lake Kinneret into the Israeli National Water Carrier, *Water Research* **33**, 1634-1644.

Porat, R., Teltsch, B., Dubinsky, Z. and Walsby, A.E. (2000) Effects of light and pressure on gas vesicle formation and buoyancy in *Aphanizomenon ovalisporum* Forti (Cyanobacteria) from Lake Kinneret, Israel, *Archiv für Hydrobiologie, Special Issues, Advances in Limnology* **55**, 333-348.

Porat, R., Teltsch, B., Perelman, A. and Dubinsky, Z. (2001) Diel buoyancy changes by the cyanobacterium *Aphanizomenon ovalisporum* from a shallow reservoir, *Journal of Plankton Research* **23**, 753-763.

Powles, S.B. (1984) Photoinhibition of photosynthesis induced by visible light, *Annual Review of Plant Physiology* **35**, 15-44.

Preston, T., Stewart, W.D.P. and Reynolds, C.S. (1980) Bloom-forming cyanobacterium *Microcystis aeruginosa* overwinters on the sediment, *Nature* **288**, 365-367.

Rabouille, S., Thébault, J.M. and Salençon, M.J. (2003) Simulation of carbon reserve dynamics in *Microcystis* and its influence on vertical migration with Yoyo model, *Comptes Rendus Biologies* **326**, 349-361.

Reynolds, C.S. (1976) Succession and vertical distribution of phytoplankton in response to thermal stratification in a lowland mere, with special reference to nutrient availability, *Journal of Ecology* **64**, 529-551.

Reynolds, C.S. (1984) *The Ecology of Freshwater Phytoplankton,* Cambridge University Press, Cambridge, United Kingdom.

Reynolds, C.S. (1987) Cyanobacterial water-blooms, *Advances in Botanical Research* **13**, 67-143.

Reynolds, C.S. (1991) Toxic blue-green algae: the " problem" in a perspective, *Freshwater Forum* **1**, 29-39.

Reynolds, C.S. (1997) *Vegetation Processes in the Pelagic: a Model for Ecosystem Theory*, Ecology Institute, Oldendorf, Germany.

Reynolds, C.S. and Walsby, A.E. (1975) Water-blooms, *Biological Reviews* **50**, 437-481.

Reynolds, C.S. and Rogers, D.A. (1976) Seasonal variations in the vertical distribution and buoyancy of *Microcystis aeruginosa* Kutz. emend. Elenkin in Rostherne Mere, England, *Hydrobiologia* **48**, 17-23.

Reynolds, C.S., Jaworski, G.H.M., Cmeich, H.A. and Leedale, G.F. (1981) On the annual cycle of the blue-green alga *Microcystis aeruginosa* Kutz. emend. Elenkin, *Philosophical Transactions of the Royal Society B* **293**, 419-477.

Reynolds, C.S., Wiseman, S.W., Godfrey, B.M. and Butterwick, C. (1983) Some effects of artificial mixing on the dynamics of phytoplankton populations in large limnetic enclosures, *Journal of Plankton Research* **5**, 203-232.

Reynolds, C.S. and Bellinger, E.G. (1992) Patterns of abundance and dominance of the phytoplankton of Rostherne Mere, England: evidence from an 18-year data set, *Aquatic Sciences* **54**, 10-36.

Rippka, R., Deruelles, J.B., Waterbury, J.B., Herdman, M. and Stanier, R.Y. (1979) Generic assignments, strain histories and properties of pure cultures of cyanobacteria, *Journal of General Microbiology* **111**, 1-61.

Robarts, R.D. and Zohary, T. (1984) *Microcystis aeruginosa* and underwater light attenuation in a hypertrophic lake (Hartbeespoort Dam, South Africa), *Journal of Ecology* **72**, 1001-1017.

Rohrlack, T., Henning, M. and Kohl, J.G. (1999a) Does the toxic effect of *Microcystis aeruginosa* on *Daphnia galeata* depend on microcystin ingestion rate? *Archiv für Hydrobiologie* **146**, 385-395.

Rohrlack, T., Henning, M. and Kohl, J.G. (1999b) Mechanisms of the inhibitory effect of the cyanobacterium *Microcystis aeruginosa* on *Daphnia galeata*'s ingestion rate, *Journal of Plankton Research* **21**, 1489-1500.

Rohrlack, T., Dittmann, E., Henning, M., Börner, T. and Kohl, J.G. (1999c) Role of microcystins in poisoning and food ingestion inhibition of *Daphnia galeata* caused by the cyanobacterium *Microcystis aeruginosa*, *Applied and Environmental Microbiology* **65**, 737-739.

Rohrlack, T., Dittmann, E., Börner, T. and Christoffersen, K. (2001) Effects of cell-bound microcystins on survival and feeding of *Daphnia* spp., *Applied and Environmental Microbiology* **67**, 3523-3529.

Rohrlack, T., Christoffersen, K., Hansen, P.E., Zhang, W., Czarnecki, O., Henning, M., Fastner, J., Erhard, M., Neilan, B.A. and Kaebernick, M. (2003) Isolation, characterization, and quantitative analysis of microviridin J, a new *Microcystis* metabolite toxic to *Daphnia*, *Journal of Chemical Ecology* **29**, 1757-1770.

Sas, H., Ed. (1989) *Lake Restoration by Reduction of Nutrient Loading: Expectations, Experience, Extrapolations*, Academia Verlag Richarz, St. Augustin.

Scheffer, M., Rinaldi, S., Gragnani, A., Mur, L.R. and van Nes, E.H. (1997) On the dominance of filamentous cyanobacteria in shallow, turbid lakes, *Ecology* **78**, 272-282.

Shapiro, J. (1997) The role of carbon dioxide in the initiation and maintenance of blue-green dominance in lakes, *Freshwater Biology* **37**, 307-323.

Smith, V.H. (1983) Low nitrogen to phosphorus ratios favor dominance by blue-green algae in lake phytoplankton, *Science* **221**, 669-671.

Symons, J.M., Irwin, W.H., Robinson, E.L. and Robeck, G.G. (1967) Impoundment destratification for raw water quality control using either mechanical or diffused-air-pumping, *Journal of the American Water Works Association* **59**, 1268-1291.

Symons, J.M., Carswell, J.K. and Robeck, G.G. (1970) Mixing of water supply reservoirs for quality control, *Journal of the American Water Works Association* **62**, 322-334.

Takamura, N., Yasuno, M. and Sugahara, K. (1984) Overwintering of *Microcystis aeruginosa* Kutz. in a shallow lake, *Journal of Plankton Research* **6**, 1019-1029.

Talling, J.F. (1976) The depletion of carbon dioxide from lake water by phytoplankton, *Journal of Ecology* **64**, 79-121.

Ten Winkel, E.H. and Davids, C. (1982) Food selection by *Dreissena polymorpha* Pallas (Mollusca: Bivalva), *Freshwater Biology* **12**, 553-558.

Thomas, R.H. and Walsby, A.E. (1985) Buoyancy regulation in a strain of *Microcystis*, *Journal of General Microbiology* **131**, 799-809.

Thomas, R.H. and Walsby, A.E. (1986) The effect of temperature on recovery of buoyancy by *Microcystis*, *Journal of General Microbiology* **132**, 1665-1672.

Thostrup, L. and Christoffersen, K. (1999) Accumulation of microcystin in *Daphnia magna* feeding on toxic *Microcystis*, *Archiv für Hydrobiologie* **145**, 447-467.

Toetz, D.W. (1981) Effects of whole lake mixing on water quality and phytoplankton, *Water Research* **15**, 1205-1210.

Trimbee, A.M. and Harris, G.P. (1984) Phytoplankton population dynamics of a small reservoir: effect of intermittent mixing on phytoplankton succession and the growth of blue-green algae, *Journal of Plankton Research* **6**, 699-713.

Trimbee, A.M. and Prepas, E.E. (1988) The effect of oxygen depletion on the timing and magnitude of blue-green algal blooms, *Verhandlungen der Internationalen Vereinigung für Theoretische und Angewandte Limnologie* **23**, 220-226.

Tsujimura, S., Tsukada, H., Nakahara, H., Nakajima, T. and Nishino, M. (2000) Seasonal variations of *Microcystis* populations in sediments of Lake Biwa, Japan, *Hydrobiologia* **434**, 183-192.

Utkilen, H.C., Oliver, R.L. and Walsby, A.E. (1985) Buoyancy regulation in a red *Oscillatoria* unable to collapse gas vacuoles by turgor pressure, *Archiv für Hydrobiologie* **102**, 319-329.

Vanderploeg, H.A., Liebig, J.R., Carmichael, W.W., Agy, M.A., Johengen, T.H., Fahnenstiel, G.L. and Nalepa, T.F. (2001) Zebra mussel (*Dreissena polymorpha*) selective filtration promoted toxic *Microcystis* blooms in Saginaw Bay (Lake Huron) and Lake Erie, *Canadian Journal of Fisheries and Aquatic Sciences* **58**, 1208-1221.

Van Liere, L. and Mur, L.R. (1980) Occurrence of *Oscillatoria agardhii* and some related species, a survey, in J. Barica and L.R. Mur (eds.), *Hypertrophic Ecosystems*, Junk Publishers, The Hague, The Netherlands, pp. 67-77.

Van Rijn, J. and Shilo, M. (1985) Carbohydrate fluctuations, and vertical migration of scum-forming cyanobacteria in fishponds, *Limnology and Oceanography* **30**, 1219-1228.

Verspagen, J.M.H., Snelder, E.O.F.M., Visser, P.M., Huisman, J., Mur, L.R. and Ibelings, B.W. (2004) Recruitment of benthic *Microcystis* (Cyanophyceae) to the water column: internal buoyancy changes or resuspension? *Journal of Phycology* **40,** 260-270.

Visser, P.M., Ibelings, B.W. and Mur, L.R. (1995) Autumnal sedimentation of *Microcystis* spp. as result of an increase in carbohydrate ballast at reduced temperature, *Journal of Plankton Research* **17**, 919-933.

Visser, P.M., Ibelings, B.W., Van der Veer, B., Koedood, J. and Mur, L.R. (1996a) Artificial mixing prevents nuisance blooms of the cyanobacterium *Microcystis* in Lake Nieuwe Meer, the Netherlands, *Freshwater Biology* **36,** 435-450.

Visser, P.M., Ketelaars, H.A.M., Van Breemen, L.W.C.A. and Mur, L.R. (1996b) Diurnal buoyancy changes of *Microcystis* in an artificially mixed storage reservoir, *Hydrobiologia* **331**, 131-141.

Visser, P.M., Massaut, L., Huisman, J. and Mur, L.R. (1996c) Sedimentation losses of *Scenedesmus* in relation to mixing depth, *Archiv für Hydrobiologie* **136**, 289-308.

Visser, P.M., Passarge, J. and Mur, L.R. (1997) Modelling vertical migration of the cyanobacterium *Microcystis*, *Hydrobiologia* **349**, 99-109.

Vos, R.J., Hakvoort, J.H.M., Jordans, R.W.J. and Ibelings B.W. (2003) Multiplatform optical monitoring of eutrophication in temporally and spatially variable lakes, *Science of the Total Environment* **312**, 221-243.

Wallace, B.B. and Hamilton, D.P. (1999) The effect of variations in irradiance on buoyancy regulation in *Microcystis aeruginosa*, *Limnology and Oceanography* **44**, 273-281.

Walsby, A.E. (1968) An alga's buoyancy bag, *New Scientist* **40**, 436-437.

Walsby, A.E. (1969) The permeability of blue-green algal gas-vacuole membranes to gas, *Proceedings of the Royal Society of London* B **173**, 235-255.

Walsby, A.E. (1970) The nuisance algae: curiosities in the biology of planktonic blue-green algae, *Water Treatment and Examination* **19**, 359-373.

Walsby, A.E. (1992) The control of gas-vacuolate cyanobacteria, in D.W. Sutcliffe and J. Gwynfryn Jones (eds.), *Eutrophication: Research and Application to Water Supply*, Freshwater Biological Association, Ambleside, United Kingdom, pp. 150-162.

Walsby, A.E. (1994) Gas vesicles, *Microbiological Reviews* **58**, 94-144.

Walsby, A.E. (1997) Numerical integration of phytoplankton photosynthesis through depth and time in a water column, *New Phytologist* **136**, 189-209.

Walsby, A.E. and Booker, M.J. (1980) Changes in buoyancy of a planktonic blue-green alga in response to light intensity, *British Phycological Journal* **15**, 311-319.

Walsby, A.E., Utkilen, H.C. and Johnsen, I.J. (1983) Buoyancy changes of a red coloured *Oscillatoria agardhii* in Lake Gjersjøen, Norway, *Archiv für Hydrobiologie* **97**, 18-38.

Walsby, A.E. and McAllister, G.K. (1987) Buoyancy regulation by *Microcystis* in Lake Okaro, *New Zealand Journal of Marine and Freshwater Research* **21**, 521-524.

Walsby, A.E. and Bleything, A. (1988) The dimensions of cyanobacterial gas vesicles in relation to their efficiency in providing buoyancy and withstanding pressure, *Journal of General Microbiology* **134**, 2635-2645.

Walsby, A.E., Reynolds, C.S., Oliver, R.L. and Kromkamp, J. (1989) The role of gas vacuoles and carbohydrate content in the buoyancy and vertical distribution of *Anabaena minutissima* in Lake Rotongaio, New Zealand, *Archiv für Hydrobiologie/ Ergebnisse der Limnologie* **32**, 1-25.

Walsby, A.E., Hayes, P.K., Boje, R. and Stal, L.J. (1997) The selective advantage of buoyancy provided by gas vesicles for planktonic cyanobacteria in the Baltic Sea, *New Phytologist* **136**, 407-417.

Walsby, A.E., Ng, G., Dunn, C. and Davis, P.A. (2004) Comparison of the depth where *Planktothrix rubescens* stratifies and the depth where the daily insolation supports its neutral buoyancy, *New Phytologist* **162**, 133-145.

Watanabe, M.F., Oishi, S., Harada, K.I., Matsuura, K., Kawai, H. and Suzuki, M. (1988) Toxins contained in *Microcystis* species of cyanobacteria (blue- green algae), *Toxicon* **26**, 1017-1025.

Whitelam, G.C. and Codd, G.A. (1983) Photoinhibition of photosynthesis in the cyanobacterium *Microcystis aeruginosa*, *Planta* **157**, 561-566.

Xie, L.Q., Xie, P. and Tang, H.J. (2003) Enhancement of dissolved phosphorus release from sediment to lake water by *Microcystis* blooms - an enclosure experiment in a hyper-eutrophic, subtropical Chinese lake, *Environmental Pollution* **122**, 391-399.

Zohary, T. and Robarts, R.D. (1989) Diurnal mixed layers and the long-term dominance of *Microcystis aeruginosa*, *Journal of Plankton Research* **11**, 25-40.

Zohary, T. and Pais-Madeira, A.M. (1990) Structural, physical and chemical characteristics of *Microcystis aeruginosa* from a hypertrophic lake, *Freshwater Biology* **23**, 339-352.

Zohary, T. and Robarts, R.D. (1990) Hyperscums and the population dynamics of *Microcystis aeruginosa*, *Journal of Plankton Research* **12**, 423-432.

CHAPTER 7

POPULATION DYNAMICS OF HARMFUL CYANOBACTERIA

Factors affecting species composition

Jef HUISMAN & Florence D. HULOT

1 Introduction

The growth potential of cyanobacteria is impressive. Under ideal circumstances, cyanobacteria typically double once or twice per day. In 40 days, a single cyanobacterium that doubles twice a day will have multiplied to $2^{80} \approx 10^{24}$ cells. This huge number is sufficient to cover an area of 1.000.000 km^2 in which the upper meter of the water column is populated by an extremely dense population of 1.000.000 cyanobacterial cells per millilitre of water. Adding another 20 days would suffice to cover the entire planet Earth with a dense surface bloom. A frightful thought! Why, then, are most lakes and oceans of our planet not covered by immense cyanobacterial scums?

This chapter deals with the population dynamics of cyanobacteria. We review how nutrients may limit the development of cyanobacterial populations, how mixing processes drive the expansion or demise of cyanobacterial blooms, how competition and predation affect the species composition of these blooms, and how careful water management strategies can help to counter the development of harmful cyanobacteria.

2 Nutrient limitation

2.1 THE BASICS

In reality, exponential growth of cyanobacteria cannot be sustained for a prolonged time because, at some point, nutrients become exhausted, the light energy becomes insufficient to support a larger population, or predators and parasites keep further population growth in check. Phosphorus and/or nitrogen limitation of phytoplankton growth is commonly observed (Schindler, 1974; Sommer, 1989; Elser et al., 1990; Sterner, 1994). In several marine ecosystems, iron is an important limiting nutrient

J. Huisman, H.C.P. Matthijs and P.M. Visser (eds.), Harmful Cyanobacteria, 143–176.

(Martin and Fitzwater, 1988; De Baar et al., 1995; Behrenfeld et al., 1996). Occasional limitations of trace metals like cadmium, cobalt, manganese, and zinc have also been found (Coale, 1991; Morel et al., 1994; Lee et al., 1995). In soft waters, inorganic carbon may limit population growth in dense cyanobacterial blooms (Klemer et al., 1995; Ibelings and Maberly, 1998).

To better understand how nutrients may limit the growth of cyanobacterial populations, consider a simple model that couples the growth of a cyanobacterial population to the dynamics of a potentially limiting nutrient. Let N denote the population density of the cyanobacteria, and let R denote the concentration of the limiting nutrient. We assume that the population dynamics of the cyanobacteria are driven by growth and mortality. Furthermore, we assume that the nutrient dynamics are driven by the supply of nutrients (for instance, via mineralization or by nutrient inputs from discharged water), losses of nutrients from the system (e.g., washout, denitrification), and consumption of nutrients by the cyanobacteria. The model then reads:

$$\frac{dN}{dt} = \mu(R)N - mN \tag{1a}$$

$$\frac{dR}{dt} = S - DR - c\mu(R)N \tag{1b}$$

Here, $\mu(R)$ is the specific growth rate of the cyanobacteria as a function of nutrient availability R, m is the specific mortality rate of the cyanobacteria, S is the rate at which new nutrients are supplied, D is the specific loss rate of nutrients from the system, and c is the amount of nutrient that must be consumed to produce one cyanobacterial cell.

We assume that the specific growth rate is an increasing function of nutrient availability that asymptotically approaches the maximal specific growth rate (as in Fig. 1A,B). This function can be described by, for instance, Monod's (1950) equation:

$$\mu(R) = \frac{\mu_{\max} R}{H + R} \tag{2}$$

Here, $\mu_{\max}$ is the maximal specific growth rate, and H is the half-saturation constant of nutrient-limited growth. The latter indicates the nutrient availability at which the species grows at half its maximal specific growth rate. The Monod equation appears valid, as a first approximation, for many species and potentially limiting nutrients.

The model specified by Eqs.1a,b predicts that there is a critical nutrient availability, which we shall call R^*, at which the specific growth rate exactly equals the specific mortality rate. If nutrient availability exceeds this critical level R^*, growth exceeds mortality, and the phytoplankton population will increase. The growing population will consume more and more nutrients, and as a consequence nutrient availability diminishes, until the nutrient availability has been depleted to the level R^*. At this point, specific growth rate equals specific mortality rate, and the phytoplankton population stabilizes.

2.2 COMPETITION AND THE R^* RULE

Phytoplankton species that consume the same nutrients reduce the availability of these nutrients both for themselves and for competing species. As a result, competition for limiting nutrients may have a major impact on the species composition of phytoplankton communities. An example is given in Figure 1. Here, the maximum specific growth rate of species A is higher than the maximum specific growth rate of species B (compare Fig. 1A and Fig. 1B). However, the mortality rate of species B is lower than the mortality rate of species A. There is a certain critical nutrient availability for species A, which we shall call R_A^*, at which the specific production rate of species A exactly equals its mortality rate. Similarly, there is a critical nutrient availability for species B, which we shall call R_B^*. It turns out that R_B^* is lower than R_A^* (compare Fig. 1A and Fig. 1B). Thus, species B can survive at a lower nutrient availability than species A. The model predicts that, initially, when nutrient availability is high, both species will increase. Thereby, the two species consume nutrients, and nutrient availability thus diminishes. Species A starts to decrease when the nutrient availability has been depleted below the level R_A^*. Species B, however, continues to increase as long as the nutrient availability exceeds R_B^*. As a consequence, owing to nutrient depletion, in the end species B wins (Fig. 1C).

The example of Figure 1 can be generalised to any number of species. When a number of species compete for a single nutrient in a homogeneous, well-mixed and constant environment, the winner of competition is the species with the lowest R^* value (Hsu et al., 1977; Armstrong and McGehee, 1980; Tilman, 1982). Hence, suppose that there are 5 species, and that their R^* values are ranked as follows:

$$R_A^* < R_B^* < R_C^* < R_D^* < R_E^* \tag{3}$$

Here, theory predicts that species A will outcompete all other species. Species B will be second-best, species C ranks third, and so on. Thus, when species compete for a single nutrient, the species that can survive at the lowest nutrient level will prevail. This result, known as the R^* rule, is a fundamental principle in competition theory. The R^* rule has been confirmed by numerous competition experiments (Tilman, 1977; Holm and Armstrong, 1981; Sommer, 1986; Grover, 1989; Van Donk and Kilham, 1990; Rothhaupt, 1996; De Nobel et al., 1997; Ducobu et al., 1998).

We emphasize that in a strict sense the R^* rule applies only under ideal conditions. It assumes spatial homogeneity ensured by complete mixing and a constant environment. Also, it assumes that species interact via nutrient consumption only. Deviations from the R^* rule may occur when the water column is not completely mixed or under temporally fluctuating conditions. Furthermore, deviations from the R^* rule may occur when species interact not only via consumption of nutrients but also via other mechanisms, for instance via the production and release of toxins. Hence, the R^* rule can best be interpreted as a basic null model in competition theory, to which further complexities can be added.

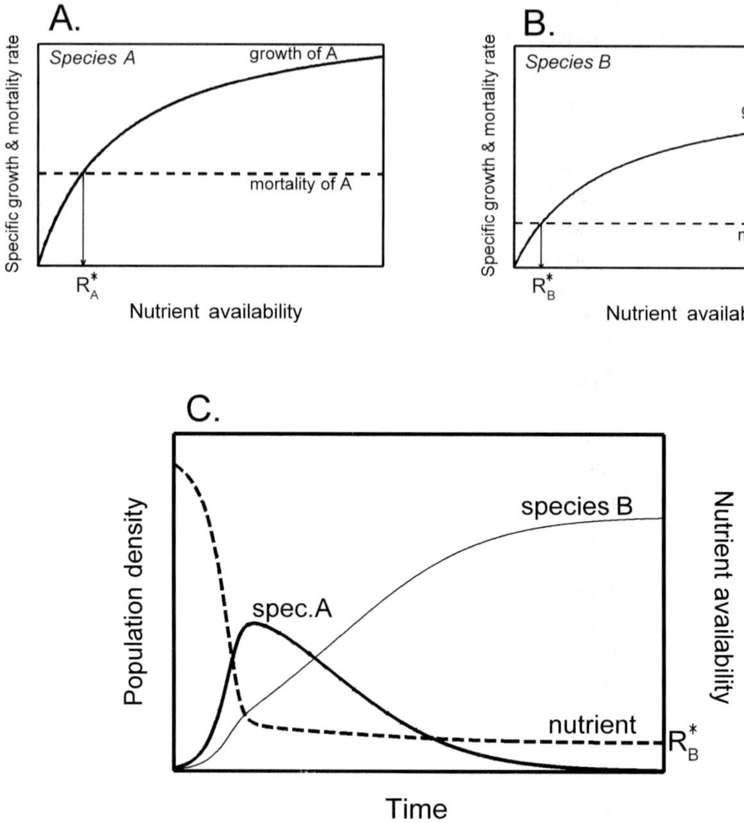

Figure 1. Competition for a single nutrient in a constant environment. (A) and (B) show the specific growth rates and specific mortality rates of species A and species B, respectively, as functions of nutrient availability. Note that species B has a lower R^* value than species A. (C) The time course of competition. The species with lowest R^* wins.

2.3 COMPETITION FOR PHOSPHORUS

Phosphorus limits the growth rate and total biomass of phytoplankton in a wide variety of freshwater ecosystems (Schindler, 1974; Sommer, 1989; Elser et al., 1990; Sterner, 1994) and in several marine ecosystems as well (e.g., Thingstad et al., 1998; Ammerman et al., 2003). Since phytoplankton species can differ considerably in their phosphorus requirements, phosphorus limitation has major implications for phytoplankton competition. As an example, we discuss the work of Ducobu et al. (1998), who studied competition for phosphorus between two freshwater phytoplankton species: the potentially toxic cyanobacterium *Planktothrix agardhii* and the prochlorophyte *Prochlorothrix hollandica*. Ducobu et al. (1998) measured in monoculture experiments that *Planktothrix* has a higher maximal production rate than *Prochlorothrix*. *Prochlorothrix*, however, has a much higher affinity for phosphorus than *Planktothrix*. As a result, *Prochlorothrix* has a lower R^* value for phosphorus than *Planktothrix*. In addition to its high affinity for phosphorus, *Prochlorothrix* is a true phosphorus storage specialist as well. It can store phosphorus in the form of polyphosphates, even to the extent that these polyphosphates can make up more than 10% of its total cellular dry weight (Ducobu et al., 1998). This internal storage capacity allows *Prochlorothrix* to survive during prolonged periods of low external phosphorus availability. Thus, competition theory predicts that the prochlorophyte *Prochlorothrix* should be a better competitor for phosphorus than the cyanobacterium *Planktothrix*. Indeed, laboratory competition experiments confirmed that *Prochlorothrix* competitively displaces *Planktothrix* under both constant and fluctuating phosphorus conditions (Fig. 2).

Several other studies also investigated the competitive abilities of cyanobacteria for phosphorus (Holm and Armstrong, 1981; Tilman et al., 1986; Grover, 1989; Hu and Zhang, 1993; De Nobel et al., 1997; Fujimoto et al., 1997). These studies revealed that, generally speaking, cyanobacteria are poor competitors for phosphorus in comparison to diatoms, prochlorophytes, and green algae. This is in line with the common observation that cyanobacteria seldom reach dominance in oligotrophic waters where phosphorus is a major limiting factor (Duarte and Canfield, 1992; Downing et al., 2001).

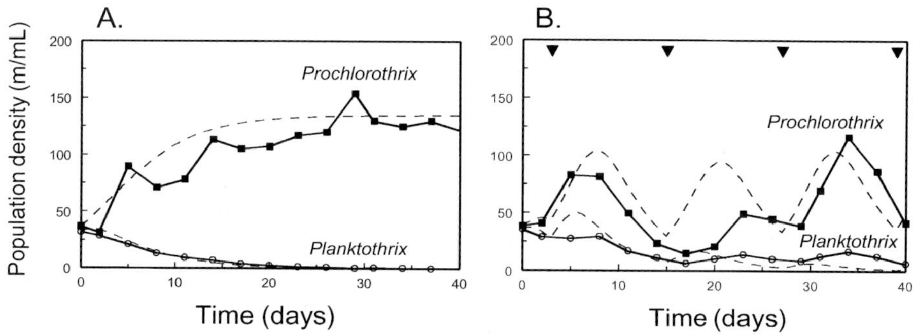

Figure 2. Competition for phosphorus between the cyanobacterium Planktothrix agardhii *and the prochlorophyte* Prochlorothrix hollandica. *(A) Phosphorus is supplied at a constant rate. (B) Phosphorus is delivered in pulses added once every 12 days. The inverted triangles in B indicate the timing of the phosphorus additions. The symbols and solid lines indicate the experimental data, whereas the dashed lines indicate the predictions of the competition model. Population densities are expressed in terms of total filament length per species (in m mL^{-1}). (Redrawn after Ducobu et al., 1998, with permission from Blackwell Publishing).*

There are exceptions to this broad generalization, though. In particular, small pico-cyanobacteria like *Synechococcus* and *Synechocystis* are very efficient competitors for phosphorus (Moutin et al., 2002; J. Passarge, unpublished results). Pico-cyanobacteria dominate the phytoplankton in some lakes (Postius and Ernst, 1999) and large parts of the oligotrophic ocean (Agawin et al., 2000; Irigoien et al., 2004; Chapter 5 in this book). Furthermore, several cyanobacteria, including species from the genera *Planktolyngbya*, *Anabaena*, and *Aphanizomenon*, appear to be strong competitors for phosphorus in lakes with high levels of organically bound phosphorus. These cyanobacteria produce a special enzyme, alkaline phosphatase, to release phosphate from small organic phosphorus compounds (Jansson et al., 1988; Dignum et al., 2004; Chapter 4 in this book).

2.4 NITROGEN FIXATION AND N:P RATIOS

In addition to phosphorus, nitrogen is an important element that may limit the growth of phytoplankton in both freshwater and marine ecosystems. Many cyanobacteria are capable to fix nitrogen gas (Fay, 1992; Bergman et al., 1997; Gallon, 2001), including heterocyst-forming species like *Anabaena*, *Nodularia*, and *Nostoc* as well as non-heterocystous species like *Gloeocapsa*, *Trichodesmium*, and some *Synechococcus* species. Although nitrogen fixation is quite costly from a physiological perspective, nitrogen fixation is advantageous when concentrations of dissolved nitrate and ammonium are low.

Suppose that a nitrogen-fixing cyanobacterium and a phytoplankton species incapable of nitrogen fixation compete for phosphorus and nitrogen. This competitive interaction can be analysed graphically, using the isocline approach developed by Tilman (1982). The graph consists of zero net growth isoclines, consumption vectors, and nutrient supply points (Fig. 3). The zero net growth isocline (ZNGI) of a species depicts the nitrogen and phosphorus availabilities at which this species would remain in equilibrium. Assuming Von Liebig's (1840) 'Law of the Minimum', the ZNGI of the non-nitrogen fixer (species B in Fig. 3A) can be plotted as two arms at a right-angle corner. The positions of the two arms are given by the R^* values of this species for dissolved phosphorus and dissolved nitrogen, respectively. In contrast, the nitrogen-fixer (species A) does not rely on dissolved nitrogen because it can use atmospheric nitrogen. Hence, the ZNGI of the nitrogen-fixer is a vertical line, given by the R^* value of the nitrogen-fixer for dissolved phosphorus. Generally speaking, nitrogen-fixing cyanobacteria are poor competitors for phosphorus. We therefore assume that the R^* value for phosphorus of the nitrogen-fixer is higher than that of the non-nitrogen-fixer. As a consequence, the ZNGIs of the two species intersect (Fig. 3A). The consumption vector of a species is given by the ratio at which this species consumes dissolved nitrogen and phosphorus. Whereas the consumption vector of the non-nitrogen fixer is directed diagonally, the consumption vector of the nitrogen-fixer is directed horizontally since it does not rely on dissolved nitrogen (Fig. 3A). The final ingredient in this graphical approach is the nutrient supply point, which indicates the rates at which nitrogen and phosphorus are supplied.

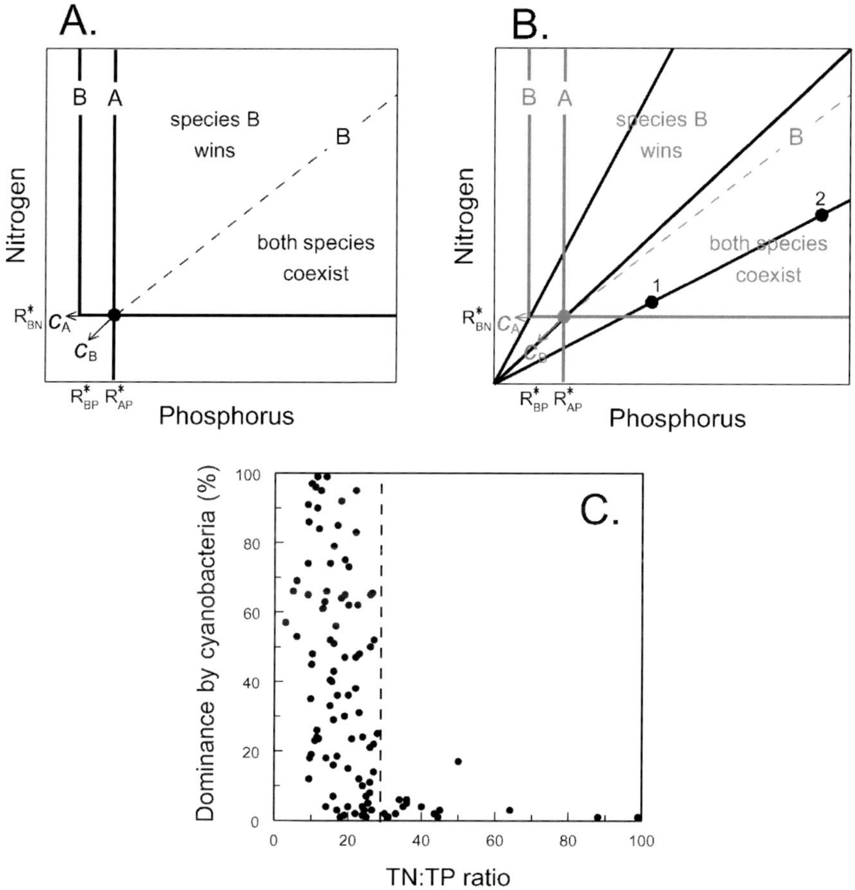

Figure 3. Competition for nitrogen and phosphorus between a nitrogen-fixing cyanobacterium (species A) and a non-nitrogen-fixing phytoplankton species (species B). (A) Theory. The solid lines show the zero net growth isoclines of the two species plotted in the plane of nitrogen and phosphorus availability. The vectors C_A and C_B are the consumption vectors of the two species. (B) The three lines represent three different N:P supply ratios. The two nutrient supply points, 1 and 2, fall on the same line and hence have the same N:P supply ratio. (C) Observed relation between the dominance of cyanobacteria and the ratio of total nitrogen (TN) to total phosphorus (TP) in 17 lakes. Each data point represents one lake season. The vertical dashed line indicates TN:TP=29. (C) is redrawn after Smith (1983), with permission from Science. © 1983 AAAS.

For all supply points that are located above the dashed line in Figure 3A, where the supply of dissolved phosphorus is relatively low compared to the supply of dissolved nitrogen, both species are limited by phosphorus. Hence, in this parameter region, the best phosphorus competitor (i.e., the non-nitrogen fixer) wins. In contrast, for all supply points that fall below the dashed line the supply of dissolved nitrogen is relatively low. Here, the non-nitrogen fixer is limited by the low availability of dissolved nitrogen, but the nitrogen-fixer is still limited by the availability of dissolved phosphorus. Hence, neither of the species can competitively displace the other. As a result, for all supply points below the dashed line, the two species stably coexist (Fig. 3A).

A key prediction of this competition model is that the species composition depends on the ratio of nitrogen supply to phosphorus supply. In Figure 3B, we have drawn three lines representing three different N:P supply ratios. This shows that, as the N:P supply ratio increases, the species composition shifts from coexistence of the species at low N:P ratios to a system dominated by non-nitrogen fixing phytoplankton at high N:P ratios.

To test the theoretical hypothesis that low N:P ratios promote dominance by nitrogen-fixing cyanobacteria, Smith (1983) compiled data from 17 lakes. The ratio of total nitrogen concentration (TN) to total phosphorus concentration (TP) in the water column was used as a surrogate for the N:P supply ratio. The data show that cyanobacteria were rare when TN:TP ratios exceeded 29 by weight, whereas cyanobacteria could become dominant when TN:TP ratios remained below 29 by weight (Fig. 3C). This result indeed supports the hypothesis that N:P ratios are an important determinant of cyanobacterial dominance. These findings are of major practical importance from a water management perspective, because N:P ratios can be manipulated in many lakes, for instance, by sewage diversion, phosphorus removal, or nitrogen addition.

Several later studies, however, have criticized the emphasis on N:P ratios to explain cyanobacterial dominance. Statistical analyses of more extensive data sets revealed that, even though cyanobacterial dominance in lakes correlated with TN:TP ratios, it correlated even better with absolute levels of TN and TP (Trimbee and Prepas, 1987; Downing et al., 2001). There are a variety of reasons that may explain these observations. In particular, whereas the theory described above focuses on nitrogen fixation, not all cyanobacteria are nitrogen fixers. A flaw in the studies of Smith (1983), Trimbee and Prepas (1987), and Downing et al. (2001) is that they did not distinguish between nitrogen-fixing cyanobacteria and non-nitrogen-fixing cyanobacteria. Also, as described in the previous section, some cyanobacterial species are very efficient P competitors, and these will not be negatively affected by a high N:P ratio. Finally, crucial to the hypothesis that N:P ratios determine cyanobacterial dominance is the underlying assumption that N and P are the only limiting factors. In several oligotrophic parts of the oceans, low iron availability may impede nitrogen fixation, thus preventing the dominance of nitrogen-fixing cyanobacteria (Falkowski, 2000). In eutrophic lakes with high nitrogen and phosphorus levels, factors like light or inorganic carbon may limit cyanobacterial growth (Mur et al., 1977; Ibelings and Maberly, 1998).

3 Competition for light in well-mixed waters

3.1 THE BASICS

Light intensity decreases with depth, because photons are absorbed by water, dissolved organic matter, phytoplankton, clay particles, and many other light-absorbing substances. In the context of phytoplankton growth, it is generally useful to distinguish between light absorption caused by the phytoplankton themselves and the background turbidity caused by the non-phytoplankton components. More specifically, the decrease of light intensity with depth can be described by Lambert-Beer's law. Let z denote the depth of the water column, where z ranges from z=0 at the top to z=z_m at the bottom of the water column. Furthermore, let $I(z,t)$ denote the light intensity at depth z and time t. Hence, according to Lambert-Beer's Law, the light gradient can be described as:

$$I(z,t) = I_{in} \exp(-K_{bg}z - kNz) \tag{4}$$

where I_{in} is the incident light intensity at the top of the water column, K_{bg} is the background turbidity caused by non-phytoplankton components, and k is the specific light attenuation coefficient of the phytoplankton. We note, from this equation, that the light gradient is dynamic. An increased phytoplankton population absorbs more light, and hence leads to a more turbid water column.

Since light intensity decreases with depth, photosynthesis varies with depth as well. Accordingly, if cyanobacteria are uniformly distributed throughout the water column by complete mixing, the population dynamics of light-limited cyanobacteria can be described as (Huisman and Weissing, 1994):

$$\frac{dN}{dt} = \frac{1}{z_m}\int_0^{z_m} \mu(I(z))N dz - mN \tag{5}$$

Here, $\mu(I)$ is the specific growth rate of the cyanobacteria as an increasing saturating function of light intensity, and m is the specific mortality rate of the cyanobacteria. There are a variety of mathematical expressions to describe the dependence of the specific growth rate on light intensity (Jassby and Platt, 1976; Henley, 1993). Here we again use the Monod equation (Eq.2).

The population dynamics described by this light-limited model lead to an equilibrium. In fact, Eq.5 predicts that the population dynamics of light-limited phytoplankton will always equilibrate at a particular value of the light penetration at the bottom of the water column. This equilibrium light penetration has been called the 'critical light intensity' for phytoplankton growth (Huisman and Weissing, 1994), and is denoted by I^*_{out}. The model predicts that a species will increase if the light penetration through the water column exceeds its critical light intensity. Conversely, a species will decline if light penetration is less than its critical light intensity. The critical light intensity is species specific. Its value depends on the growth characteristics and loss characteristics of the species concerned. As a result, the critical light intensity of a species also depends on environmental factors that affect these growth and loss characteristics, such as incident light intensity, temperature, pH, and so on.

Owing to the complexity of Eq.5, it is not possible to derive a simple analytical equation for the critical light intensity of a species. However, the value of the critical light intensity can be estimated in two different ways. Firstly, the value of the critical light intensity can be measured directly, using a light-limited chemostat. It is simply the light penetration through the chemostat culture measured at steady state. Secondly, the value of the critical light intensity can be calculated numerically, by estimating all components of Eq.5 and subsequent numerical simulation of the model until an equilibrium is reached. Laboratory experiments show that, generally speaking, these two different methods yield similar estimates of the critical light intensity (Huisman et al., 1999a).

The equilibrium population density can be expressed as a function of the critical light intensity. That is, according to Lambert-Beer's law (see Eq.4), the equilibrium population density, N^*, can be written as:

$$N^* = \frac{\ln(I_{in}) - \ln(I^*_{out})}{k\, z_m} - \frac{K_{bg}}{k} \tag{6}$$

One interesting prediction of this equation is that the equilibrium population density should be inversely proportional to water-column depth. To test this prediction, Huisman (1999) ran light-limited chemostats with the green alga *Chlorella vulgaris* using four different water-column depths. Figures 4A-D show the time courses of these experiments. The experiments demonstrate that, despite large differences in water-column depth and population density, light penetration in all four chemostat experiments equilibrated at the same steady-state value (ca. 5 μmol photons $m^{-2}\ s^{-1}$). This confirms the model prediction that the critical light intensity is independent of water-column depth. Furthermore, the chemostat experiments demonstrate that population densities are much higher in shallow systems than in deep systems, fully in line with the predictions of Eq.6. The explanation is straightforward. Light energy is supplied as a flux, and a given photon flux per unit surface area can sustain only a given production per unit surface area. In deep systems, this same production per unit surface area is divided over a much larger volume than in shallow systems. Hence, the steady-state population density in well-mixed waters is inversely related to water-column depth (Fig. 4E).

Our experiments were performed in small laboratory chemostats with mixing depths ranging from 3 to 20 cm. At a larger scale, in mesocosms, similar scaling arguments were found by Petersen et al. (1997). They observed that primary production per unit volume was inversely proportional to water-column depth in light-limited mesoscosms with mixing depths of 45-215 cm. At again a larger scale, in field experiments, Diehl et al.

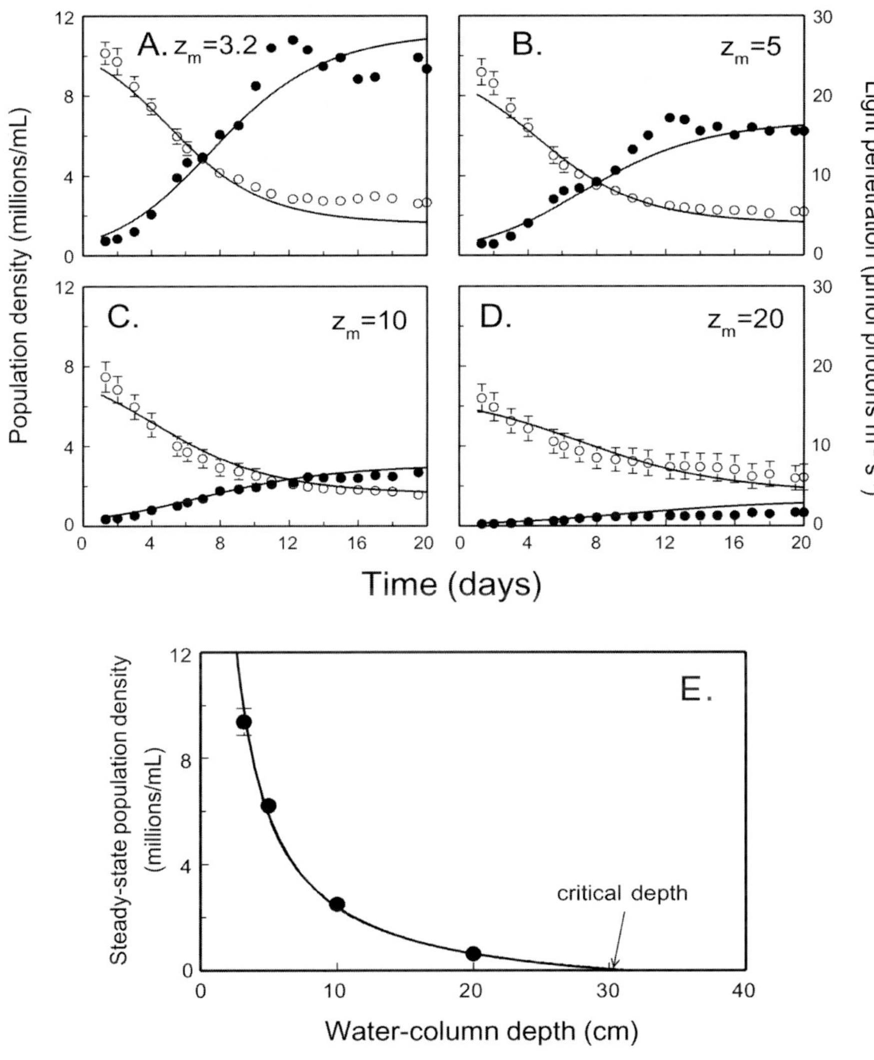

Figure 4. Light-limited growth in well-mixed waters. (A-D) Time course of the population density of Chlorella vulgaris *(closed circles) and light penetration (open circles) in four light-limited chemostats, each with a different mixing depth: (A) z_m = 3.2 cm, (B) z_m = 5 cm, (C) z_m = 10 cm, (D) z_m = 20 cm. Solid lines show the time courses predicted by the model. (E) The steady-state population densities obtained in the experiments (A-D), plotted against water-column depth. The solid line shows the steady-state population density predicted by Eq.6. Beyond a critical mixing depth of 30 cm, the* Chlorella *population would be washed out. (Redrawn after Huisman, 1999, with permission from the Ecological Society of America).*

(2002) report the same inverse relation between primary production per unit surface area and water-column depth, using water-column depths of 2-15 m. Hence, this scaling rule of population density versus water-column depth seems to work well over several orders of magnitude, provided that the phytoplankton is mixed over the entire depth of the water column. As a result, all else being equal, population densities under well-mixed, light-limited conditions are predicted to be roughly 10 times higher in a shallow lake of 2 meter deep than in a deep lake of 20 m depth.

Extrapolating the trend in Figure 4E shows that the curve will intersect the x-axis. Hence, there is a maximum water-column depth beyond which the background turbidity absorbs so much light that a phytoplankton population cannot be sustained. Following Sverdrup's (1953) terminology, we shall call this depth the 'critical depth'. The critical depth, z_{cr}, can be obtained by solving Eq.6 for $N^*=0$. This yields:

$$z_{cr} = \frac{\ln(I_{in}) - \ln(I^*_{out})}{K_{bg}} \tag{7}$$

This equation shows that the critical depth for a phytoplankton population can be calculated from its critical light intensity.

3.2 ARE CYANOBACTERIA STRONG COMPETITORS FOR LIGHT ?

Theory predicts that the outcome of competition for light in well-mixed waters depends on the 'critical light intensities' of the species. More precisely, the species with the lowest critical light intensity is predicted to be the superior competitor for light (Huisman and Weissing, 1994). To test this theoretical prediction, Huisman et al. (1999a) studied competition for light between the green alga *Chlorella vulgaris*, the cyanobacterium *Aphanizomenon flos-aquae* and the cyanobacterium *Microcystis aeruginosa*. The species were grown in the laboratory, using continuous culture systems specifically designed to study light-limited phytoplankton. Monoculture experiments showed that the critical light intensity of *Chlorella* was lower than the critical light intensities of the two cyanobacterial species. In a competition experiment between *Chlorella* and *Aphanizomenon*, in which both species were inoculated with a low initial population density, both *Chlorella* and *Aphanizomenon* initially increased (Fig. 5A). Owing to light absorption by the two phytoplankton species, light penetration through the water column decreased with increasing population densities. As soon as the light penetration was reduced below the critical light intensity of *Aphanizomenon*, *Aphanizomenon* started to decline whereas *Chlorella* continued to increase. Hence, in the end, *Chlorella* competitively displaced *Aphanizomenon* (Fig. 5A). In a similar fashion, in the other competition experiment, *Chlorella* competitively displaced *Microcystis* (Fig. 5B). Thus, as predicted by theory, the species with the lowest critical light intensity was indeed the better competitor for light in well-mixed waters.

Are cyanobacteria strong competitors for light? The early competition experiments of Mur et al. (1977) revealed that the cyanobacterium *Planktothrix agardhii* is a much better competitor for light than the green alga *Scenedesmus protuberans*. At the time, this finding led to the suggestion that cyanobacteria are generally better adapted to low light conditions, and hence better competitors for light, than green algae and diatoms. This suggestion received further support from the observation that prokaryotic cyanobacteria have lower maintenance costs than their eukaryotic competitors (van Liere and Mur, 1979; Gons and Mur, 1980). In contrast, in the competition experiments of Huisman et al. (1999a) the cyanobacteria *Aphanizomenon* and *Microcystis* were both competitively displaced by the green alga *Chlorella*. In line with still unpublished work from our lab, we therefore conjecture that cyanobacteria may differ widely in their critical light intensities. The filamentous cyanobacterium *Planktothrix agardhii* seems to have a very low critical light intensity. Hence, its dominance in various shallow eutrophic lakes may indeed be explained by its low light requirements compared to other phytoplankton species, consistent with the experiments of Mur et al. (1977). The buoyant cyanobacteria *Aphanizomenon* and *Microcystis* have higher light requirements than *Planktothrix*, however. Hence, *Aphanizomenon* and especially *Microcystis* are relatively poor competitors for light in well-mixed waters.

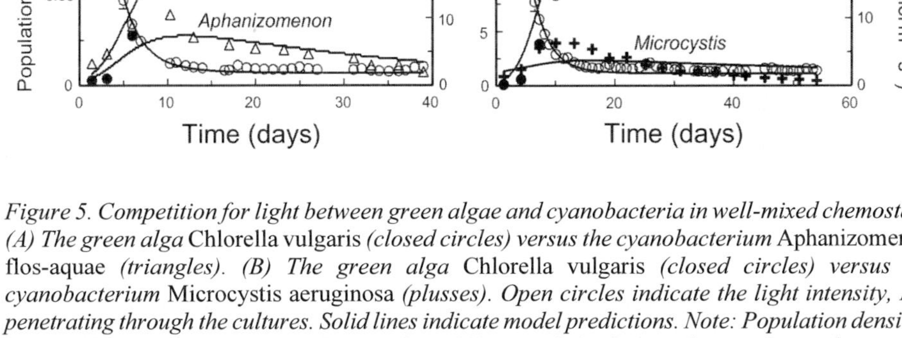

Figure 5. Competition for light between green algae and cyanobacteria in well-mixed chemostats. (A) The green alga Chlorella vulgaris *(closed circles) versus the cyanobacterium* Aphanizomenon flos-aquae *(triangles). (B) The green alga* Chlorella vulgaris *(closed circles) versus the cyanobacterium* Microcystis aeruginosa *(plusses). Open circles indicate the light intensity, I_{out}, penetrating through the cultures. Solid lines indicate model predictions. Note: Population densities of* Aphanizomenon *were counted as number of filaments/ml, whereas the population densities of* Chlorella *and* Microcystis *were counted as number of cells/ml. (Redrawn after Huisman et al., 1999a, with permission from the Ecological Society of America).*

4 Hydrodynamics and surface blooms

4.1 THE BASICS

Physical mixing processes have a profound impact on the population dynamics of phytoplankton (Riley et al., 1949; Hutchinson, 1967; Spigel and Imberger, 1987; Reynolds, 1997). Vertical mixing, in particular, is a major determinant of the development of surface blooms of harmful cyanobacteria. On the one hand, cyanobacteria require light for photosynthesis. Hence, cyanobacteria will proliferate only if they remain in the euphotic zone. On the other hand, most cellular components have a slightly higher density than water. As a result, many cyanobacterial species tend to sink into the deep (Smayda, 1970). There are also a number of buoyant cyanobacteria, however, including several potentially toxic species like *Microcystis*, *Aphanizomenon*, and *Anabaena*. These buoyant cyanobacteria produce gas vesicles, giving their cells a lower density than water, so that they may float upwards (Walsby, 1994; Chapter 6 in this book).

During quiet days with little wind mixing, buoyant cyanobacteria that float upwards may form surface blooms. These surface blooms can reach very high population densities, even if cyanobacteria were initially present in the water column at low concentrations only. This is because the development of a surface bloom concentrates all buoyant cells in a very thin surface layer. For instance, consider a lake of 5 m deep with buoyant cyanobacteria uniformly mixed throughout the water column at a low concentration of 1000 cells mL^{-1}. If all these cells would float upward to form a surface bloom of 1 cm thickness, the concentration in this surface bloom would be 500.000 cells mL^{-1}. If, a few hours later, the wind picks up and blows the surface bloom towards the shore, the surface blooms may even be concentrated further and may pile up in dense scums against the lakeshore.

This section describes how the hydrodynamics of aquatic ecosystems interact with phytoplankton competition and the formation of surface blooms.

4.2 TURBULENT DIFFUSION

The word plankton stems from the Greek πλαγκτοσ, which means roaming or wandering (Hutchinson, 1974). Most plankton species wander passively through the water column, carried by the parcel of water in which they are suspended. These water parcels may flow in a specific direction, like the flow of a river. They may also drift around, by the random motion generated by turbulence. Turbulent motion is caused by different-sized whirls of water, known as eddies, created by wind and waves, by local density differences in the water column, by the frictional forces of streaming water, or by the wake of swimming objects. Eddies are intertwined, and may come and go as eddies have a transient nature. The size of eddies spans a broad spectrum, from a few millimetres to several meters and sometimes even several kilometres. The random motion generated by these slowly moving water parcels is generally referred to as turbulent diffusion, or eddy diffusion.

Consider the transport of particles caused by turbulent diffusion. For convenience, we consider transport in one dimension, along the x-axis only. We assume, for the moment, that the population of particles does not increase or decrease in numbers. Instead of a plankton population, it might be a population of small clay particles, for instance. Let N denote the concentration of particles. Furthermore, let J denote the flux of particles (i.e., the number of particles that crosses a unit area perpendicular to the x-axis). An increase of the flux of particles along the x-axis implies a decrease of the local particle concentration. More precisely,

$$\frac{\partial N}{\partial t} = -\frac{\partial J}{\partial x} \tag{8}$$

Analogous to molecular diffusion, the net flux of particles caused by turbulent diffusion is proportional to the concentration gradient:

$$J = -D\frac{\partial N}{\partial x} \tag{9}$$

Here, the minus sign indicates that the net flux of particles proceeds from high to low particle concentrations. The constant of proportionality, D, is known as the turbulent diffusion coefficient or eddy diffusivity.

Substituting Eq.9 into Eq.8, the change in particle concentration caused by turbulent diffusion can be described as:

$$\frac{\partial N}{\partial t} = D\frac{\partial^2 N}{\partial x^2} \tag{10}$$

The effect of turbulent diffusion is comparable to the effect of molecular diffusion by Brownian motion. The random motion of eddies causes a net transport of particles from high to low concentrations, and thereby it tends to homogenize the concentration of entrained particles. Turbulent diffusion, however, is much faster and therefore moves over a much larger scale than molecular diffusion.

4.3 BUOYANT VERSUS SINKING PHYTOPLANKTON

If mixing is weak, the vertical positions of cyanobacterial species are determined by their growth and loss characteristics, their sinking or floating velocity, and the random motion caused by turbulent diffusion. The dynamics of the vertical population density distribution can be described by a partial differential equation:

$$\frac{\partial N}{\partial t} = \mu(I(z))N - mN + v\frac{\partial N}{\partial z} + D\frac{\partial^2 N}{\partial z^2} \qquad (11)$$

Here, v is the vertical velocity of the phytoplankton by flotation or sinking (with $v > 0$ for buoyant species, and $v < 0$ for sinking species), and D is the vertical turbulent diffusion coefficient. As boundary conditions, we assume that there is no influx or efflux of phytoplankton, neither at the top nor at the bottom of the water column:

$$J(z,t) = 0 \qquad at\ z = 0\ and\ z = z_m \qquad (12)$$

Numerical simulation of this model is computationally quite demanding, because shading between phytoplankton species introduces a nonlocal term in the partial differential equation. An advanced simulation technique that tackles this computational problem is described in detail in Huisman and Sommeijer (2002). We simulated the model many times, for a wide range of different vertical turbulent diffusivities and water-column depths. For each simulation, we monitored whether a cyanobacterial bloom could develop. The results are shown in Figure 6. Note the log scales of the axes: the graphs span the entire spectrum from very shallow and quiescent lakes in the lower left corner of the graph to extremely deep and turbulent lakes in the upper right corner of the graph. Figure 6A applies to buoyant cyanobacteria. If the depth of the water column exceeds a critical depth (sensu Sverdrup, 1953) and turbulent mixing exceeds a critical turbulence (sensu Huisman et al., 1999b), the cyanobacteria will be mixed to great depths and the light conditions thus experienced are so dark that losses exceed production. In contrast, if turbulent mixing is less than the critical turbulence, buoyant cyanobacteria escape vertical mixing and float upwards to suitable light conditions. Hence, buoyant cyanobacteria can develop a bloom if water-column depth is less than a critical depth or if turbulent mixing is less than a critical turbulence. The exact values of the critical depth and critical turbulence depend on the species, and also depend on environmental conditions like the incident light intensity and the background turbidity of the water column. For instance, in murky waters with high background turbidity, the critical depth will be shallow and the critical turbulence will be low, since high turbidity prevents growth beyond a few meters depth. In contrast, in very clear waters where light penetrates more than 100 m deep, the critical depth may exceed several hundred meters and the critical turbulence will be high.

Figure 6B applies to sinking phytoplankton. Similar to buoyant cyanobacteria, sinking phytoplankton cannot develop a bloom if water-column depth exceeds a critical depth and turbulent mixing exceeds a maximal turbulence. Furthermore, sinking phytoplankton cannot be entrained in waters with a very low turbulence. That is, if turbulent mixing is less than a minimal turbulence, sinking phytoplankters will vanish into the deep (Riley et al., 1949; Shigesada and Okubo, 1981). Thus, there is a turbulence window for sinking phytoplankton. Sinking phytoplankton species can develop blooms in deep waters only if turbulent mixing takes an intermediate value, between a minimal and a maximal turbulence (Huisman et al., 2002; Ghosal and Mandre, 2003).

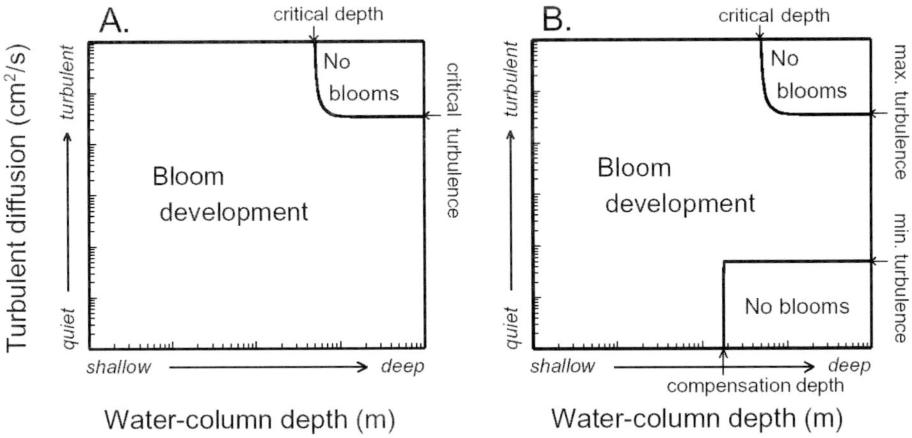

Figure 6. Combinations of water-column depth and vertical turbulent diffusion that allow bloom development for (A) buoyant phytoplankton, and (B) sinking phytoplankton. (Redrawn after Huisman et al., 1999b, 2002, with permission from (A) the American Society of Limnology and Oceanography, Inc, and (B) the University of Chicago Press).

4.4 WEAK MIXING AND COMPETITION FOR LIGHT

Changes in turbulent mixing may shift the competitive balance between buoyant and sinking phytoplankton species. For example, Huisman et al. (2004) modeled the impact of weak mixing on competition for light between the buoyant cyanobacterium *Microcystis* and sinking diatom species, using parameter estimates based on laboratory studies and numerical simulation techniques tailored to solve the system of coupled integro-partial differential equations. According to laboratory experiments, diatoms generally have lower light requirements and hence a lower critical light intensity than *Microcystis* (e.g., Reynolds, 1997; Huisman, 1999). As a consequence, diatoms should be superior competitors for light in well-mixed waters, at least compared to *Microcystis*. What happens, however, if mixing is weak? This scenario is illustrated in Figure 7A. Initially, the water column is relatively clear, and the species start with a low uniform population density distribution. Mixing is weak, just sufficient to entrain the sinking diatoms but insufficient to entrain the buoyant *Microcystis*. During the first weeks, the diatoms and *Microcystis* both increase, as there is sufficient light available for both species. The buoyant *Microcystis* colonies float upwards, however, and gradually form a dense surface bloom with population densities locally exceeding 400.000 cells mL^{-1}. As a result of the development of this dense surface bloom, nearly all light is absorbed in the upper meter of the water column. The diatoms are thus shaded by *Microcystis*, and start to decline. In the end, *Microcystis* wins (Fig. 7A).

Huisman et al. (2004) ran numerous simulations with the competition model, for a wide range of different water-column depths and vertical turbulent diffusivities. For each simulation, they documented the final outcome of competition. The results are plotted in Figure 7B. The diatoms win in turbulent waters. Conversely, *Microcystis* wins in quiet waters. In a narrow region at intermediate diffusivities, the two species coexist.

Strikingly, the pattern in Figure 7B is relatively simple. The boundary between the region where the diatoms win and the region where *Microcystis* wins is a narrow diagonal band with a slope of 1 when plotted on a log-log scale. The position of this boundary can be explained by one of the classic dimensionless numbers in hydrodynamics, the Péclet number. The Péclet number is defined as the ratio between the time scale of turbulent mixing and the time scale of vertical velocity (e.g., Lucas et al., 1998; Condie, 1999). The time scale of turbulent mixing indicates the time required for a phytoplankter to travel through the water column by means of turbulent mixing. It is given by $\tau_{mix}=(z_m)^2/D$. The time scale of the vertical velocity of a species indicates the time required for a phytoplankter to travel through the water column by means of its sinking rate or floating rate. It is given by $\tau_v=z_m/v$. Hence, the Péclet number, *Pe*, is defined as:

$$Pe = \frac{\tau_{mix}}{\tau_v} = \frac{z_m v}{D} \tag{13}$$

When this equation is plotted in a log-log graph of D versus z_m, for a constant value of the Péclet number, it reveals a straight line with slope 1 (as in Fig. 7B). We note that the Péclet number is species specific, because the vertical velocity is different for different phytoplankton species. Here, we focus on the Péclet number of *Microcystis*, because the simulations indicate that *Microcystis* wins the competition when it is able to overtop the diatoms. The diagonal boundary line between '*Microcystis* wins' and 'diatoms win', in Figure 7B, is described by a Péclet number of Pe=7. That is, if the rate of turbulent mixing exceeds the vertical floating rate of *Microcystis* (i.e., $Pe < 1$), both *Microcystis* and the diatoms will be entrained in the water column, *Microcystis* cannot form a surface bloom, and the diatoms will win. Conversely, if the vertical floating rate of *Microcystis* exceeds the rate of turbulent mixing (i.e., $Pe > 10$), *Microcystis* escapes from entrainment, forms a surface bloom, and hence *Microcystis* wins. The prediction, in Figure 7B, that *Microcystis* is favoured more in deep waters than in shallow waters is consistent with field observations (see Chapter 6), and can also be explained by the Péclet number. According to Eq.13, the Péclet number is proportional to water-column depth. That is, *Microcystis* can escape from entrainment more easily in deep waters than in shallow waters.

In conclusion, the model illustrates that buoyant cyanobacteria have a competitive advantage during periods of weak mixing, and especially so in deep waters. Although dimensionless numbers provide at best only a simple approximation of the complete dynamics of a system, the Péclet number may provide a useful rule of thumb for water management purposes, to predict whether buoyant cyanobacteria will be able to form surface blooms.

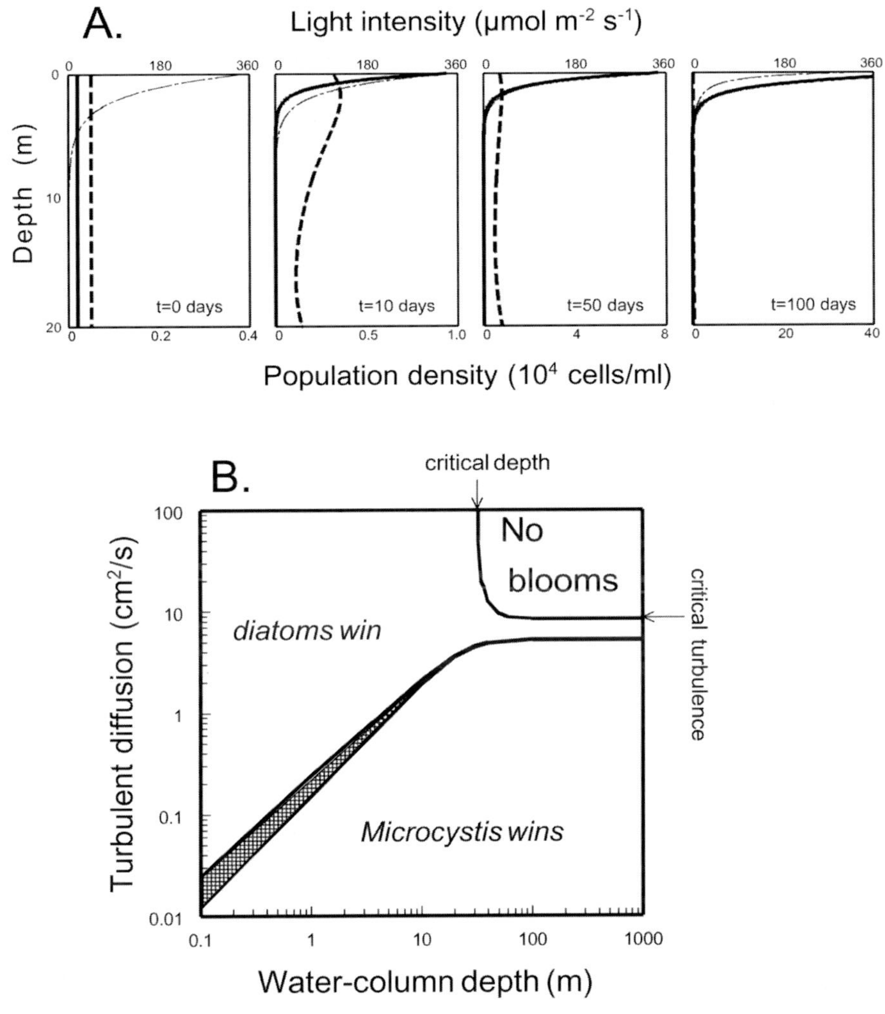

Figure 7. Competition for light between the buoyant cyanobacterium Microcystis *and sinking diatoms. (A) A model simulation assuming weak vertical mixing. The bold dashed line indicates the population density distribution of the diatoms, and the bold solid line indicates* Microcystis. *The thin dashed line indicates the light gradient. In the end,* Microcystis *wins. (B) The final outcome of competition, plotted for different combinations of water-column depth and vertical turbulent diffusion. In the narrow hatched area, the cyanobacterium and the diatoms coexist. (Redrawn after Huisman et al., 2004, with permission from the Ecological Society of America).*

4.5 FIELD STUDIES ON THE ROLE OF HYDRODYNAMICS

The prediction that changes in turbulent mixing may shift the competitive dominance between sinking and buoyant phytoplankton species is consistent with field observations. In many aquatic ecosystems shifts from dominance by green algae and/or diatoms during periods of intense vertical mixing to dominance by buoyant cyanobacteria during weak vertical mixing have been observed (Reynolds et al., 1984; Harris and Baxter, 1996; Walsby et al., 1997; Bormans and Condie, 1998; Sherman et al., 1998; McCausland, 2001). A well-studied example is Lake Nieuwe Meer, a recreational lake in the city of Amsterdam, The Netherlands (Visser, 1995; Visser et al., 1996). Lake Nieuwe Meer is a hypertrophic lake, connected to the canals of Amsterdam. To the despair of the local water management authorities, surface blooms of the buoyant cyanobacterium *Microcystis aeruginosa* dominated the lake during the summer months for many years. Die-off of dense surface blooms of *Microcystis* in autumn spread a disgusting smell and caused anoxia. Moreover, many *Microcystis* strains produce microcystins, which can damage the liver of fish, birds, and mammals, including humans. Thereby, *Microcystis* blooms provide a serious threat for water quality, fisheries, and human health (Chorus and Bartram, 1999; Codd et al., 1999; Chapter 1 in this book). In 1993, the water authority responsible for Lake Nieuwe Meer started artificial mixing of the lake by means of air bubbling, using a system of perforated air tubes installed just above the lake sediment. Artificial mixing led to major changes in phytoplankton species composition. Surface blooms of the buoyant cyanobacterium *Microcystis* were replaced by a mixture of sinking phytoplankton species, most notably diatoms from the genera *Cyclotella* and *Stephanodiscus* and large green algae from the *Scenedesmus* genus (Fig. 8). Detailed measurements of the temperature microstructure, using autonomous microstructure profilers (Kocsis et al., 1999; Sharples et al., 2001), recently allowed estimation of the vertical turbulent diffusivities in the lake, both with and without artificial mixing. This demonstrated that the predictions of the competition model are consistent with the observed changes in phytoplankton species composition: diatoms and green algae dominated for $Pe < 1$, whereas *Microcystis* dominated for $Pe > 10$ (Huisman et al., 2004). Thus, Lake Nieuwe Meer is a clear example where changes in turbulent mixing shifted the competitive balance between buoyant and sinking species.

Whereas turbulent mixing was changed artificially in Lake Nieuwe Meer, in most aquatic ecosystems changes in turbulent mixing are driven by climatic factors. Heat exchange and wind action, in particular, are major determinants of the turbulence structure of natural waters. In natural waters, a shift from cloudy weather with occasional storms to sunny weather with weak wind mixing may easily reduce the vertical turbulent diffusivity by two or three orders of magnitude, which will tend to favour the development of surface blooms. It is therefore tempting to speculate that climate change, through its impact on mixing processes, will have a major impact on the development of surface blooms of buoyant cyanobacteria, including harmful species like *Microcystis*, *Anabaena*, and *Aphanizomenon*.

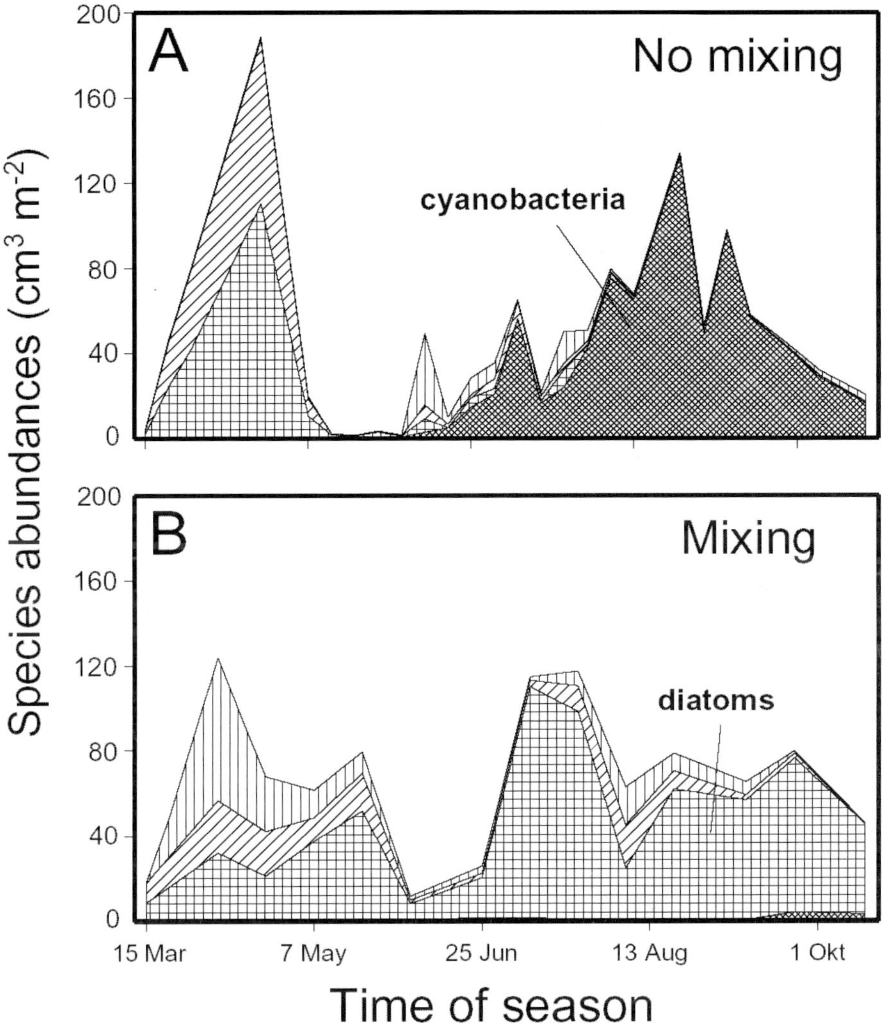

Figure 8. Changes in the phytoplankton species composition observed in Lake Nieuwe Meer in response to artificial mixing. The graph in (A) shows the seasonal pattern of phytoplankton during a year without artificial mixing (1990), whereas (B) shows the seasonal pattern of phytoplankton during a year with artificial mixing (1993). Species abundances are expressed in biovolumes per unit surface area of the lake. Dark hatched area = cyanobacteria (mostly Microcystis*), boxes = diatoms, diagonal lines = green algae, vertical lines = small flagellates. (Redrawn after Huisman et al., 2004, with permission from the Ecological Society of America).*

5 Allelopathic interactions

Cyanobacteria may gain dominance not only by competition for nutrients and light, but also by releasing toxic compounds against their competitors (Keating, 1977; Flores and Wolk, 1986; Gross, 2003). For instance, the cyanobacteria *Scytonema hofmanni* and *Fischerella muscicola* produce compounds that inhibit the photosynthetic electron transport of their competitors (Gleason and Baxa, 1986; Gross et al., 1991). The cyanobacterium *Anabaena flos-aquae* produces a toxin that paralyzes the motile green alga *Chlamydomonas* (Kearns and Hunter, 2001). The cyanobacterium *Microcystis* can inhibit carbon fixation of competing phytoplankton species (Sukenik et al., 2002). These interactions, in which a species directly affects competing species through the release of a toxic chemical, are known as allelopathic interactions.

5.1 WHO WINS? THE ROLE OF INITIAL CONDITIONS

Allelopathic interactions can be analysed by means of simple competition models (Levin, 1988; Frank, 1994; Durret and Levin, 1997). Hulot and Huisman (2004) consider a toxic cyanobacterium and a sensitive phytoplankton species that compete for a limiting nutrient. Moreover, the cyanobacterium produces a toxin that has a negative effect on the sensitive phytoplankton species. The predictions of this allelopathic interaction can again be captured by a graphical isocline approach. In Figure 9A, the zero net growth isoclines (ZNGIs) of the toxic cyanobacterium and the sensitive phytoplankton species are plotted as a function of the environmental concentrations of the nutrient and the toxin. Because the toxin has no effect on the cyanobacterium, the ZNGI of the cyanobacterium is a line parallel to the y-axis (toxin concentration), which intercepts the x-axis (nutrient availability) at the critical value R^*_{NT} . The growth rate of the sensitive phytoplankton species is positively affected by nutrient availability but negatively affected by the toxin. Therefore, its ZNGI is a curve that intercepts the x-axis at the critical value R^*_{NS} and then increases monotonically with increasing nutrient availability. This curved ZNGI shows that at higher nutrient availability a higher toxin concentration is required to suppress net growth of the sensitive phytoplankton species. The two ZNGIs intersect if the toxin-sensitive species is a better competitor for nutrients than the cyanobacterium (i.e., if $R^*_{NS} < R^*_{NT}$). In this case, the model has three equilibria: a coexistence equilibrium at the intersection point of the two ZNGIs, a monoculture equilibrium with the cyanobacterium only, and a monoculture equilibrium with the sensitive phytoplankton species only. Which equilibrium will be reached depends on the 'impact' vectors and the supply point. The impact vector of the sensitive species is directed horizontally, because the species consumes nutrient only. The impact vector of the cyanobacterium is directed diagonally because it consumes nutrients and produces toxin.

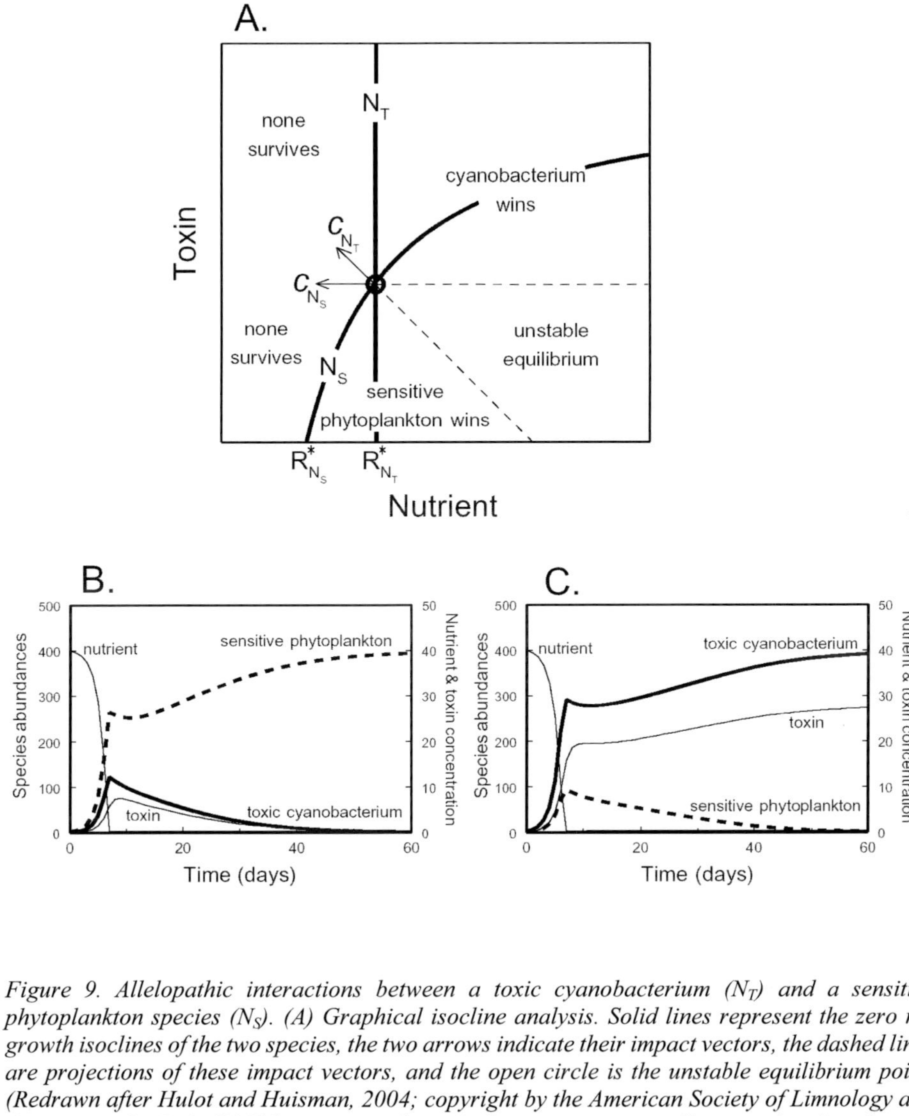

Figure 9. Allelopathic interactions between a toxic cyanobacterium (N_T) and a sensitive phytoplankton species (N_S). (A) Graphical isocline analysis. Solid lines represent the zero net growth isoclines of the two species, the two arrows indicate their impact vectors, the dashed lines are projections of these impact vectors, and the open circle is the unstable equilibrium point. (Redrawn after Hulot and Huisman, 2004; copyright by the American Society of Limnology and Oceanography, Inc). (B) If the sensitive phytoplankton species is initially more abundant than the toxic cyanobacterium, the sensitive phytoplankton species wins. (C) If the toxic cyanobacterium is initially more abundant than the sensitive phytoplankton species, the toxic cyanobacterium wins.

The ZNGIs and the projection of the species' impact vectors define four regions in the nutrient-toxin plane (Fig. 9A). If the supply point lies left of both ZNGIs, neither species can survive because the nutrient availability is below their minimal nutrient requirements. If the supply point lies in the region 'cyanobacterium wins', toxin concentrations are so high that the sensitive species is competitively excluded by the cyanobacterium. If the supply point lies in the region 'sensitive species wins', toxin concentrations are low but nutrient availability is depleted to very low levels. Hence, in this region, the cyanobacterium is competitively excluded by the sensitive species. Finally, and most interestingly, if the supply point lies in the region 'unstable equilibrium', the coexistence equilibrium exists, but it is, indeed, unstable. That is, when the sensitive phytoplankton species has a head start, it will deplete nutrient availability below the minimal nutrient requirements of the cyanobacterium, and hence the sensitive species wins (Fig. 9B). Conversely, when the cyanobacterium has a head start, it will raise toxin levels above the tolerance limits of the sensitive species, and hence the cyanobacterium becomes dominant (Fig. 9C). Thus, either the toxic cyanobacterium or the sensitive phytoplankton species will win. The winner depends on the initial conditions. This impact of the initial conditions on the final outcome of competition is related to the initial quantity of toxins produced. A higher toxin dosage has a larger effect on the sensitive species. As a result, toxic cyanobacteria must exceed a threshold abundance before they can take full advantage of the production of allelopathic compounds.

5.2 HYDRODYNAMICS AND ALLELOPATHIC INTERACTIONS

Further theory and experiments have revealed that the impact of allelopathic interactions greatly depends on mixing intensity (Chao and Levin, 1981; Durrett and Levin, 1997; Kerr et al., 2002; Hulot and Huisman, 2004). Therefore, it is useful to distinguish between aquatic ecosystems where cyanobacteria are intensely mixed and aquatic ecosystems where cyanobacteria are locally concentrated (in surface blooms or biofilms).

5.2.1 Intense mixing

In the model of Hulot and Huisman (2004), allelopathic interactions are embedded in a microbial foodweb. The model considers a well-mixed water column, in which a toxic cyanobacterium and a sensitive phytoplankton species compete for nutrients and produce detritus. The cyanobacterium also produces a toxin. The detritus is broken down by two species of heterotrophic bacteria. Only one of these two heterotrophic bacteria is capable to break down the toxin. The heterotrophic bacteria release inorganic nutrients. The inorganic nutrients, in turn, become available for the toxic cyanobacterium and the sensitive phytoplankton species.

The model was simulated for many different parameter settings, assuming intense mixing. The simulation results are plotted in Figure 10A, as a function of the toxicity of the cyanobacterial toxin and the initial abundance of the cyanobacterium. In line with the previous section, the winner of competition may indeed depend on the initial conditions.

If the toxicity remains below a certain critical threshold value (0.26 in Fig. 10A), the toxin is not sufficiently powerful to poison the sensitive phytoplankton species. In this case, the sensitive species always wins. If the toxicity exceeds this threshold, however, the toxic cyanobacterium wins if its initial abundance is high whereas the sensitive phytoplankton species wins if the initial abundance of the toxic cyanobacterium is low (Fig. 10A).

5.2.2 Weak mixing

To explore the implications of weak mixing on allelopathic interactions, Hulot and Huisman (2004) simulated the same microbial foodweb using a cellular automata model. In cellular automata models, spatial processes are explicitly described in a two-dimensional grid of cells, thus mimicking a surface scum or biofilm. Each cell in the grid supports at most one individual primary producer and at most one individual heterotrophic bacterium. Furthermore, each grid cell supports inorganic nutrients, detritus and toxin at different concentrations. Each grid cell interacts with its four neighbouring cells. At each time step, the primary producers and heterotrophic bacteria may reproduce, with a probability that depends on the resource and toxin availability within the grid cell, analogous to the model formulation for well-mixed waters. When a primary producer or heterotrophic bacterium is allowed to reproduce, it sends its offspring to an adjoining cell, provided that this adjoining cell is not already occupied by an individual of the same functional group. Furthermore, at each time step primary producers and heterotrophic bacteria may move from one cell to an adjoining cell of the grid by means of horizontal turbulent diffusion. Finally, at each time step, nutrients, detritus, and toxin may diffuse from one cell to the neighbouring cells.

Qualitatively, computer simulations with the cellular automata model (Fig. 10B) show similar results as the model simulations for well-mixed systems (Fig. 10A). Depending on the toxicity and initial conditions, either the cyanobacterium wins or the sensitive phytoplankton species wins. Quantitatively, however, there is a huge difference between the simulations with the cellular automata and the simulations for well-mixed systems. In the cellular automata model, the toxic cyanobacterium becomes dominant at a toxicity and an initial abundance that are much lower than the toxicity and initial abundance required for dominance in well-mixed systems (compare the axis scales in Fig. 10A and Fig. 10B). Thus, theory predicts that toxic cyanobacteria are favoured by weak mixing. There are two mechanisms that underlie this prediction. Firstly, during weak mixing, cyanotoxins are not diluted but will accumulate locally, and may thereby reach lethal concentrations for neighbouring sensitive phytoplankton. This makes toxic cyanobacteria locally successful. These local victors may gradually invade the entire system, thus excluding the sensitive species. Secondly, degradation of the toxin by heterotrophic bacteria is less effective during weak mixing, because it then takes longer before toxin-degrading bacteria have arrived at the site where the toxin is produced, and during that timespan other heterotrophic bacteria may competitively displace the toxin-degrading bacteria at all sites where the toxin does not occur. In conclusion, weak mixing will favour species that produce allelopathic toxins.

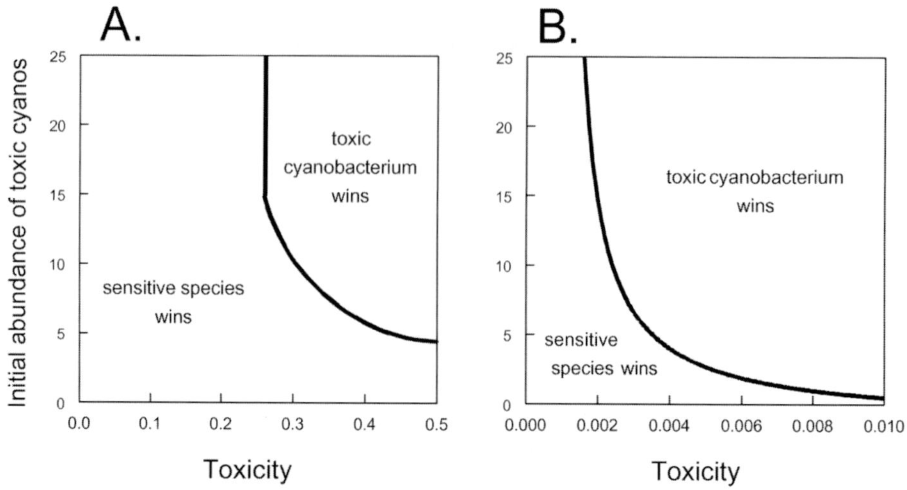

Figure 10. Model predictions illustrating the effects of mixing intensity on allelopathic interactions between a toxic cyanobacterium and a sensitive phytoplankton species. The winner of competition is plotted as a function of toxicity and the initial abundance of the toxic cyanobacterium. (A) assumes intense mixing, whereas (B) assumes weak mixing. (Redrawn after Hulot and Huisman, 2004; copyright by the American Society for Limnology and Oceanography).

5.3 EXPERIMENTAL STUDIES ON ALLELOPATHY

Experimental studies are consistent with the model predictions. A first prediction is that toxic cyanobacteria must exceed a threshold abundance before they can take full advantage of their toxic compounds. A toxin called cyanobacterin, produced by the cyanobacterium *Scytonema hofmanni*, only slows the growth rate of *Synechococcus* at a concentration of 1 $\mu g\ mL^{-1}$ (Gleason and Paulson, 1984) while it fully inhibits growth of *Synechococcus* at a concentration of 1.5 $\mu g\ mL^{-1}$ (Mason et al., 1982). Von Elert and Jüttner (1996) observed negative effects of the toxic cyanobacterium *Trichormus doliolum* on its competitor *Anabaena variabilis* only when *Trichormus* represented at least 25% of the total biomass of the mixed culture. Sukenik et al. (2002) observed that *Microcystis* has a stronger inhibitory effect on the growth rate of *Peridinium* at higher initial population densities of *Microcystis*. These studies confirm, not unexpectedly, that higher population densities of toxic species have a larger allelopathic effect.

A second prediction, related to the first, is that the outcome of allelopathic interactions depends on the initial abundances of the toxic species. We are not aware of studies that investigated this aspect for cyanobacteria, but it has been studied for heterotrophic bacteria. Chao and Levin (1981) studied two strains of the heterotrophic bacterium *Escherichia coli*. One of the strains produced an allelopathic toxin, known as colicin. The other strain was sensitive to colicin. They ran a large number of competition experiments, with different initial abundances of the colicin-producing strain. This revealed, indeed, that the sensitive strain became dominant only if the initial abundance of the colicin-producing strain was below a critical threshold value, whereas the colicin-producing strain became dominant when its initial abundance exceeded this critical threshold value.

A third prediction is that weak mixing favours allelopathic interactions, because weak mixing allows local accumulation of high toxin concentrations. A nice experimental example is again provided by studies on *E. coli*. Kerr et al. (2002) threw three *E. coli* strains into the competitive arena. They used a colicin-producing strain, a colicin-sensitive strain, and they also added a colicin-resistant strain that doesn't produce the colicin itself. In well-mixed liquid culture, the resistant strain rapidly displaced the toxin-producing strain and the toxin-sensitive strain. However, on agar plates, where the species were not well mixed, all three species coexisted. Hence, these *E. coli* studies confirm the prediction that species producing allelopathic toxins are favoured by incomplete mixing. We are not aware of similar studies with cyanobacteria. Given these *E. coli* results, however, it is tempting to speculate that dense surface blooms of buoyant cyanobacteria, which generally develop during weak wind mixing, are ideal breeding grounds for the evolution of allelopathic interactions.

6 Grazing by herbivores

Cyanobacteria, as well as other phytoplankton species, are potential resources for herbivores such as zooplankton, molluscs (e.g., zebra mussels), and planktivorous fish. Several traits of harmful cyanobacteria contribute to their relatively low edibility.

Size and morphology. Most herbivores consume only a certain size range within the full spectrum of food particles. For filter-feeding herbivores, the lower size limits of food particles are set by the mesh size of their filtration appendages, whereas the upper size limits are set by their ability to handle and ingest the food item (Burns, 1968; Geller and Müller, 1981; Gliwicz, 2003). These size limits usually depend on body size, such that microzooplankton generally feed on small phytoplankton species whereas mesozooplankton can consume larger phytoplankton species. The grazing pressure on large colonial cyanobacteria (e.g., *Microcystis*, *Trichodesmium*) is often quite low, because their colonies are often beyond the upper size limit of food particles that can be handled by zooplankton. Filamentous cyanobacteria (e.g., *Planktothrix*) are also difficult to handle for many zooplankton species, because long filaments may clog the filtration apparatus, thus reducing the efficiency of filter feeding (Gliwicz and Lampert, 1990; DeMott et al., 2001). Therefore, compared to most other phytoplankton species, colonial and filamentous cyanobacteria appear well defended by their size and shape against zooplankton species. The situation might be different for filter-feeding molluscs, like the zebra mussels, which can handle larger particle sizes than zooplankton (see Chapter 6).

Nutritional quality. A variety of minerals, vitamins, and fatty acids are essential components in the natural diets of most animal species. One key element is phosphorus.

The phosphorus content of phytoplankton can be quite variable, as the C:P ratio of phytoplankton often tracks the environmental availability of phosphorus and light. As a result, the growth rates of herbivores can become limited by phosphorus (Andersen and Hessen, 1991; DeMott and Gulati, 1999; Sterner and Elser, 2002). Low phosphorus contents are not specific for cyanobacteria, however. Thus far, there is no strong evidence that phosphorus levels make cyanobacteria less edible than other phytoplankton species. In contrast, cyanobacteria generally have very low contents of poly-unsaturated fatty acids in comparison to most other phytoplankton groups. Owing to their low levels of poly-unsaturated fatty acids, the nutritional quality of cyanobacteria is relatively low, and zooplankton species do not grow well on pure diets of cyanobacteria (Ahlgren et al., 1990; Weers and Gulati, 1997; Müller-Navarra et al., 2004).

Toxin production. Cyanobacteria may defend themselves against predation using chemical defences. Laboratory experiments show that cyanotoxins may inhibit and sometimes even kill cladocerans (DeMott, 1999; Rohrlack et al., 1999; Ferrão-Filho et al., 2000), copepods (Koski et al., 1999; Reinikainen et al., 2002), and rotifers (Gilbert, 1994). Sometimes, the chemical defences of cyanobacteria are induced by herbivores. Jang et al. (2003) studied the production of microcystin by several strains of *Microcystis* exposed to three species of cladocerans. They showed that the direct or indirect (filtered zooplankton media) exposure of the cyanobacteria to zooplankton was followed by an increase of the mass-specific microcystin production. This illustrates the intricacies in the chemical warfare between phytoplankton and zooplankton, where chemical cues released by zooplankton are detected by cyanobacteria to induce a chemical defence.

Thus, as a broad generalization, harmful cyanobacteria are of relatively low nutritional quality, can be toxic for herbivores, and are relatively inedible when compared to other phytoplankton species (Ghadouani et al., 2003). What are the implications for the population dynamics of harmful cyanobacteria? Mathematical models have analysed the effects of edible and inedible prey species on plant-herbivore interactions (Grover, 1995; Leibold, 1996). These models predict, in general terms, that edible and inedible prey species may coexist, if the edible prey species is a better competitor for limiting resources than the inedible prey species. Nutrient enrichment, however, is predicted to benefit inedible species more than edible species, because edible species are stronger suppressed by top-down control. Accordingly, nutrient enrichment favours a shift in species composition, from efficient nutrient competitors at low nutrient levels to predator-resistant species at high nutrient levels (Leibold, 1996; Irigoien et al., 2004).

It has been suggested that harmful cyanobacteria can be suppressed by means of biomanipulation. Results of biomanipulation experiments are controversial, however (DeMelo et al., 1992; Gulati and Van Donk, 2002; Drenner and Hambright, 2002). In shallow lakes, biomanipulation appears quite successful, especially when it creates windows of opportunity for the establishment of macrophytes like *Chara*, *Elodea* or *Potamogeton*. Macrophytes can monopolize limiting resources, thereby preventing the development of further phytoplankton blooms (Scheffer et al., 1993; Van Donk et al., 1993). In deep lakes, biomanipulation is usually directed at top-down control of the pelagic foodweb, which seems a less successful strategy. The idea is that reduction of planktivorous fish cascades downwards in the foodweb, favouring herbivorous zooplankton, which in turn suppresses the phytoplankton. In many cases, however, such top-down effects gradually fade away from the top towards the bottom of the foodweb (McQueen et al., 1986; DeMelo et al., 1992; Drenner and Hambright, 2002). In fact, even opposite effects may occur, since high grazing pressures by zooplankton will tend to select for inedible phytoplankton species, including cyanobacteria (Leibold, 1996; Vanderploeg et al., 2001; Gliwicz, 2003). It is this selection mechanism that will often impede top-down control of harmful cyanobacteria.

7 Discussion

The population dynamics of harmful cyanobacteria can be governed by a myriad of ecological factors. For instance, several cyanobacterial species are poor competitors for phosphorus that proliferate mainly in eutrophic waters, whereas other cyanobacteria are strong phosphorus competitors favoured in oligotrophic systems with low phosphorus availability. Some cyanobacteria are capable of nitrogen fixation and are favoured under nitrogen-limited conditions, whereas other cyanobacteria are incapable of nitrogen fixation and require high nitrogen levels to produce their nitrogen-demanding phycobili pigments. Buoyant cyanobacteria can form surface blooms during weak wind mixing, thus accumulating in high densities near the water surface. Other cyanobacteria lack buoyancy, however, and may require dispersal throughout the entire water column. Some cyanobacteria produce powerful toxins that can be used against competing phytoplankton, other cyanobacterial toxins are used as a defence against herbivores, and for several cyanotoxins no clear function has been identified yet. Cyanobacteria form a very diverse group of organisms, and as a result there are a wide variety of different environmental conditions that may determine their population dynamics. A single-factor explanation for 'the dominance of cyanobacteria' is thus bound to fail.

As a consequence, a single remedy to prevent harmful cyanobacteria does not exist. Instead, different measures may be required for different cyanobacterial species in different aquatic ecosystems. The success of water management strategies to combat harmful cyanobacteria hinges on a proper identification of the cyanobacterial species involved and the ecosystem processes that govern their population dynamics.

8 Acknowledgements

We thank Bas Ibelings, Petra Visser, and an anonymous reviewer for their helpful comments on this chapter. The research of JH was supported by the Earth and Life Sciences Foundation (ALW), which is subsidized by the Netherlands Organization for Scientific Research (NWO). The research of FDH was supported by a Marie Curie fellowship of the European Community program "Human Potential" under contract number HPMF-CT-2002-01751.

9 References

Agawin, N.S.R., Duarte, C.M. and Agustí, S. (2000) Nutrient and temperature control of the contribution of picoplankton to phytoplankton biomass and production, *Limnology and Oceanography* **45**, 591-600.

Ahlgren, G., Lundstedt, L., Brett, M. and Forsberg, C. (1990) Lipid composition and food quality of some freshwater phytoplankton for cladoceran zooplankters, *Journal of Plankton Research* **12**, 809-818.

Ammerman, J., Hood, R., Case, D. and Cotner, J. (2003) Phosphorous deficiency in the Atlantic: an emerging paradigm in oceanography, *EOS* **84**, 165-170.

Andersen, T. and Hessen, D.O. (1991) Carbon, nitrogen, and phosphorus content of freshwater zooplankton, *Limnology and Oceanography* **36**, 807-814.

Armstrong, R.A. and McGehee, R. (1980) Competitive exclusion, *American Naturalist* **115**, 151-170.

Behrenfeld, M.J., Bale, A.J., Kolber, Z.S., Aiken, J. and Falkowski, P.G. (1996) Confirmation of iron limitation of phytoplankton photosynthesis in the Equatorial Pacific Ocean, *Nature* **383**, 508-511.

Bergman, B., Gallon, J.R., Rai, A.N. and Stal, L.J. (1997) N_2 fixation by non-heterocystous cyanobacteria, *FEMS Microbiology Reviews* **19**, 139-185.

Bormans, M. and Condie, S.A. (1998) Modelling the distribution of *Anabaena* and *Melosira* in a stratified river weir pool, *Hydrobiologia* **364**, 3-13.

Burns, C.W. (1968) The relationship between body size of filter-feeding Cladocera and the maximum size of particle ingested, *Limnology and Oceanography* **13**, 675-678.

Chao, L. and Levin, B.R. (1981) Structured habitats and the evolution of anticompetitor toxins in bacteria, *Proceedings of the National Academy of Sciences (USA)* **78**, 6324-6328.

Chorus, I. and Bartram, J., eds. (1999) *Toxic Cyanobacteria in Water: A Guide to their Public Health Consequences, Monitoring and Management*, Spon, London, United Kingdom.

Coale, K.H. (1991) Effects of iron, manganese, copper, and zinc enrichments on productivity and biomass in the Subarctic Pacific, *Limnology and Oceanography* **36**, 1851-1864.

Codd, G.A., Bell, S.G., Kaya, K., Ward, C.J., Beattie, K.A. and Metcalf, J.S. (1999) Cyanobacterial toxins, exposure routes and human health, *European Journal of Phycology* **34**, 405-415.

Condie, S.A. (1999) Settling regimes for non-motile particles in stratified waters, *Deep-Sea Research I* **46**, 681-699.

De Baar, H.J.W., De Jong, T.M., Bakker, D.C.E., Loscher, B.M., Veth, C., Bathmann, U. and Smetacek, V. (1995) Importance of iron for plankton blooms and carbon dioxide drawdown in the Southern Ocean, *Nature* **373**, 412-415.

DeMelo, R., France, R. and McQueen, D.J. (1992) Biomanipulation: hit or myth? *Limnology and Oceanography* **37**, 192-207.

DeMott, W.R. (1999) Foraging strategies and growth inhibition in five daphnids feeding on mixtures of a toxic cyanobacterium and a green alga, *Freshwater Biology* **42**, 263-274.

DeMott, W.R. and Gulati, R.D. (1999) Phosphorus limitation in *Daphnia*: evidence from a long-term study of three hypertrophic Dutch lakes, *Limnology and Oceanography* **44**, 1557-1564.

DeMott, W.R., Gulati, R.D. and van Donk, E. (2001) *Daphnia* food limitation in three hypereutrophic Dutch lakes: evidence for exclusion of large-bodied species by interfering filaments of cyanobacteria, *Limnology and Oceanography* **46**, 2054-2060.

De Nobel, W.T., Huisman, J., Snoep, J.L. and Mur, L.R. (1997) Competition for phosphorus between the nitrogen-fixing cyanobacteria *Anabaena* and *Aphanizomenon*, *FEMS Microbiology Ecology* **24**, 259-267.

Diehl, S., Berger, S., Ptacnik, R. and Wild, A. (2002) Phytoplankton, light, and nutrients in a gradient of mixing depths: field experiments, *Ecology* **83**, 399-411.

Dignum, M., Hoogveld, H.L., Matthijs, H.C.P., Laanbroek, H.J. and Pel, R. (2004) Detecting the phosphate status of phytoplankton by enzyme-labelled fluorescence and flow cytometry, *FEMS Microbiology Ecology* **48**, 29-38.

Downing, J.A., Watson, S.B. and McCauley, E. (2001) Predicting cyanobacteria dominance in lakes, *Canadian Journal of Fisheries and Aquatic Sciences* **58**, 1905-1908.

Drenner, R.W. and Hambright, K.D. (2002) Piscivores, trophic cascades, and lake management, *The Scientific World* **2**, 284-307.
Duarte, C.M. and Canfield, D.E. (1992) Patterns in phytoplankton community structure in Florida lakes, *Limnology and Oceanography* **37**, 155-161.
Ducobu, H., Huisman, J., Jonker, R.R. and Mur, L.R. (1998) Competition between a prochlorophyte and a cyanobacterium under various phosphorus regimes: comparison with the Droop model, *Journal of Phycology* **34**, 467-476.
Durrett, R. and Levin, S.A. (1997) Allelopathy in spatially distributed populations, *Journal of Theoretical Biology* **185**, 165-171.
Elser, J.J., Marzolf, E.R. and Goldman, C.R. (1990) Phosphorus and nitrogen limitation of phytoplankton growth in the freshwaters of North America: a review and critique of experimental enrichments, *Canadian Journal of Fisheries and Aquatic Sciences* **47**, 1468-1477.
Falkowski, P.G. (2000) Rationalizing elemental ratios in unicellular algae, *Journal of Phycology* **36**, 3-6.
Fay, P. (1992) Oxygen relations of nitrogen fixation in cyanobacteria, *Microbiological Reviews* **56**, 340-373.
Ferrão-Filho, A.S., Azevedo, S.M.F.O. and DeMott, W.R. (2000) Effects of toxic and non-toxic cyanobacteria on the life history of tropical and temperate cladocerans, *Freshwater Biology* **45**, 1-19.
Flores, E. and Wolk, C.P. (1986) Production, by filamentous, nitrogen-fixing cyanobacteria, of a bacteriocin and of other antibiotics that kill related strains, *Archives of Microbiology* **145**, 215-219.
Frank, S.A. (1994) Spatial polymorphism of bacteriocins and other allelopathic traits, *Evolutionary Ecology* **8**, 369-386.
Fujimoto, N., Sudo, R., Sugiura, N. and Inamori, Y. (1997) Nutrient-limited growth of *Microcystis aeruginosa* and *Phormidium tenue* and competition under various N:P supply ratios and temperatures, *Limnology and Oceanography* **42**, 250-256.
Gallon, J.R. (2001) N_2 fixation in phototrophs: adaptations to a specialized way of life, *Plant & Soil* **230**, 39-48.
Geller, W. and Müller, H. (1981) The filtration apparatus of *Cladocera*: filter mesh sizes and their implications on food selectivity, *Oecologia* **49**, 316-321.
Ghadouani, A., Pinel-Alloul, B. and Prepas, E.E. (2003) Effects of experimentally induced cyanobacterial blooms on crustacean zooplankton communities, *Freshwater Biology* **48**, 363-381.
Ghosal, S. and Mandre, S. (2003) A simple model illustrating the role of turbulence on phytoplankton blooms, *Journal of Mathematical Biology* **46**, 333-346.
Gilbert, J.J. (1994) Susceptibility of planktonic rotifers to a toxic strain of *Anabaena flos-aquae*, *Limnology and Oceanography* **39**, 1286-1297.
Gleason, F.K. and Paulson, J.L. (1984) Site of action of the natural algicide, cyanobacterin, in the blue-green alga, *Synechococcus* sp., *Archives of Microbiology* **138**, 273-277.
Gleason, F.K. and Baxa, C.A. (1986) Activity of the natural algicide, cyanobacterin, on eukaryotic microorganisms, *FEMS Microbiology Letters* **33**, 85-88.
Gliwicz, Z.M. (2003) *Between Hazards of Starvation and Risk of Predation: The Ecology of Offshore Animals*, Ecology Institute, Oldendorf, Germany.
Gliwicz, Z.M. and Lampert, W. (1990) Food thresholds in *Daphnia* species in the absence and presence of blue-green filaments, *Ecology* **71**, 691-702.
Gons, H.J. and Mur, L.R. (1980) Energy requirements for growth and maintenance of *Scenedesmus protuberans* Fritsch in light-limited continuous cultures, *Archives of Microbiology* **125**, 9-17.
Gross, E.M. (2003) Allelopathy of aquatic autotrophs, *Critical Reviews in Plant Sciences* **22**, 313-339.
Gross, E.M., Wolk, C.P. and Jüttner, F. (1991) Fischerellin, a new allelochemical from the freshwater cyanobacterium *Fischerella muscicola*, *Journal of Phycology* **27**, 686-692.
Grover, J.P. (1989) Influence of cell shape and size on algal competitive ability, *Journal of Phycology* **25**, 402-405.
Grover, J.P. (1995) Competition, herbivory, and enrichment: nutrient-based models for edible and inedible plants, *American Naturalist* **145**, 746-774.
Gulati, R.D. and Van Donk E. (2002) Lakes in the Netherlands, their origin, eutrophication and restoration: state-of-the-art review, *Hydrobiologia* **478**, 73-106.
Harris, G.P. and Baxter, G. (1996) Interannual variability in phytoplankton biomass and species composition in a subtropical reservoir, *Freshwater Biology* **35**, 545-560.
Henley, W.J. (1993) Measurement and interpretation of photosynthetic light-response curves in algae in the context of photoinhibition and diel changes, *Journal of Phycology* **29**, 729-739.

Holm, N.P. and Armstrong, D.E. (1981) Role of nutrient limitation and competition in controlling the populations of *Asterionella formosa* and *Microcystis aeruginosa* in semicontinuous culture, *Limnology and Oceanography* **26**, 622-634.

Hsu, S.B., Hubbell, S.P. and Waltman, P. (1977) A mathematical theory of single-nutrient competition in continuous cultures of micro-organisms, *SIAM Journal of Applied Mathematics* **32**, 366-383.

Hu, S. and Zhang, D.Y. (1993) The effects of initial population density on the competition for limiting nutrients in two freshwater algae, *Oecologia* **96**, 569-574.

Huisman, J. (1999) Population dynamics of light-limited phytoplankton: microcosm experiments, *Ecology* **80**, 202-210.

Huisman, J. and Weissing, F.J. (1994) Light-limited growth and competition for light in well-mixed aquatic environments: an elementary model, *Ecology* **75**, 507-520.

Huisman, J., Jonker, R.R., Zonneveld, C. and Weissing, F.J. (1999a) Competition for light between phytoplankton species: experimental tests of mechanistic theory, *Ecology* **80**, 211-222.

Huisman, J., van Oostveen, P. and Weissing, F.J. (1999b) Critical depth and critical turbulence: two different mechanisms for the development of phytoplankton blooms, *Limnology and Oceanography* **44**, 1781-1788.

Huisman, J., Arrayás, M., Ebert, U. and Sommeijer, B. (2002) How do sinking phytoplankton species manage to persist? *American Naturalist* **159**, 245-254.

Huisman, J. and Sommeijer, B. (2002) Population dynamics of sinking phytoplankton in light-limited environments: simulation techniques and critical parameters, *Journal of Sea Research* **48**, 83-96.

Huisman, J., Sharples, J., Stroom, J., Visser, P.M., Kardinaal, W.E.A., Verspagen, J.M.H. and Sommeijer, B. (2004) Changes in turbulent mixing shift competition for light between phytoplankton species, *Ecology* **85**, 2960-2970.

Hulot, F.D. and Huisman, J. (2004) Allelopathic interactions between phytoplankton species: the roles of heterotrophic bacteria and mixing intensity, *Limnology and Oceanography* **49**, 1424-1434.

Hutchinson, G.E. (1967) *A Treatise on Limnology. II. Introduction to Lake Biology and the Limnoplankton*, Wiley, New York.

Hutchinson, G.E. (1974) De rebus planktonicis, *Limnology and Oceanography* **19**, 360-361.

Ibelings, B.W. and Maberly, S.C. (1998) Photoinhibition and the availability of inorganic carbon restrict photosynthesis by surface blooms of cyanobacteria, *Limnology and Oceanography* **43**, 408-419.

Irigoien, X., Huisman, J. and Harris, R.P. (2004) Global biodiversity patterns of marine phytoplankton and zooplankton, *Nature* **429**, 863-867.

Jang, M.H., Ha, K., Joo, G.J. and Takamura, N. (2003) Toxin production of cyanobacteria is increased by exposure to zooplankton, *Freshwater Biology* **48**, 1540-1550.

Jansson, M., Olsson, H. and Petterson, K. (1988) Phosphatases: origin, characteristics and function in lakes, *Hydrobiologia* **170**, 157-175.

Jassby, A.D. and Platt, T. (1976) Mathematical formulation of the relationship between photosynthesis and light for phytoplankton, *Limnology and Oceanography* **21**, 540-547.

Kearns, K.D. and Hunter, M.D. (2001) Toxin-producing *Anabaena flos-aquae* induces settling of *Chlamydomonas reinhardtii*, a competing motile alga, *Microbial Ecology* **42**, 80-86.

Keating, K.I. (1977) Allelopathic influences on blue-green bloom sequence in a eutrophic lake, *Science* **296**, 885-887.

Kerr, B., Riley, M.A., Feldman, M.W. and Bohannan, B.J.M. (2002) Local dispersal promotes biodiversity in a real-life game of rock-paper-scissors, *Nature* **418**, 171-174.

Klemer, A.R., Hendzel, L.L., Findlay, D.L., Hedin, R.A., Mageau, M.T. and Konopka, A. (1995) Carbon availability and the pattern of cyanobacterial dominance in enriched low-carbon waters, *Journal of Phycology* **31**, 735-744.

Kocsis, O., Prandke, H., Stips, A., Simon, A. and Wüest, A. (1999) Comparison of dissipation of turbulent kinetic energy determined from shear and temperature microstructure, *Journal of Marine Systems* **21**, 67-84.

Koski, M., Engström, J. and Viitasalo, M. (1999) Reproduction and survival of the calanoid copepod *Eurytemora affinis* fed with toxic and non-toxic cyanobacteria, *Marine Ecology Progress Series* **186**, 187-197.

Lee, J.G., Roberts, S.B. and Morel, F.M.M. (1995) Cadmium: a nutrient for the marine diatom *Thalassiosira weissflogii*, *Limnology and Oceanography* **40**, 1056-1063.

Leibold, M.A. (1996) A graphical model of keystone predators in food webs: trophic regulation of abundance, incidence, and diversity patterns in communities, *American Naturalist* **147**, 784-812.

Levin, B.R. (1988) Frequency dependent selection in bacterial populations, *Philosophical Transactions of the Royal Society of London B* **319**, 459-472.

Lucas, L.V., Cloern, J.E., Koseff, J.R., Monismith, S.G. and Thompson, J.K. (1998) Does the Sverdrup critical depth model explain bloom dynamics in estuaries? *Journal of Marine Research* **56**, 375-415.

Martin, J.H. and Fitzwater, S.E. (1988) Iron-deficiency limits phytoplankton growth in the Northeast Pacific Subarctic, *Nature* **331**, 341-343.

Mason, C.P., Edwards, K.R., Carlson, R.E., Pignatello, J., Gleason, F.K. and Wood, J.M. (1982) Isolation of chlorine-containing antibiotic from the freshwater cyanobacterium *Scytonema hofmanni*, *Science* **215**, 400-402.

McCausland, M.A., Thompson, P.A. and Blackburn, S.I. (2001) The effect of changes in light availability caused by mixing on the growth of *Anabaena circinalis* (Nostocales, Cyanobacteria) and *Aulacoseira* sp. (Centrales, Bacillariophyceae), *Phycologia* **40**, 530-541.

McQueen, D.J., Post, J.R. and Mills, E.L. (1986) Trophic relationships in freshwater pelagic ecosystems, *Canadian Journal of Fisheries and Aquatic Sciences* **43**, 1571-1581.

Monod, J. (1950) La technique de culture continue, théorie et applications, *Annales de l'Institut Pasteur* **79**, 390-410.

Morel, F.M.M., Reinfelder, J.R., Roberts, S.B., Chamberlain, C.P., Lee, J.G. and Yee, D. (1994) Zinc and carbon co-limitation of marine phytoplankton, *Nature* **369**, 740-742.

Moutin, T., Thingstad, T.F., van Wambeke, F., Marie, D., Slayk, G., Raimbault, P. and Claustre, H. (2002) Does competition for nanomolar phosphate supply explain the predominance of the cyanobacterium *Synechococcus*? *Limnology and Oceanography* **47**, 1562-1567.

Müller-Navarra, D.C., Brett, M.T., Park, S., Chandra, S., Ballantyne, A.P., Zorita, E. and Goldman, C.R. (2004) Unsaturated fatty acid content in seston and tropho-dynamic coupling in lakes, *Nature* **427**, 69-72.

Mur, L.R., Gons, H.J. and Van Liere, L. (1977) Some experiments on the competition between green algae and blue-green bacteria in light-limited environments, *FEMS Microbiology Letters* **1**, 335-338.

Petersen, J.E., Chen, C.C. and Kemp, W.M. (1997) Scaling aquatic primary productivity: experiments under nutrient- and light-limited conditions, *Ecology* **78**, 2326-2338.

Postius, C. and Ernst, A. (1999) Mechanisms of dominance: coexistence of picocyanobacterial genotypes in a freshwater ecosystem, *Archives of Microbiology* **172**, 69-75.

Reinikainen, M., Lindvall, F., Meriluoto, J.A.O., Repka, S., Sivonen, K., Spoof, L. and Wahlsten, M. (2002) Effects of dissolved cyanobacterial toxins on the survival and egg hatching of estuarine calanoid copepods, *Marine Biology* **140**, 577-583.

Reynolds, C.S. (1997) *Vegetation Processes in the Pelagic: A Model for Ecosystem Theory*, Ecology Institute, Oldendorf, Germany.

Reynolds, C.S., Wiseman, S.W. and Clarke, M.J.O. (1984) Growth- and loss-rate responses of phytoplankton to intermittent artificial mixing and their potential application to the control of planktonic algal biomass, *Journal of Applied Ecology* **21**, 11-39.

Riley, G.A., Stommel, H. and Bumpus, D.F. (1949) Quantitative ecology of the plankton of the western North Atlantic, *Bulletin of the Bingham Oceanographic Collection* **12**, 1-169.

Rohrlack, T., Dittmann, E., Henning, M., Börner, T. and Kohl, J.G. (1999) Role of microcystin in poisoning and food ingestion inhibition of *Daphnia galeata* caused by the cyanobacterium *Microcystis aeruginosa*, *Applied and Environmental Microbiology* **65**, 737-739.

Rothhaupt, K.O. (1996) Laboratory experiments with a mixotrophic chrysophyte and obligately phagotrophic and phototrophic competitors, *Ecology* **77**, 716-724.

Scheffer, M., Hosper, S.H., Meijer, M.L., Moss, B. and Jeppesen, E. (1993) Alternative equilibria in shallow lakes, *Trends in Ecology and Evolution* **8**, 275-279.

Schindler, D.W. (1974) Eutrophication and recovery in experimental lakes: implications for lake management, *Science* **184**, 897-899.

Sharples, J., Moore, C.M. and Abraham, E.R. (2001) Internal tide dissipation, mixing, and vertical nitrate flux at the shelf edge of NE New Zealand, *Journal of Geophysical Research* **106**, 14069-14081.

Sherman, B.S., Webster, I.T., Jones, G.J. and Oliver, R.L. (1998) Transitions between *Aulacoseira* and *Anabaena* dominance in a turbid river weir pool, *Limnology and Oceanography* **43**, 1902-1915.

Shigesada, N. and Okubo, A. (1981) Analysis of the self-shading effect on algal vertical distribution in natural waters, *Journal of Mathematical Biology* **12**, 311-326.

Smayda, T.J. (1970) The suspension and sinking of phytoplankton in the sea, *Oceanography and Marine Biology Annual Review* **8**, 353-414.

Smith, V.H. (1983) Low nitrogen to phosphorus ratios favor dominance by blue-green algae in lake phytoplankton, *Science* **221**, 669-671.

Sommer, U. (1986) Nitrate- and silicate competition among Antarctic phytoplankton, *Marine Biology* **91**, 345-351.

Sommer, U. (1989) Nutrient status and nutrient competition of phytoplankton in a shallow, hypertrophic lake, *Limnology and Oceanography* **34**, 1162-1173.

Spigel, R.H. and Imberger, J. (1987) Mixing processes relevant to phytoplankton dynamics in lakes, *New Zealand Journal of Marine and Freshwater Research* **21**, 361-377.

Sterner, R.W. (1994) Seasonal and spatial patterns in macro- and micronutrient limitation in Joe Pool Lake, Texas, *Limnology and Oceanography* **39**, 535-550.

Sterner, R.W. and Elser, J.J. (2002) *Ecological Stoichiometry: The Biology of Elements from Molecules to the Biosphere*, Princeton University Press, Princeton, USA.

Sukenik, A., Eshkol, R., Livne, A., Hadas, O., Rom, M., Tchernov, D., Vardi, A. and Kaplan, A. (2002) Inhibition of growth and photosynthesis of the dinoflagellate *Peridinium gatunense* by *Microcystis* sp. (Cyanobacteria): a novel allelopathic mechanism, *Limnology and Oceanography* **47**, 1656-1663.

Sverdrup, H.U. (1953) On conditions for the vernal blooming of phytoplankton, *Journal du Conseil Conseil Permanent International pour l'Exploration de la Mer* **18**, 287-295.

Thingstad, T.F., Zweifel, U.L. and Rassoulzadegan, F. (1998) P limitation of heterotrophic bacteria and phytoplankton in the northwest Mediterranean, *Limnology and Oceanography* **43**, 88-94.

Tilman, D. (1977) Resource competition between planktonic algae: an experimental and theoretical approach, *Ecology* **58**, 338-348.

Tilman, D. (1982) *Resource Competition and Community Structure*, Princeton University Press, Princeton, USA.

Tilman, D., Kiesling, R., Sterner, R.W., Kilham, S.S. and Johnson, F.A. (1986) Green, bluegreen and diatom algae: taxonomic differences in competitive ability for phosphorus, silicon and nitrogen, *Archiv für Hydrobiologie* **4**, 473-485.

Trimbee, A.M. and Prepas, E.E. (1987) Evaluation of total phosphorus as a predictor of the relative biomass of blue-green algae with emphasis on Alberta lakes, *Canadian Journal of Fisheries and Aquatic Sciences* **44**, 1337-1342.

Vanderploeg, H.A., Liebig, J.R., Carmichael, W.W., Agy, M.A., Johengen, T.H., Fahnenstiel, G.L. and Nalepa, T.F. (2001) Zebra mussel (*Dreissena polymorpha*) selective filtration promoted toxic *Microcystis* blooms in Saginaw Bay (Lake Huron) and Lake Erie, *Canadian Journal of Fisheries and Aquatic Sciences* **58**, 1208-1221.

Van Donk, E. and Kilham, S.S. (1990) Temperature effects on silicon- and phosphorus-limited growth and competitive interactions among three diatoms, *Journal of Phycology* **26**, 40-50.

Van Donk, E., Gulati, R.D., Iedema, A. and Meulemans, J.T. (1993) Macrophyte-related shifts in the nitrogen and phosphorus contents of the different trophic levels in a biomanipulated shallow lake, *Hydrobiologia* **251**, 19-26.

Van Liere, L. and Mur, L.R. (1979) Growth kinetics of *Oscillatoria agardhii* Gomont in continuous culture, limited in its growth by the light energy supply, *Journal of General Microbiology* **115**, 153-160.

Visser, P.M. (1995) Growth and vertical movement of the cyanobacterium *Microcystis* in stable and artificially mixed water columns, PhD thesis, University of Amsterdam, The Netherlands.

Visser, P.M., Ibelings, B.W., van der Veer, B., Koedood, J. and Mur, L.R. (1996) Artificial mixing prevents nuisance blooms of the cyanobacterium *Microcystis* in Lake Nieuwe Meer, The Netherlands, *Freshwater Biology* **36**, 435-450.

Von Elert, E. and Jüttner, F. (1996) Factors influencing the allelopathic activity of the planktonic cyanobacterium *Trichormus doliolum*, *Phycologia* **36**, 68-73.

Von Liebig, J. (1840) *Die Organische Chemie in ihrer Anwendung auf Agrikultur und Physiologie*, Vieweg, Braunschweig, Germany.

Walsby, A.E. (1994) Gas vesicles, *Microbiological Reviews* **58**, 94-144.

Walsby, A.E., Hayes, P.K., Boje, R. and Stal, L.J. (1997) The selective advantage of buoyancy provided by gas vesicles for planktonic cyanobacteria in the Baltic Sea, *New Phytologist* **136**, 407-417.

Weers, P.M.M. and Gulati, R.D. (1997) Effect of the addition of polyunsaturated fatty acids to the diet on the growth and fecundity of *Daphnia galeata*, *Freshwater Biology* **38**, 721-729.

CHAPTER 8

OPTICAL DETECTION OF CYANOBACTERIAL BLOOMS

Shipboard observation and remote sensing

Herman J. GONS, Hans HAKVOORT,
Steef W.M. PETERS & Stefan G.H. SIMIS

1 Introduction

Planktonic cyanobacteria exhibit buoyancy regulation and morphologic adaptations by which population losses due to sinking and grazing are minimised. This may lead to rapid bloom formation, given favourable light and nutrient availability. Blooms will initiate at sites best offering these conditions – 'hot spots' or 'fronts' of productivity – and spread to other areas depending on the basin's morphometry and hydrology (Oliver and Ganf, 2000). By their buoyancy control several species are able to dwell in particular depth intervals. Colonies of *Microcystis* species exhibit vertical migration, and are known to form surface scums, which can be wind-swept to other lake parts and accumulate on the shore. Due to these factors, temporal and spatial distribution of cyanobacterial biomass is highly complex.

Because monitoring of the spatial distribution of plankton is costly and therefore sparse, spatio-temporal patterns of bloom formation in large inland waters are poorly known. Availability of rapid and cost-effective methods to acquire the distribution of biomass and ecophysiological characteristics is important from scientific and water management points of view. Worldwide, there are more than 2000 lakes with a surface area exceeding 10 km^2 (Wetzel, 1990) for which assessments of the spatial distribution of plankton biomass could be greatly improved by remote sensing. Advances in hydrologic optics (Kirk, 1994), and the advent of satellite-based digital imaging spectrometers with a less than 1-km spatial resolution and suitable spectral band width now enable mapping of biomass distribution in many of these inland waters. The gap between measurements *in situ* and satellite imagery can be bridged by shipboard and airborne spectrometry (Kirk, 1994; Bukata et al., 1995). Moreover, the data from the different platforms are complementary (Vos et al., 2003). The methods used by these platforms are here collectively referred to as optical teledetection, which can be broadly defined as macroscopic observation of a system's properties by means of the spectral composition of the emergent light flux.

J. Huisman, H.C.P. Matthijs and P.M. Visser (eds.), Harmful Cyanobacteria, 177–199.

The fraction of the incident light that is reflected from a particular material varies with wavelength. Thus, a sandy beach appears brighter than seawater, and also has a very different hue. Also, when water bodies are compared, their appearance in terms of brightness and hue may vary widely (Fig. 1). Phytoplankton pigments have a great colour effect, as illustrated by changes from the oligotrophic Keweenaw Bay to the mesotrophic and eutrophic Otisco Lake basins (upper row in Fig. 1). Also, other components in the water affect its colour, as illustrated by the River IJssel, with high loads of organic silt, and Lake Loosdrecht, stained by humic substances (lower row in Fig. 1). As will be detailed later, cyanobacterial abundance can be deduced from the spectral composition of the emergent light flux. Spectral valleys around 670 and 620 nm are due to light absorption by the photosynthetic pigments chlorophyll *a* (Chl *a*) and phycocyanin. The latter pigment occurs in cyanobacteria but not in green algae and diatoms.

Indeed, irrespective of the platform, obtaining spectral reflectance data is a *conditio sine qua non* for the identification and separation of substances in water by optical teledetection, because the spectral reflectance can be directly related to inherent optical properties of the aquatic medium. Concentrations of photosynthetic pigments and other substances in the water column may be retrieved by application of retrieval algorithms, i.e., models that estimate these concentrations from the spectral composition of the emergent light flux. The water should be so deep that reflection from benthic organisms and sediment is insignificant. For the whole band of photosynthetically available radiation, about 90% of the light emerges from a depth range given by the reciprocal of the vertical light attenuation coefficient, $1/K_d$, which will be < 0.5 m for most eutrophic inland waters. Another prerequisite is full mixing of this near-surface layer. For full mixing from surface to bottom, the estimated pigment concentration for the near-surface layer applies to the whole water column. Obviously, column-integrated pigment concentration could be considerably overestimated in case of surface blooms. However, compact *Microcystis* scums can be recognised by high near-infrared to red reflectance ratio, and water covered by *Microcystis* streaks exhibits lower 440 to 380 nm reflectance ratios than well-mixed water (Gons, 1999).

Retrieval algorithms are based on mathematical models relating spectral reflectance to the inherent optical properties of waterbodies, namely absorption and backscattering (Gordon et al., 1975). Given the spectral distributions of absorption and backscattering coefficients of, e.g., algae, cyanobacteria, and dissolved humic substances, concentrations of these materials can be resolved by fitting modelled spectral reflectance to observed spectral reflectance. These fits can be achieved by various numerical methods including matrix inversion (Hoge and Lyon, 1996; Hakvoort et al., 2002) or neural networks (Schiller and Doerffer, 1999). If the optical properties are incompletely known, semi-empirical algorithms may be derived by calibration of coefficients in a simplified model. These algorithms may apply to many waterbodies with similar concentration ranges as in the calibration. Fully empirical algorithms are not based on optical theory, but are usually derived from multiple regression analysis. In general the predictive power is limited, but these empirical algorithms can nevertheless be useful in mapping concentrations provided 'ground truth' for the relevant water is available.

Figure 1. Natural water colours. Upper row from left to right: Keweenaw Bay of Lake Superior (USA/Canada), northern Otisco Lake, and southern Otisco Lake (New York, USA). Lower row from left to right: River IJssel, Lake IJsselmeer, and Lake Loosdrecht (The Netherlands).

Since the operation of the Coastal Zone Colour Scanner (CZCS) aboard the Nimbus-7 satellite (from 1978 until 1986) the usefulness of remote sensing in oceanography has been recognised widely. Based on data from this instrument the global distribution of oceanic chlorophyll could be mapped, paving the way for global production estimates (Antoine et al., 1995; Longhurst et al., 1995; Behrenfeld and Falkowski, 1997a). The next step is remote sensing of the share of particular plankton groups in the total primary production. Regarding eutrophic inland waters rapid progress can be expected from satellite sensors involving more spectral bands with improved spatial resolution compared to CZCS. The Medium Resolution Imaging Spectrometer (MERIS) aboard the Envisat-1 satellite, launched in 2002, can be programmed to relay data from 15 channels in the visible and near-infrared wavebands for 300-m pixels (Rast et al., 1999). The spatial resolution makes this new sensor applicable to many inland waters, and the spectral resolution facilitates recognition of specific photosynthetic pigments. Among plankton groups, the cyanobacteria stand out by extensive use of phycobilin pigments. The pigments phycocyanin and phycoerythrin provide distinct spectral signatures, which are picked up by an instrument like MERIS. Owing to these features, MERIS is one of the first satellite sensors that can be applied to map cyanobacterial distributions in inland waters. Based on retrieval algorithms to estimate phycocyanin, MERIS data can produce synoptic views of cyanobacterial bloom formation and dispersal.

In this chapter, we discuss these recent advances in optical methods to study the spatial and temporal distribution of cyanobacteria in large inland and coastal waters. Our review is illustrated by detailed observations on cyanobacterial blooms in Lake IJsselmeer, The Netherlands.

2 Spectral reflectance – the intrinsic water colour

2.1 RELATIONSHIP WITH INHERENT OPTICAL PROPERTIES

Due to light scattering processes, part of the downwelling light penetrating into water is redirected upward (Kirk, 1994). This is called the subsurface irradiance reflectance. For measurement using a cosine-corrected sensor with 180° field of view the spectral subsurface irradiance reflectance at null depth, $R(0,\lambda)$, is defined as

$$R(0,\lambda) = \frac{E_u(0,\lambda)}{E_d(0,\lambda)} \tag{1}$$

where $E_u(0,\lambda)$ and $E_d(0,\lambda)$ are upward and downward irradiance immediately below the water surface, respectively. From Monte Carlo simulations Gordon et al. (1975) obtained the approximate relationship

$$R(0,\lambda) = C\frac{b_b(\lambda)}{a(\lambda) + b_b(\lambda)} \tag{2}$$

where the scaling factor C can be computed for known sun and skylight. The absorption coefficient $a(\lambda)$ and backscatter coefficient $b_b(\lambda)$ can be partitioned into partial coefficients relating to water constituents

$$a(\lambda) = a_p(\lambda) + a_t(\lambda) + a_g(\lambda) + a_w(\lambda) \tag{3}$$

$$b_b(\lambda) = b_{bp}(\lambda) + b_{bt}(\lambda) + b_{bw}(\lambda) \tag{4}$$

where the subscripts p, t, g and w denote biomass of pigmented plankton (cyanobacteria and algae), tripton (inanimate particles), gilvin (dissolved humic substances) and water, respectively. Backscattering by gilvin is considered insignificant and thus omitted. In many cases $b_b(\lambda) \ll a(\lambda)$ so that $R(0,\lambda)$ varies simply in proportion to $b_b(\lambda)/a(\lambda)$. Compared to absorption by photosynthetic pigments, the total backscattering exhibits a flat spectrum without marked peaks and valleys. The spectral shape of $R(0,\lambda)$ therefore mainly depends on the relative contributions of the partial absorption coefficients (Fig. 2), which can be estimated by spectrophotometry in the laboratory. Of these coefficients $a_w(\lambda)$ is accurately known (Buiteveld et al., 1994), $a_p(\lambda)$ and $a_t(\lambda)$ can be estimated by using the quantitative filter method (Roesler, 1998; Tassan and Allali, 2002) in combination with pigment bleaching (Kishino et al., 1985), and $a_g(\lambda)$ can be measured after removing the particles by filtration. In the example of Figure 2, for λ in the range 400-700 nm, $a(\lambda)$ was mainly due to photosynthetic pigments. The blue and red absorption peaks of Chl *a* were mirrored by reflectance valleys. The presence of cyanobacteria was indicated by an absorption peak attributable to phycocyanin around 620 nm, with a corresponding reflectance valley at this wavelength.

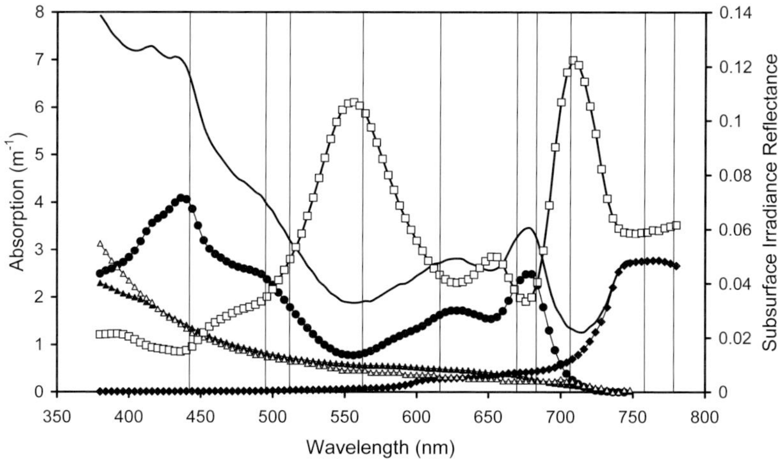

Figure 2. Partitioning of the absorption coefficient (line without symbols) among plankton biomass (solid circles), tripton (solid triangles), gilvin (open triangles) and water (solid diamonds). Also shown is the subsurface irradiance reflectance (open squares) from above-water radiance measurements. The observations were made on southern Lake IJsselmeer, July 29, 1997. Vertical lines indicate wavebands of the Medium Resolution Imaging Spectrometer (MERIS).

2.2 ABOVE-WATER MEASUREMENTS

In principle, $R(0,\lambda)$ as measured *in situ* represents the benchmark of optical teledetection. In the real world with water surface roughness and clouds, $R(0,\lambda)$ is not easily obtained. For the eutrophic waters considered here, another problem is sensor shading at the measurement of the upward irradiance (Gordon and Ding, 1992). Indeed, above-water spectrometry may yield more accurate estimation of $R(0,\lambda)$ and is much more rapid than measurement *in situ*. Several instruments suitable for routine optical teledetection are commercially available. Because the direction of view is of paramount importance, the sensors measure radiance $L(\theta,\varphi,\lambda)$, *i.e.*, radiant flux per unit area and per unit solid angle. Let us consider above-water measurement from the deck of a research vessel (Fig. 3). The direction of view is specified by angle θ_a, *i.e.*, the angle to the upward vertical, and azimuth angle φ, *i.e.*, the angle to the vertical plane of the sun. It is convenient to also express an angle θ_w, *i.e.*, the angle of subsurface upward radiance (L_u) to the downward vertical. The water-leaving radiance (L_w) is less than L_u because part of the latter light is reflected back into the water column, and due to refraction at the water-air interface:

$$L_w(\theta_a,\varphi,\lambda) = \frac{1-\rho(\theta_w)}{n^2} L_u(\theta_w,\varphi,\lambda) \tag{5}$$

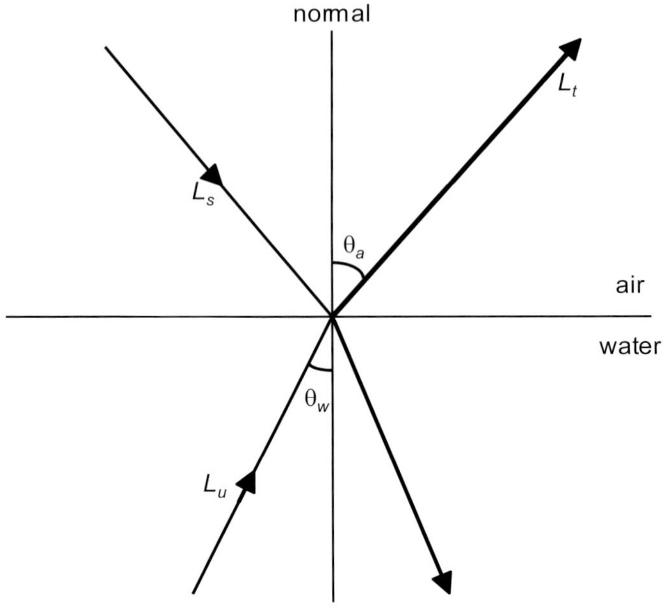

Figure 3. Schematic representation of the geometry of subsurface upward radiance (L_u) and sky radiance (L_s) contributing to the radiance (L_t) observed at fixed angle θ_a for a flat water surface.

where $\rho(\theta_w)$ is the Fresnel reflectance for internal reflection of L_u and n is the relative refraction index of water. When using a hand-held spectroradiometer, shipboard measurements must be carried out at an angle to the vertical in order to avoid shade and reflection from the boat. Measurement at azimuth angle $\varphi = 90^o$ or 270° avoids sunglint as well. The sensor of a spectroradiometer cannot directly measure L_w, but instead measures total radiance L_t (Fig. 3), which for still water is given by

$$L_t(\theta_a, \varphi, \lambda) = L_w(\theta_a, \varphi, \lambda) + \rho(\theta_a) L_s(\theta_a, \varphi, \lambda) \tag{6}$$

where L_s is the sky radiance of which the angle of incidence equals the angle of observation θ_a, and $\rho(\theta_a)$ is the Fresnel reflectance for this angle of incidence. Ideally, θ_a is sufficiently large while a favourable ratio between emergent and reflected skylight is maintained. Fresnel reflectance of the air-water interface is small for angles of incidence $< 50^o$. Because L_u is nearly constant in the $\leq 30^o$ nadir angle interval (Kirk, 1994; Morel and Gentili, 1993), *i.e.*, at angles $\leq 42^o$ for L_w, a choice of $\theta_a = 42^o$ appears to be optimal. From equation (5) and equation (6) L_w and thus L_u are obtained

$$L_u(\theta_w,\varphi,\lambda) = \frac{(L_t(\theta_a,\varphi,\lambda) - \rho(\theta_a)L_s(\theta_a,\varphi,\lambda))\,n^2}{1-\rho(\theta_w)} \tag{7}$$

For specified measurement geometry under condition of still water the upward irradiance just beneath the surface is proportional to the upward radiance

$$E_u(0,\lambda) = Q(\theta_w,\varphi)L_u(\theta_w,\varphi,\lambda) \tag{8}$$

where Q is a factor for the conversion of upward radiance to irradiance, which applies to the geometry of the light field. For turbid water the value of Q mainly depends on the angular distribution of the incident light (Morel and Gentili, 1993) and is expected to be close to π for intermediate sun elevation (Gons, 1999). Note that Q has been defined so as to apply to any direction of L_u instead of the vertical upward direction only.

The downward irradiance immediately below the water surface $E_d(0,\lambda)$ consists of refracted sunlight and diffuse light, and the upwelling light reflected downward again at the water-air interface. The latter quantity is not negligible in turbid water, and equals about half of $E_u(0,\lambda)$. The precise downward reflectance cannot be given without knowledge of the angular distribution of the upwelling light. The value of $E_d(0,\lambda)$ can be estimated from the spectral downward irradiance above the water surface $E_{ad}(\lambda)$

$$E_d(0,\lambda) = E_{ad}(\lambda)\left[1-\rho(\theta_o)(1-F(\lambda)) - r_d F(\lambda)\right] + 0.5Q(\theta_w,\varphi)L_u(\theta_w,\varphi,\lambda) \tag{9}$$

where $F(\lambda)$ is the fraction of diffuse light in $E_{ad}(\lambda)$, and $\rho(\theta_0)$ and r_d are the Fresnel reflectance for the solar beam at the relevant sun zenith angle θ_0 and the average reflectance for the diffuse light, respectively. Values of r_d depend on the assumed sky radiance distribution (Jerlov, 1968), but a mean value of 0.06 is sufficiently accurate. Using the same spectrometer as for recording L_t and L_s, E_{ad} and its diffuse fraction F can be conveniently obtained by measuring the radiance from a sunlight exposed and next shaded reference diffuser.

In the real world, the water surface generally will not be flat but show considerable roughness due to wind-driven surface waves. Theory for simple waves provides a basis for evaluating potential error due to oscillations in skylight contribution to L_t (equation 6). For basins with known bathymetry at steady-state wind, wave characteristics can be predicted for given wind speed and fetch. Regarding shipboard measurements, 'all-weather' skylight reflectance is expected to be about 12% greater than the value of $\rho(\theta_a)$ for still water (Gons, 1999; Mobley, 1999).

The subsurface irradiance reflectance $R(0,\lambda)$ is defined by equation (1) as the quotient of upward and downward irradiance, which can be estimated according to equation (8) and equation (9), respectively. The customary quantity in case of airborne and satellite remote sensing is a form of radiance reflectance known as the remote sensing reflectance R_{rs}

$$R_{rs} = \frac{L_w(\theta_a,\varphi,\lambda)}{E_{ad}(\lambda)} \tag{10}$$

For known angular distribution of the incident light, $R(0,\lambda)$ can be derived from R_{rs}. Estimation of R_{rs} from remote platforms requires correction for atmospheric scattering and absorption (Thomas and Stamnes, 1999), which are not considered in this chapter.

3 Retrieval algorithms

Given the inherent optical properties of a waterbody, retrieval algorithms can be based on the matrix inversion method (MIM). The adequacy of this approach has been demonstrated for mapping of total suspended matter and Chl *a* from both airborne and satellite imagery (Hakvoort et al., 2002; Vos et al., 2003), but has not yet proved to be stable for retrieval of phycocyanin. Applicability of MIM critically depends on precise estimation of $R(0,\lambda)$, and hence requires precise atmospheric correction. For the shipboard observations and MERIS imagery shown in section 4, semi-empirical algorithms were applied instead. The band ratioing makes these algorithms less sensitive to error in $R(0,\lambda)$. Especially the Chl-*a* algorithm has been rigorously tested in the field to establish its temporal and spatial validity.

3.1 RETRIEVAL OF CHLOROPHYLL

Colour of clear oceanic waters mainly changes with plankton biomass and covarying detritus. Satisfactory estimates of the concentration of Chl *a* in oceanic waters has been based on reflectance ratios for the blue and green wavebands. Similar detection of Chl *a* appeared impossible for turbid coastal and inland waters due to interference by dissolved humic substances and particulate matter from terrestrial and sediment inputs (Morel and Prieur, 1977). A major problem is the spectral overlap of the strong blue-light absorption by these materials with absorption by photosynthetic pigments. Studies on eutrophic inland waters (Dekker et al., 1991; Mittenzwey et al., 1992) indicated the potential use of a red to near-infrared band ratio, i.e., for the wavelengths around and near the red Chl-*a* peak, where absorption by gilvin and tripton are small compared to their absorption values in the blue region (Fig. 2). Shipboard measurements of reflectance on the IJssel Lagoon have shown that Chl *a* can be adequately estimated by high-resolution measurement around 672 nm in the red waveband, and around 704 and 776 nm in the near-infrared band (Gons, 1999). Assuming that absorption at 776 nm is entirely attributable to water, equation (2) can be written as $R(0,776) = C\, b_b/(a_w(776)+b_b)$. The spectral absorption of water is accurately known (Buiteveld et al., 1994). For known angular distribution of the incident light, the scaling factor C can be calculated, and hence the value for the backscatter coefficient b_b in this spectral region can be estimated. Chl *a* can then be retrieved from the reflectance band ratio for λ = 704 and 672 nm by adopting equation (2) for insignificant absorption by gilvin and tripton, and for spectrally neutral b_b. The algorithm proved to be so robust as to accommodate adequate Chl-*a* estimates in many other inland as well as coastal waters (Gons, 1999; Gons et al., 2000). Moreover, the algorithm worked equally well using the 664-nm band instead of the 672-nm band, and hence proved to be applicable with the MERIS band setting (Gons et al., 2002). Thus, we obtained the following algorithm:

$$[Chl] = \frac{R_{Chl}\{a_w(704) + b_b\} - a_w(664) - b_b{}^p}{a^*_{Chl}(664)} \tag{11}$$

where $[Chl]$ is the Chl-*a* concentration in mg m^{-3}, R_{Chl} is the reflectance band ratio $R(0,704)/R(0,664)$, p is an empirically derived coefficient which has a value of 1.06, and $a^*_{Chl}(664)$ is a mean Chl-*a*-specific absorption coefficient which has a value of 0.0146 m^2 (mg of Chl *a*)$^{-1}$. The values of p and $a^*_{Chl}(664)$ were calibrated for the IJssel Lagoon data from 1993 until 1996, and proved to hold for later observations on the same water as well as other water bodies.

3.2 RETRIEVAL OF PHYCOCYANIN

Phycocyanin (PC) belongs to the water-soluble biliprotein pigments occurring in cyanobacteria, red algae and cryptomonads (MacColl, 1998). The pigment exhibits interspecific variability of the absorption spectra with peaks near 620 nm. The signature of PC is a valley in $R(0,\lambda)$ next to that due to the red absorption peak of Chl *a* (Fig. 2). To date no PC retrieval algorithm has been published that proved to be widely applicable. A major problem is significant Chl-*a* absorption around the PC peak. For the application to MERIS imagery as shown below, the reflectance band ratio for λ = 620 nm and 705 nm is used in a similar way as for the Chl-*a* retrieval according to equation (11) (Simis et al., 2004). Also in this PC algorithm, b_b is derived from near-infrared reflectance at 776 nm, and it is assumed that at 620 nm absorption by gilvin and tripton can be neglected, whereas correction for absorption by Chl *a* at 620 nm is important. The relative contributions of PC and Chl *a* to the total absorption are expected to vary markedly depending on plankton species composition and nutrient status. This variability can be corrected for, because the Chl-*a* absorption can be derived independently (equation 11), and next be substituted into the algorithm for PC retrieval

$$[PC] = \frac{R_{PC}(a_w(705) + b_b) - b_b - a_w(620) - a^*_{Chl}(620)[Chl]}{a^*_{PC}(620)} \tag{12}$$

where $[PC]$ is the PC concentration in mg m^{-3}, R_{PC} is the MERIS reflectance band ratio $R(0,705)/R(0,620)$, $a_w(705)$ and $a_w(620)$ are the absorption coefficients of water for the MERIS bands centered at 705 nm and 620 nm, respectively, $[Chl]$ is the concentration of Chl *a* in mg m^{-3} estimated according to equation (11), $a^*_{Chl}(620)$ is the Chl-*a*-specific and $a^*_{PC}(620)$ is the PC-specific absorption coefficient for the 620-nm MERIS band, with units m^2 (mg of pigment)$^{-1}$. This PC-retrieval algorithm proved to be applicable for the markedly different lakes Loosdrecht and IJsselmeer (Simis et al., 2004), but still needs testing for other water bodies.

3.3 RETRIEVAL OF THE VERTICAL LIGHT ATTENUATION COEFFICIENT

Shipboard optical teledetection also provides reasonably accurate estimation of the vertical attenuation coefficient K_d for photosynthetically available radiation. For various inland waters, with K_d in the range 0.7-5.4 m^{-1}, the value of K_d could be estimated with a standard error of 0.3 m^{-1} according to (Gons et al., 1998)

$$K_d = 3.7R_1 + 1.2b_b + 0.08R_2 + 1.5R_3 - 1.6 \quad (13)$$

where R_1, R_2 and R_3 are the reflectance band ratios $R(0{,}700)/R(0{,}572)$, $R(0{,}672)/R(0{,}776)$, and $R(0{,}440)/R(0{,}572)$, respectively.

4 Patterns of biomass distribution in a freshwater lagoon

4.1 STUDY AREA

The River IJssel is the northernmost branch of the River Rhine, discharging 10 to 25% of the Rhine water entering the Netherlands. The river water passes Lake Ketelmeer (area 35 km^2; mean depth 3 m) before flowing into Lake IJsselmeer (area 1190 km^2; mean and maximum depth 4.5 m and 9 m, respectively). The IJssel Lagoon – formerly known as Zuyder Sea – came into existence as a freshwater reservoir by the construction of a 32 km long dam, called the Afsluitdijk, isolating it from the Wadden Sea in 1932 (Fig. 4). Large areas were reclaimed for land use, and the waterbodies now known as lakes IJsselmeer and Markermeer were separated by a second dam. By surface area this coastal lake system is the largest freshwater body in Western and Central Europe. The water table is controlled by discharges to the Wadden Sea, through sluices in the Afsluitdijk. Water retention times of lakes Ketelmeer and IJsselmeer are about 1 week and 0.4 year, respectively.

Parts of Lake IJsselmeer in summer showed depth-averaged Chl a > 180 mg m^{-3} during blooms of *Microcystis* and *Aphanizomenon* species (Gons et al., 1998). Biological data include the distributions of cyanobacteria, algae, crustaceans, molluscs, fish and birds. Zebra mussels are on the diet of several bird species, and may affect the cyanobacterial distribution in the lake (Lammens and Hosper, 1998).

4.2 SHIPBOARD OBSERVATIONS

Cruises of the r/v *Luctor* (NIOO, Yerseke) comprising the mouth of the River IJssel, Lake Ketelmeer, and Lake IJsselmeer, have been made each spring and summer from 1992 onwards. Observations since 1993 established temporal and spatial variation in $R(0,\lambda)$ as well as optical conditions in *situ*. More recently, cruises were made in conjunction with airborne and satellite remote sensing. The 2003 surveys served algorithm calibration for the MERIS imagery presented in section 4.4.

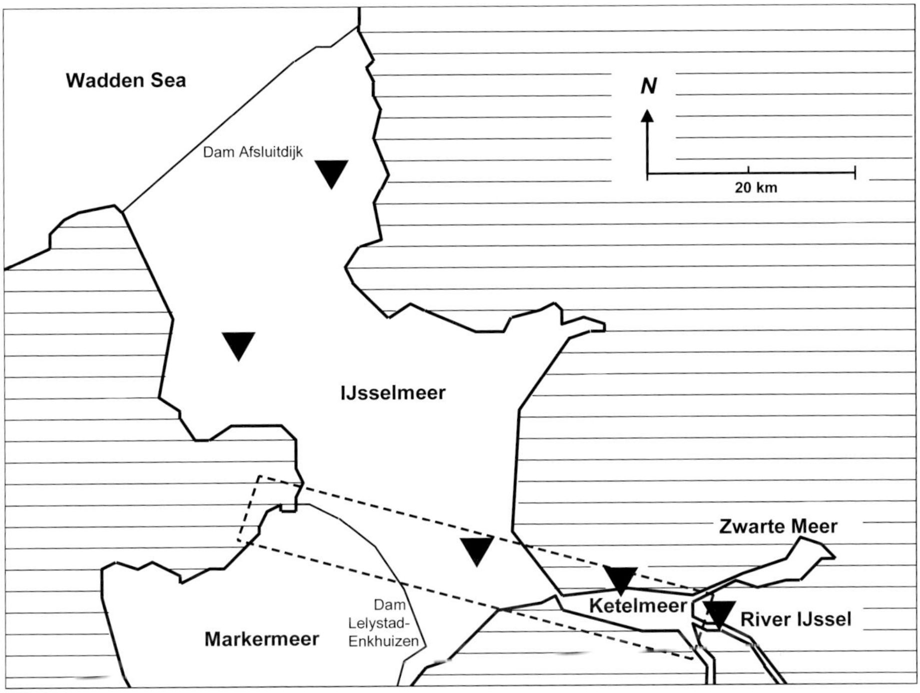

Figure 4. Schematic map showing the IJssel Lagoon. Land surface is indicated by the hatched area. Sampling stations of r/v Luctor cruises and the trajectory covered by the EPS-A flight (Fig. 8) are indicated by triangles and the dashed rectangle, respectively.

In these surveys, optical teledetection at cruising speed by a PR-650 SpectraColorimeter (PhotoResearch, Chatsworth, CA, USA) complemented the data for 5 sampling stations. The PR-650 instrument recorded continuous spectra from 380 to 780 nm at 4-nm steps (Gons 1999). The data from the mid-summer cruise exhibited a marked horizontal gradient in Chl-*a* concentration from the River IJssel to Lake IJsselmeer (Figs. 5-7). Two different types of spectra were acquired, namely for 'river water' as occurred from the mouth of River IJssel through Lake Ketelmeer until the south of Lake IJsselmeer (Fig. 5A), and for 'plankton blooms' in Lake IJsselmeer (Fig. 5B-C). In the 'plankton bloom' spectra, the valley at 620 nm attributed to PC indicated an abundant presence of cyanobacteria. The river water contained high concentrations of mineral particles with low population densities of centric diatoms, chlorococcal algae and flagellates. Because of the short water retention time, the composition of the water did not change much in Lake Ketelmeer. As indicated by the slightly decreasing K_d in Lake Ketelmeer (Fig. 6), particle deposition occurred before the river water mixed with the IJsselmeer water proper. Whereas in most surveys the water was sufficiently wind-mixed to prevent stratification in Lake IJsselmeer, vertical temperature gradients up to 3 °C and near-surface accumulation of cyanobacterial biomass were observed during this cruise.

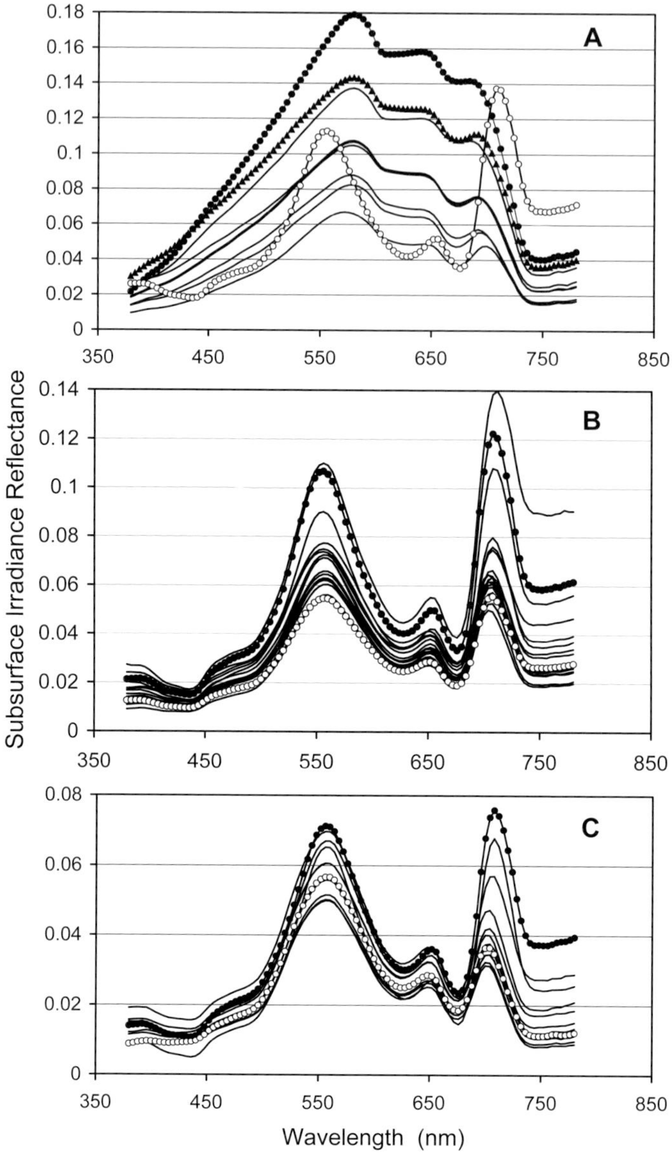

Figure 5. Subsurface irradiance reflectance spectra derived during a survey of the IJssel Lagoon on July 29, 1997. (A) Transect from 0 to 21 km, from the River IJssel (solid circles) via central Lake Ketelmeer (solid triangles) to southern Lake IJsselmeer (open circles), (B) from 23 to 52 km, from southern (solid circles) to northwestern Lake IJsselmeer (open circles), and (C) from 54 to 69 km, from northwestern (solid circles) to northeastern Lake IJsselmeer (open circles) along the transect. Note Y-axis scale differences.

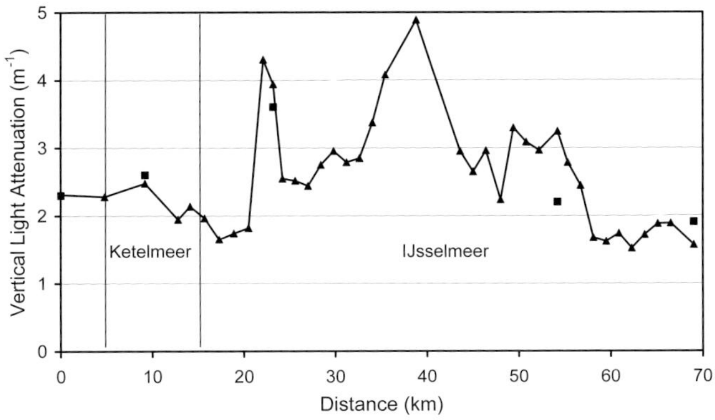

Figure 6. The vertical light attenuation coefficient, K_d, along the IJssel Lagoon transect on July 29, 1997 (optical teledetection: triangles; underwater measurements: squares).

The cyanobacteria were mainly *Aphanizomenon flos-aquae* in the southern and western parts of Lake IJsselmeer, and *Microcystis* sp. in the northeastern part of the lake. Their surface blooms were apparent from differences in near-surface and depth-averaged Chl *a*. At the sampling stations, K_d estimated from above-water spectrometry (equation 13) was consistent with measured values of K_d (Fig. 6). Also the Chl-*a* concentration in the near-surface sample was quite accurately predicted (equation 11), thus validating the data derived from optical teledetection (Fig. 7).

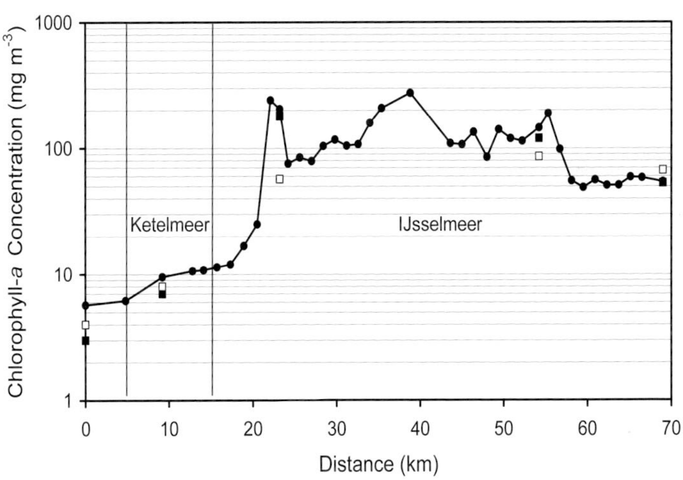

Figure 7. Chlorophyll a *concentrations along the IJssel Lagoon transect on July 29, 1997 (optical teledetection: circles; surface samples: solid squares; column-integrated samples: open squares).*

Clearly, in Lake IJsselmeer the values of Chl *a* and K_d correlated positively (Figs. 6 and 7). Peaks in the spatial distribution were at least partly attributable to formation of surface scums of buoyant cyanobacteria as could be visually observed from the ship deck. For a remote observer coincidence of very high Chl *a* and K_d alone is not conclusive evidence for accumulated cyanobacterial biomass. Work on the Chinese lake Chao Hu showed, however, that cyanobacterial biomass concentrated near or at the surface markedly enhances $R(0,\lambda)$ in the violet and NIR wavebands (Gons, 1999). The spectra for this cruise also showed these reflectance phenomena. Increase in the near-infrared to red ratio up to 2.3 occurred for positions with greatest Chl-*a* and K_d values (Fig. 5). These ratios were still considerably lower than those retrieved from compact surface scums of Lake Chao Hu.

4.3 REMOTE SENSING FROM AIRCRAFT

The mapping of detailed biomass distributions along shore lines, in river plumes and across fronts requires a higher spatial resolution than can be provided from satellites (spaceborne remote sensing; Table 1). The best satellite resolution currently available for Chl-*a* mapping is 300 m for MERIS high-resolution mode, while other satellite sensors like SeaWiFS and MODIS provide 1-km pixels for the wavebands relevant to quantify coloured water constituents. Sensors aboard airplanes (airborne remote sensing) allow for more adequate spectral and detailed spatial information to be acquired and their operation can be tuned to particular processes and events. Vos et al. (2003) made a compilation of airborne and spaceborne-derived water quality maps clearly showing the more detailed structures obtained with airborne techniques. Also, spaceborne imagery pixels adjacent to land showed erroneous results. For such coastal areas airborne remote sensing must be used for adequate mapping.

The Geo-information and ICT Department of the Netherlands Directorate-General for Public Works and Water Management guided flights of a coastal guard plane equipped with an EPS-A scanner (Hakvoort et al., 2002; Vos et al., 2003). For observations on the IJssel Lagoon in 2001 flights were made at 3 km altitude, which after pixel averaging to reduce instrumental noise gave about 10-m spatial resolution. The EPS-A waveband configuration was nearly identical to the MERIS band setting. Examples of the fine-scale mapping from this airborne platform are shown for a trajectory including the mouth of River IJssel, Lake Ketelmeer, southern Lake IJsselmeer, and northern Lake Markermeer in May 2001 (Fig. 8).

The flight showed almost reversed distribution patterns of total suspended matter and Chl *a*. In this part of Lake IJsselmeer, concentrations of total suspended matter remained below 10 g m^{-3}, whereas Chl *a* varied from 10 to 30 mg m^{-3}. Other EPS-A flights as well as the shipboard data (Fig. 7) and satellite imagery (Fig. 8) showed recurrence of the high-chlorophyll area in southern Lake IJsselmeer. In contrast, in the adjacent Lake Markermeer the total suspended matter exceeded 40 g m^{-3}, whilst Chl *a* remained below 10 mg m^{-3}. Another striking feature was the high concentrations of total suspended matter, exceeding 20 g m^{-3}, on both sides and

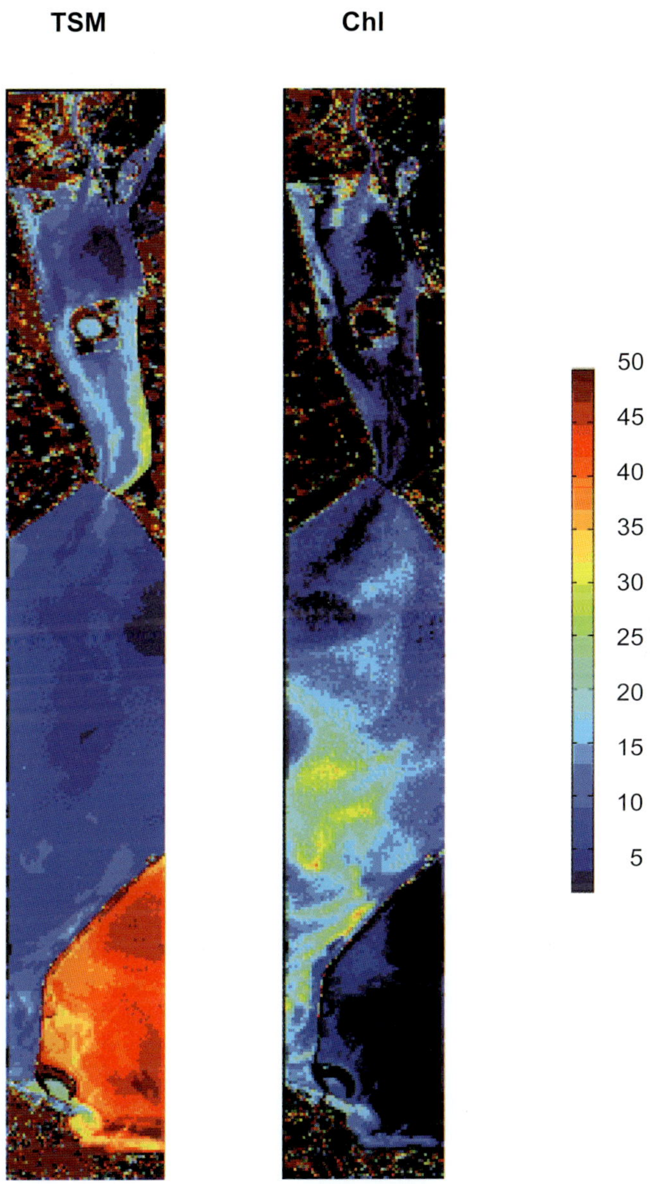

Figure 8. Distributions of total suspended matter (left panel, TSM in g *m^{-3}) and chlorophyll* a *(right panel, Chl in mg m^{-3}) along the EPS-A scanner trajectory of May 13, 2001, over Lake Ketelmeer, southern Lake IJsselmeer and northern Lake Markermeer (from top to bottom). The position of the trajectory is indicated on the map of Figure 4.*

Table 1. Current satellite-based sensors suitable for the mapping of chlorophyll-a concentrations in oceanic waters (data from the website of the International Ocean Colour Coordinating Group, IOCCG, May 2004).

Sensor	Launch Date (dd-mm-yy)	Swath (km)	Resolution (m)	Spectral Bands (number)
COCTS (China)	15-05-02	1400	1100	10
MERIS (Europe)	01-03-02	1150	300	15
MODIS-Aqua (USA)	04-05-02	2330	1000	36
MODIS-Terra (USA)	18-12-99	2330	1000	36
OCI (Japan)	27-01-99	690	825	6
OCM (India)	26-05-99	1420	350	8
OSMI (Korea)	20-12-99	800	850	6
SeaWiFS (USA)	01-08-97	2806	1100	8

extending downstream of the island in Lake Ketelmeer. This island is under construction as a depot for contaminated silt, and is so wide that the currents along the opposite lake shores became considerably stronger. The high concentrations of suspended matter are therefore attributable to erosion of sediment in those lake parts.

4.4 REMOTE SENSING FROM SATELLITES

The Coastal Zone Colour Scanner, which was in operation from 1978 until 1986, was the first satellite sensor specifically designed for mapping of chlorophyll in the sea. Thereafter, more than a decade passed before a new generation of ocean colour sensors came into orbit (Table 1). Especially the advent of SeaWiFS marked conceptual and technological leaps forward. Depending on orbit and swath, i.e., width of the image, the sensors offer a view of the entire Earth within a few days. Among the current satellite sensors with chlorophyll product, MERIS aboard the Envisat satellite features not only the best spatial resolution but also the most suitable bands to detect Chl *a* in turbid waters (Gons et al., 2002). The 68.5° field of view allows global coverage to be provided in two to three days. Three of the MERIS bands are similar to wavebands used for shipboard optical teledetection of Chl *a* (equation 11). Furthermore, the near-infrared channels can be used to detect cyanobacterial surface blooms. Adequate estimation of the vertical light attenuation coefficient is also expected. The MERIS 620-nm band is close to the absorption maximum of phycocyanin (PC).

After Envisat's launch, in March 2002, several months were needed for instrumental calibration. Therefore, no MERIS standard products were available before September 2002. For summer 2003 several clear-sky images of the IJssel

Lagoon have been acquired. Atmospherically corrected MERIS data for July 14, 2003, have been processed for mapping of Chl *a* and PC according to equations (11) and (12). Field calibration was carried out by cruises in summer 2003, during which the specific absorption coefficients of Chl *a* and PC were obtained using the quantitative filter technique with a Perkin-Elmer Lambda 800 spectrophotometer equipped with an integrating sphere (Labsphere, North Sutton, NH, USA). Concentrations of Chl *a* were determined using standard method (Gons et al., 1998). For PC analysis the particles were concentrated using high-speed centrifugation. The pigment was extracted in nine cycles of freezing and thawing, and spectrophotometrically determined (Bennett and Bogorad, 1973). The coefficients $a^*_{Chl}(620)$, $a^*_{PC}(620)$ and $a^*_{Chl}(665)$, estimated by multiple regression of pigment absorption versus concentration were 0.0082, 0.0219 and 0.0151 m^2 (mg of pigment)$^{-1}$, respectively. The latter value is similar to that derived previously in calibrating equation (11) for λ = 664 nm. It can be surmised that PC will be best predicted when cyanobacteria predominate (Simis et al., 2004). The number of observations is still too small for adopting these coefficients to the IJssel Lagoon in general, but applicability within the observation period will be a reasonable assumption. After recalculating R_{rs} of the MERIS product, mapping of Chl *a* and PC was performed according to equations (11) and (12), respectively.

The Chl-*a* concentrations varied from less than 10 mg m^{-3} in Lake Ketelmeer and southern Lake IJsselmeer up to 100 mg m^{-3} in northwestern Lake IJsselmeer (Fig. 9). A very steep gradient occurred in southern Lake IJsselmeer, where a similar biomass 'front' was mapped as in the EPS-A trajectory of May 2001 (cf. Fig. 8). Also in northern Lake IJsselmeer considerable variation in Chl *a* was mapped. The concentrations varied from 100 mg m^{-3} in the west to 40 mg m^{-3} in areas along the entire northeastern shoreline.

Broadly speaking, for PC a similar distribution was mapped as for Chl *a* (Fig. 10). PC concentrations of less than 2 mg m^{-3} occurred in Lake Ketelmeer and adjacent IJsselmeer. PC concentrations exceeding 18 mg m^{-3} were mapped for large areas in central and western Lake IJsselmeer, whilst in areas along the northeastern shoreline the values were more than three times as low. Similar to Chl *a*, in southern Lake IJsselmeer a PC 'front' appeared. The PC to Chl *a* weight ratio indicates abundance of cyanobacteria relative to algae. Mapping of this ratio (not shown) exhibited values of about 0.4 in the greater part of central and northern Lake IJsselmeer, that dropped to less than 0.2 in areas near the north-eastern shoreline of this lake and became virtually zero in Lake Ketelmeer. This spatial distribution indicates that cyanobacteria were more abundant than algae in the central and northwestern parts of the IJssel Lagoon.

The spatial patterns obtained by satellite imagery in summer 2003 are consistent with shipboard observations in the summer of 1997 (Fig. 7; see also Gons et al., 1998). Higher Chl-*a* levels in 1997 than in 2003 may reflect differences in external nutrient loading, or temporal variability associated with other factors. It may also be partly attributable to surface blooms observed in 1997. The MERIS data of 2003 lacked high near-infrared reflectance indicative of surface blooms. Wind speeds at the time of this MERIS overpass were presumably high enough for full mixing.

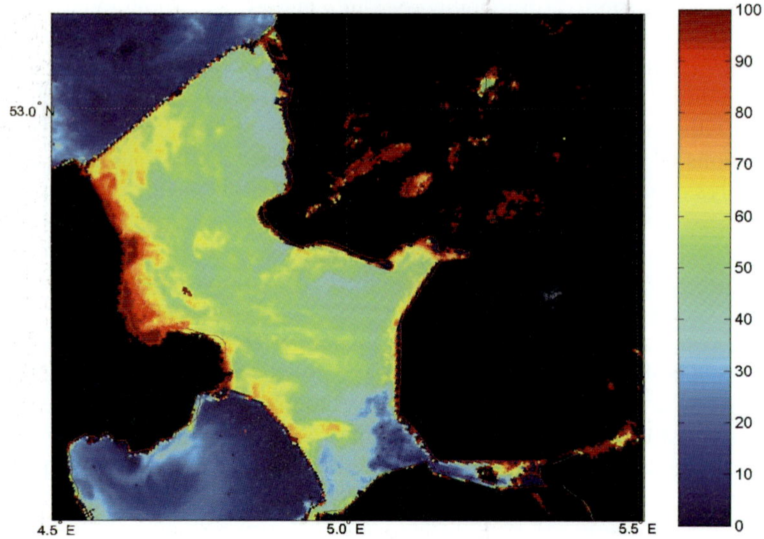

Figure 9. Chlorophyll-a distribution (in mg m^{-3}) in the IJssel Lagoon as derived from MERIS imagery, July 14, 2003.

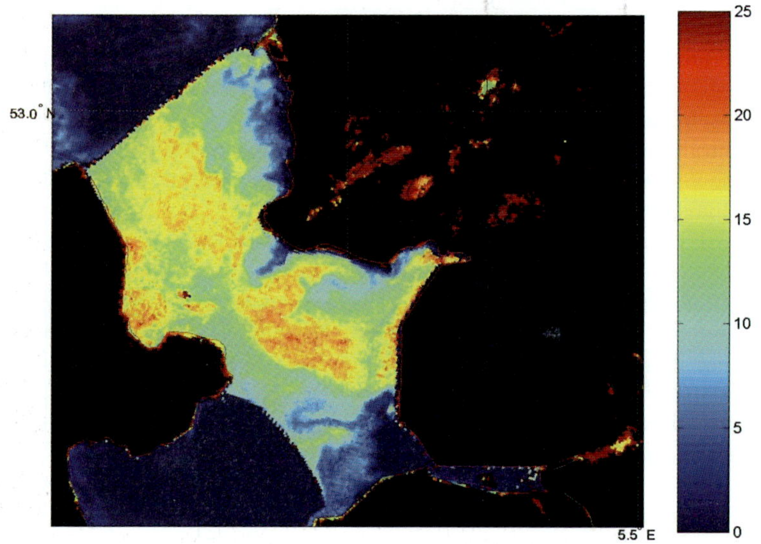

Figure 10. Phycocyanin distribution (in mg m^{-3}) in the IJssel Lagoon as derived from MERIS imagery, July 14, 2003.

5 Optical teledetection of cyanobacterial production

The previous section charted the dominant role of hydrology in structuring the biomass distribution from the River IJssel to southern Lake IJsselmeer. In order to understand spatial variation within Lake IJsselmeer – such as in total biomass and abundance of cyanobacteria relative to algae – insight into the spatial production as well as dispersal and population loss of the plankton groups must be gained. Oceanographic work has convincingly shown that information needed to assess aquatic primary productivity can be retrieved from remote sensing (Antoine et al., 1995; Behrenfeld and Falkowski, 1997a; Joint and Groom, 2000). For the IJssel Lagoon, with a typical vertical light attenuation coefficient K_d of ~3 m^{-1}, direct optical teledetection involves a layer less than 0.5 m deep. However, full vertical mixing is often the rule in this water, so that optical monitoring of the near-surface water generally applies to the whole water column. For full mixing in the IJssel Lagoon, the decrease of light irradiance with depth can be described by a single value of K_d (Gons et al., 1998).

The productivity as a function of depth can be calculated from estimates of the Chl-*a* concentration, the vertical light attenuation coefficient and the photosynthesis-irradiance relationship. Next, the daily primary production per unit surface area is obtained by integrating this productivity over the depth of the water column and the time of the photoperiod (Behrenfeld and Falkowski, 1997b). In order to establish net production, a correction must be made for respiratory losses depending on the length of the dark period and the ratio of the euphotic depth to the depth of vertical mixing. The latter depth equals the water depth in the well-mixed IJssel Lagoon.

The photosynthesis-irradiance curve is described by the initial slope and the maximum specific photosynthetic rate (P_{max}). The value of P_{max} varies depending on plankton species, their nutrient status, and ambient temperature (Falkowski and Raven, 1997; Behrenfeld et al., 2002). However, the greater part of production occurs in the well-illuminated near-surface layer from which both Chl *a* and K_d can be retrieved. The primary productivity per unit surface area (P_A), can be estimated as (Talling, 1957)

$$P_A = \frac{[Chl]P_{max}f(E_d(0))}{K_d} \tag{14}$$

where $[Chl]$ is the concentration of Chl *a*, P_{max} is the Chl-*a*-specific photosynthetic capacity, and $f(E_d(0))$ expresses the dependence on incident light level, which is determined by date, geographic position and atmospheric conditions. So it appears that once Chl *a* and K_d are obtained, plankton production can be derived for adequate empirical knowledge of P_{max} and $f(E_d(0))$. In other words, for uniform species composition and water depth, the quotient $[Chl]/K_d$ can be used as a simple index, obtained from optical methods, that scales the primary production per unit surface area across a system (Gons et al., 1998).

The species composition varies markedly over the entire IJssel Lagoon, but not so much in the 'river water' and 'plankton bloom' compartments (see section 4.2).

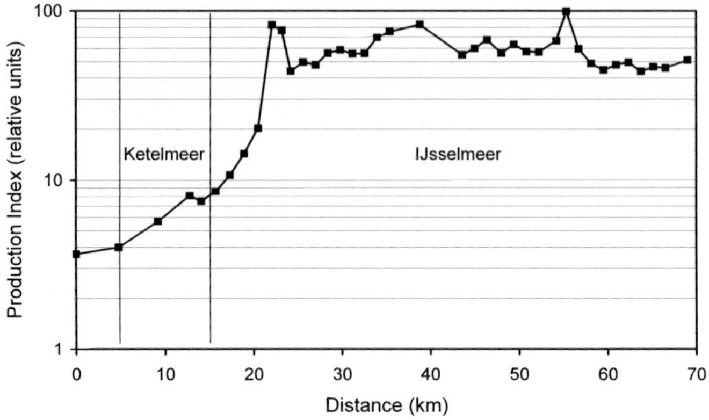

Figure 11. Primary productivity per unit surface area, scaled as the index Chl a/K_d *along the IJssel Lagoon transect on July 29, 1997.*

Proximate as the scaling may be, the index plotted for the July 1997 cruise (Fig. 11) is illustrative regarding the productivity across the IJssel Lagoon. First, the index for Lake IJsselmeer was up to 30 times as high as in the mouth of the River IJssel. Second, in Lake Ketelmeer and even more so in southern Lake IJsselmeer the index rose steeply until a peak at about 7 km from the outflow of Lake Ketelmeer. It is striking to note that the location of this peak also showed as Chl-*a* and PC 'fronts' in both aerial (Fig. 8) and satellite imagery (Figs. 9 and 10) for other dates. Third, the index was higher in the central part of Lake IJsselmeer (25-55 km along the transect) than in the northeastern part of Lake IJsselmeer (60-70 km). This pattern appeared to be recurrent (Gons et al., 1998).

The next steps are estimation of production in terms of carbon fixation rate and the share of cyanobacteria therein. Considering models such as equation (14), it seems to be a relatively simple task to link photosynthetic carbon fixation rate per unit surface area to Chl *a* and the vertical light attenuation coefficient. Following this approach, a model was derived for daily production per unit surface area from shipboard data of the NW Atlantic Ocean, which became a basis for satellite mapping of the global oceanic production (Behrenfeld and Falkowski, 1997a). Yet, considering the impacts of temperature, light history and nutrient status on the photosynthetic capacity alone (Falkowski and Raven, 1997), much work must still be done to improve the remote productivity estimates (Behrenfeld et al., 2002).

In order to estimate the cyanobacterial production, the cyanobacterial share in Chl *a* and the photosynthesis-irradiance curve of the cyanobacterial community must be known. The first parameter can be obtained through optical teledetection of the PC concentration (equation 12) for known intracellular PC to Chl *a* ratio. The photosynthetic parameters in turbid waters like the IJssel Lagoon probably cannot be estimated by optical teledetection. Accurate estimates of productivity in large water bodies in general can only be achieved after integrating optical teledetection with measurements on water samples and experimental laboratory studies. Presumably, during the next decades of research there will be a further need of shipboard observations gathering field samples that allow validation and further improvement of airborne and satellite remote sensing. Field validation will greatly benefit from flow cytometry to enumerate and sort microorganisms in natural water (Veldhuis and Kraay, 2000), and from technologies to determine physiological activities of particular groups and species. For example, variable chlorophyll fluorometry allows not only rapid detection of photosynthetic parameters in lake water (Kromkamp and Forster, 2003) but also assessments of the relative photosynthetic rates of cyanobacteria, diatoms and green algae (Schreiber, 1998). Whether the population dynamics of cyanobacteria are governed by light limitation or nutrient limitation (Chapters 6 and 7 in this book) can be assessed through molecular-biological detection of the high-affinity uptake systems in individual cells (Chapters 4 and 5 in this book). Combining such field information with remote sensing will greatly enhance the capability to predict the development of cyanobacterial blooms.

6 Conclusions

The question of the applicability of optical teledetection methods to monitor the spatial distribution of cyanobacteria can be answered affirmatively. The continued progress in spatial and spectral resolution of remote sensing allows distinction of cyanobacteria and various groups of algae on the basis of their photosynthetic pigments at a detailed spatial scale. In particular, quantification of cyanobacterial biomass and the recognition of cyanobacterial surface blooms by means of optical detection have come within reach. Systems such as the large IJssel Lagoon signify ecological changes on a regional scale. It is high time that the originally technologically driven research now focuses on the unravelling of plankton growth and dispersal in relation to land use and climate. The optical teledetection methods as described in this chapter – whether applied in shipboard monitoring or by using airborne or satellite sensors – can acquire algal and cyanobacterial biomass concentrations as well as productivity in near-real time. Indeed, these advances allow field studies at the population level (Gordon et al., 2001; Subramaniam et al., 2002). Future work may yield detailed information on biodiversity and nutrient cycling from high-resolution imagery, analysed using 'smart', hybrid algorithms integrating hydro-optics and ecological emergent properties. Obviously, monitoring data from vessels, aircraft and satellites are complementary regarding spatial and temporal information. Shipboard optical teledetection cannot provide comprehensive biomass mapping, but can be applied independent of cloud cover, which implies that about 10 days per year with satisfactory satellite images are sufficient to cover the IJssel Lagoon (Vos et al., 2003). Aircraft may underfly clouds and yield enhanced resolution compared to satellites, but for a scanner such as the EPS-A instrument decrease of altitude also implies loss of areal coverage, and thus a loss in cost-effectiveness of this very expensive method. It is a general conclusion that the three monitoring platforms need to be exploited in concerted programmes of data acquisition, validation and reporting.

7 Acknowledgements

The contribution of SGHS was funded by a grant from the User Support Programme of the National Institute for Space Research (SRON), which is subsidized by the Netherlands Organization for Scientific Research (NWO). This is publication No. 3363 of the Netherlands Institute of Ecology (NIOO-KNAW).

8 References

Antoine, D., André, J-M. and Morel, A. (1995) Oceanic primary production. 2. Estimation at global scale from satellite (coastal zone color scanner) chlorophyll, *Global Biogeochemical Cycles* **10**, 57-69.

Behrenfeld, M.J. and Falkowski, P.G. (1997a) Photosynthetic rates derived from satellite-based chlorophyll concentration, *Limnology and Oceanography* **42**, 1-20.

Behrenfeld, M.J. and Falkowski, P.G. (1997b) A consumer's guide to phytoplankton primary production models, *Limnology and Oceanography* **42**, 1479-1491.

Behrenfeld, M.J., Maranon, E., Siegel, D.A. and Hooker, S.B. (2002) Photoacclimation and nutrient-based model of light-saturated photosynthesis for quantifying oceanic primary production, *Marine Ecology Progress Series* **228**, 103-117.

Bennett, A. and Bogorad, L. (1973) Complementary chromatic adaptation in a filamentous blue-green alga, *Journal of Cell Biology* **58**, 419-435.

Buiteveld, H., Hakvoort, J.H.M. and Donze, M. (1994) The optical properties of pure water, *SPIE Proceedings on Ocean Optics XII* **2258**, 174-183.

Bukata, R.P., Jerome, J.H., Kondratyev, K.Y. and Pozdnyakov D.V. (1995) *Optical Properties and Remote Sensing of Inland and Coastal Waters*, CRC Press, Boca Raton.

Dekker, A.G., Malthus, T.J. and Seyhan, E. (1991) Quantitative modeling of inland water quality for high-resolution MSS systems, *IEEE Transactions on Geoscience and Remote Sensing* **29**, 89-95.

Falkowski, P.G. and Raven, J.A. (1997) *Aquatic Photosynthesis*, Blackwell, Malden.

Gons, H.J. (1999). Optical teledetection of chlorophyll *a* in turbid inland waters, *Environmental Science and Technology* **33**, 1127-1132.

Gons, H.J., Ebert, J. and Kromkamp, J. (1998) Optical teledetection of the vertical attenuation coefficient for downward quantum irradiance of photosynthetically available radiation in turbid inland waters, *Aquatic Ecology* **31**, 299-311.

Gons, H.J., Rijkeboer, M., Bagheri, S. and Ruddick, K.G. (2000) Optical teledetection of chlorophyll *a* in estuarine and coastal waters, *Environmental Science and Technology* **34**, 5189-5192.

Gons, H.J., Rijkeboer, M. and Ruddick, K.G. (2002) A chlorophyll-retrieval algorithm for satellite imagery (Medium Resolution Imaging Spectrometer) of inland and coastal waters, *Journal of Plankton Research* **24**, 947-951.

Gordon, H.R. and Ding, K. (1992) Self-shading of in-water optical instruments, *Limnology and Oceanography* **37**, 491-500.

Gordon, H.R., Brown, O.B. and Jacobs, M.M. (1975) Computed relationships between inherent and apparent optical properties of a flat, homogeneous ocean, *Applied Optics* **25**, 448-456.

Gordon, H.R., Boynton, G.C., Balch, W.M., Groom, S.B., Harbour, D.S. and Smyth, T.J. (2001) Retrieval of coccolithophore calcite concentration from SeaWiFS imagery, *Geophysical Research Letters* **28**, 1587-1590.

Hakvoort, H., De Haan, J., Jordans, R, Vos, R, Peters, S. and Rijkeboer, M. (2002). Towards airborne remote sensing of water quality in The Netherlands – validation and error analysis, *ISPRS Journal of Photogrammetry & Remote sensing* **57**, 171-183.

Hoge, F.E. and Lyon, P.E. (1996) Satellite retrieval of inherent optical properties by linear matrix inversion of oceanic radiance models – an analysis of radiance and measurement errors, *Journal of Geophysical Research – Oceans* **101**, 16631-16648.

Jerlov, N.G. (1968) *Optical Oceanography*, Elsevier, Amsterdam.

Joint, I. and Groom, S.B. (2000) Estimation of phytoplankton production from space: current status and future potential of satellite remote sensing, *Journal of Experimental Marine Biology and Ecology* **250**, 233-255.

Kirk, J.T.O. (1994) *Light and Photosynthesis in Aquatic Ecosystems*, 2nd Ed., Cambridge University Press, Cambridge.

Kishino, M., Takahashi, M., Okami, N. and Ichimura, S. (1985) Estimation of the spectral absorption coefficients of phytoplankton in the sea, *Bulletin of Marine Science* **37**, 634-642.

Kromkamp J.C. and Forster, R.M. (2003) The use of variable fluorescence measurements in aquatic ecosystems: differences between multiple and single turnover measuring protocols and suggested terminology, *European Journal of Phycology* **38**, 103-112.

Lammens, E.H.R.R. and Hosper, S.H. (1998) The food web of the lakes IJsselmeer and Markermeer (in Dutch), RIZA report, Lelystad.

Longhurst, A., Sathyendranath, S., Platt, T. and Caverhill, C. (1995) An estimate of global primary production in the ocean from satellite radiometer data, *Journal of Plankton Research* **17**, 1245-1271.

MacColl, R. (1998) Cyanobacterial phycobilisomes, *Journal of Structural Biology* **124**, 311-334.

Mittenzwey, K.H., Ullrich, S., Gitelson, A.A. and Kondratiev, K.Y. (1992) Determination of chlorophyll *a* of inland waters on the basis of spectral reflectance, *Limnology and Oceanography* **37**, 147-149.

Mobley, C.D. (1999) Estimation of the remote-sensing reflectance from above-surface measurements, *Applied Optics* **38**, 7442-7455.

Morel, A. and Prieur, L. (1977) Analysis of variations in ocean color, *Limnology and Oceanography* **22**, 709-722.

Morel, A. and Gentili, B. (1993) Diffuse reflectance of oceanic waters. II. Bidirectional effects, *Applied Optics* **32**, 6864-6879.

Oliver, R.L. and Ganf, G.G. (2000) Freshwater blooms, in B.A. Whitton and M. Potts (eds.), *The Ecology of Cyanobacteria*, Kluwer, Dordrecht, pp. 149-194.

Rast, M., Bézy, J.L. and Bruzzi, S. (1999) The ESA Medium Resolution Imaging Spectrometer MERIS – a review of the instrument and its mission, *International Journal of Remote Sensing* **20**, 1681-1702.

Roesler, C.S. (1998) Theoretical and experimental approaches to improve the accuracy of particulate absorption coefficients derived from the quantitative filter technique, *Limnology and Oceanography* **43**, 1649-1660.

Schiller, H. and Doerffer, R. (1999) Neural network for emulation of an inverse model operational derivation of Case II water properties from MERIS data, *International Journal of Remote Sensing* **20**, 1735-1746.

Schreiber, U. (1998) Chlorophyll fluorescence: new instruments for special applications, in G. Garab (ed.) *Photosynthesis: Mechanisms and Effects*, Volume V, Kluwer, Dordrecht, pp. 4253-4258.

Simis, S.G.H., Peters, S.W.M. and Gons, H.J. (2004) Remote sensing of the cyanobacterial pigment phycocyanin in turbid inland water, *Limnology and Oceanography*, in press.

Subramaniam, A., Brown, C.W., Hood, R.R., Carpenter, E.J. and Capone, D.G. (2002) Detecting *Trichodesmium* blooms in SeaWiFS imagery, *Deep-Sea Research II* **49**, 107-121.

Talling, J.F. (1957) The phytoplankton population as a compound photosynthetic system, *New Phytologist* **56**, 133-149.

Tassan, S. and Allali, K. (2002) Proposal for the simultaneous measurement of light absorption and backscattering by aquatic particulates, *Journal of Plankton Research* **24**, 471-479.

Thomas, G.E. and Stamnes, K. (1999). *Radiative Transfer in the Atmosphere and Ocean*, Cambridge University Press, Cambridge.

Veldhuis, M.J.W. and Kraay, G.W. (2000) Application of flow cytometry in marine phytoplankton research: current applications and future perspectives, *Scientia Marina* **64**, 121-134.

Vos, R..J., Hakvoort, J.H.M., Jordans, R.W.J. and Ibelings, B.W. (2003) Multiplatform optical monitoring of eutrophication in temporally and spatially variable lakes, *The Science of the Total Environment* **312**, 221-243.

Wetzel, R.G. (1990) Land-water interfaces: metabolic and limnological regulators, *Verhandlungen der Internationalen Vereinigung für Theoretische und Angewandte Limnologie* **24**, 6-24.

CHAPTER 9

WATER SAFETY PLANS

A better regulatory approach to prevent human exposure to harmful cyanobacteria

Ingrid CHORUS

1 Introduction

Awareness of cyanobacterial toxicity as a human health hazard has existed in local populations for a long time (see Chapter 1 in this book). This includes local 'common sense' responses to avoid exposure, reflected for instance by childhood memories of grandmother's advice '*don't swim when the water looks green*'. Research on cyanotoxins started in a small number of institutes in the 1960s and picked up momentum paralleling the increasing occurrence of cyanobacteria that accompanied rapidly increasing eutrophication. By the end of the 1980s, chemical structures of major cyanotoxins were known and their toxicity described. Some awareness of the problem existed in the scientific community, and first surveys assessing toxicity of blooms with mouse bioassays were conducted. Incidents of animal deaths and human illness attributed to toxic cyanobacteria were increasingly published (see Kuiper-Goodman et al., 1999, for an overview).

In 1992, for the first time a massive calamity – i.e. the Murray-Darling River bloom in Australia with livestock deaths ranging in the thousands – triggered responses towards regulatory approaches. These included capacity building in the form of institutionalising a national 'Algal Task Force' as well as the development of guidelines (Ressom, 1994). These were based on targeted scientific investigations, such as an epidemiological study on recreational exposure (Pilotto et al., 1997). In England, the problem received public attention after 1989, when the National River Authority published one of the first comprehensive brochures on toxic cyanobacteria following the illness of army cadets training in a lake with heavy *Microcystis* blooms (National Rivers Authority, 1990).

J. Huisman, H.C.P. Matthijs and P.M. Visser (eds.), Harmful Cyanobacteria, 201–227.

2 Current regulatory approaches and their limitations

In consequence of the events discussed above, toxic cyanobacteria were brought to the attention of the World Health Organisation (WHO). For keeping the WHO Guidelines for Drinking-water Quality up to date, a 'Chemicals Working Group' regularly assesses substances suggested for development of a WHO Drinking-water Guideline Value (GV). Criteria for including a substance on the list of WHO GV's are: that it is of significant international concern, that there is credible evidence of its occurrence in drinking-water in concentrations relevant to human health, and that adequate human health and/or toxicological information is available for derivation of at least a provisional GV. This working group decided on a provisional Guideline Value for microcystin-LR in drinking-water in 1997 (WHO 1998).

At the time, microcystin-LR was the only cyanotoxin for which enough information was available for guideline derivation. It was also perceived to be the (or at least one of the) most important cyanotoxins on the basis of the high frequency of its occurrence worldwide as well as its high toxicity. Since, microcystin-LR has been used in the sense of a 'reference microcystin', i.e. hazard assessment for other microcystins often addresses their concentrations as if these were microcystin-LR, applying the GV for microcystin-LR to them as well. This default approach is likely to result in worst-case estimates, as acute toxicity determination indicates all other microcystins to be equally or less toxic.

Some countries have set standards or national guidelines for microcystin-LR in drinking water on the basis of the provisional WHO TDI (tolerable daily intake) for microcystin-LR (e.g. Australia, Brazil, Canada, the Czech Republic, France, Poland, Spain). A proposal for a Guideline Value for cylindrospermopsin is on the working program of the Chemicals Working Group for the WHO Guidelines for Drinking-water Quality. However, discussions about legally binding regulations for cyanotoxins based on compliance to a standard for microcystin-LR show scientific discomfort with this approach for a number of reasons:

- One is the large variety of microcystins and other cyanotoxins that occur, with recent data showing that microcystin-LR often is not even the most frequent one. Using the worst-case approach of assessing all microcystins in a sample as if they were as toxic as microcystin-LR may lead to poorly justified management actions, such as closing recreational sites when a bloom is far less toxic than this approach assumes, and this would be difficult to defend against pressures to continue using the site. In some regions, e.g. northern lowland Europe with a high share of eutrophic inland water-bodies, pressures for continuing recreational use are very heavy, due to a high population density and the importance of these water-bodies for tourism.
- Furthermore, the epidemiological study addressing recreational exposure by Pilotto et al. (1997) demonstrated that the health effects observed were related to exposure time and cyanobacterial cell density, but not to the concentration of the cyanotoxins analysed. The conclusion that other, unknown cellular components must have been involved in evoking these effects is supported by results of different bioassays performed with crude extracts of cyanobacteria.

These show toxic effects and bioactivities not attributable to any of the known cyanotoxins. For example, crude extracts of *Aphanizomenon flos-aquae* have been found quite toxic on fish eggs (Oberemm et al., 1997, 2001), but the substance causing these effects is unknown and it is unclear whether these results are meaningful for oral, inhalatory or dermal exposure of humans.

- Other groups of cyanotoxins are known. Though available data suggest little chronic effects for the cyanobacterial neurotoxins (Kuiper-Goodman et al., 1999), the evidence base is insufficient to justify not addressing them in cyanotoxin risk assessment and management. Furthermore, new results for cylindrospermopsin propose a TDI in a range similar to that for microcystin-LR (Humpage and Falconer, 2002), indicating that where this cyanotoxin occurs, it definitely needs to be addressed in hazard assessment and risk management.

These observations question the justification of selecting one or two cyanotoxins as parameters for monitoring compliance to standards. Would then a more comprehensive standard be adequate, i.e. for cyanobacterial cell numbers or biomass per liter, rather than for selected toxins? This approach is taken by the WHO Guidelines for safe management practice (Falconer et al., 1999) which primarily base the guidance level on cell numbers or a measure of cyanobacterial biomass, although the corresponding microcystin concentrations were used as orientation to determine the intermediate level for restricting recreational activity.

The downside of basing standards on cell numbers is that not all cyanobacterial blooms are equally toxic. For example, *Aphanizomenon flos-aquae* forms blooms quite frequently, but these so far have not been demonstrated to contain microcystins, and appear to contain neurotoxins only rarely. Would it be justified to close a bathing site, perhaps even call of a major sports event such as an international sailing regatta, because of a bloom of *A. flos-aquae*? Or because of a *Microcystis* bloom which has been demonstrated not to contain microcystins?

Cyanobacterial toxins, however, are just one example of the shortcomings of an approach to risk management which is driven by monitoring compliance to standards. In the bigger picture, the past decades have seen an increasing number of substances discussed with respect to regulation of their occurrence in the environment, particularly in water. This increase is due to both the development of new chemical compounds and to an increased scientific understanding of human health impacts of both natural and anthropogenic substances.

Likewise, and often even more relevant for water-borne illness, the scientific understanding of the occurrence of pathogenic microorganisms in water has changed. Until recently, their regulation was based on the occurrence of indicator organisms, such as *E. coli* and coliform bacteria, whose occurrence in water reflects faecal contamination. Standards are set to determine when water is regarded to be free of these indicators, e.g. that no cells or colonies of these indicator organisms should be detectable in a 100-ml water sample. Such standard-driven regulatory approaches have substantially contributed to a dramatic improvement of drinking-water safety during the last century, but in spite of this major success story for public health, shortcomings of the approach have increasingly been causing concern during the past 2 decades. These indicator organisms do not adequately reflect the occurrence of all types of pathogens. Survival of e.g. protozoan cysts is one problem – they are much more resistant to disinfection as well as to natural die-off in rivers and reservoirs, and may still be present even if none of the indicator organisms can be detected. (In fact, disinfection may mask their presence, as the indicators react more sensitively and are not detected in disinfected water, while oocysts of Cryptosporidia or Giardia may well be present). Attenuation of viruses in the underground or in slow sand filters is another problem – they may travel much further and remain infectious much longer than the indicator bacteria which are supposed to indicate their presence.

In consequence, do these developments mean that we need to keep on expanding the list of chemicals and microorganisms to monitor in drinking-water and at recreational sites? Would we improve water safety and public health by adding cyanobacterial cells, Cryptosporidia, Giardia and some viruses to the microorganisms for which we set standards? Should we add further microcystins and other cyanotoxins (e.g. cylindrospermopsin) to the list of chemicals for which standards must be met? Conceptually, this would be in line with the well-established approach taken by many national governments, and in Europe by the European Union Drinking-water and Bathing-water Directives (98/83/EU, 76/160/EU), which currently largely focus on lists of standards that finished drinking water and bathing water must meet, and how to monitor these parameters.

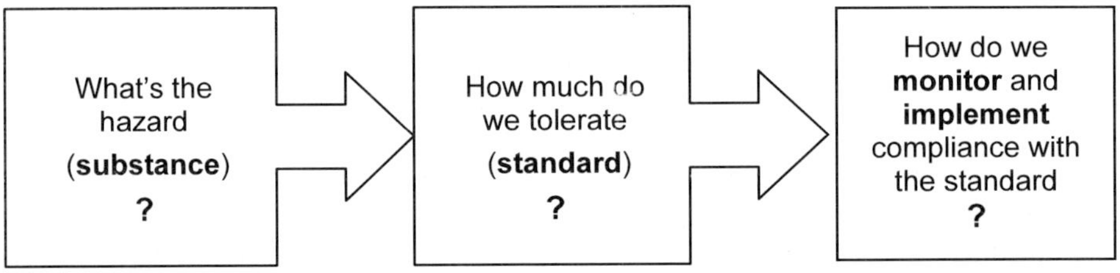

Figure 1. Approach to the safety of drinking-water driven by monitoring the end-point for compliance to standards.

This approach, depicted in Figure 1, may be characterised as driven by 'product monitoring', as it focuses on the finished drinking-water as the product of the supply process, or as driven by 'end-point monitoring', as it focuses on the end-point of the supply chain rather than on the functioning of barriers that prevent contamination. Doubtlessly, this approach has improved drinking-water quality quite effectively in many countries. In Europe, pressure towards compliance with standards is high, particularly through the risk of paying substantial penalties to the EU commission for non-compliance. Vice versa, however, this financial risk pushes public authorities towards focusing their efforts on avoiding non-compliance to standards, rather than on assessing the whole supply system and developing system-specific risk assessment and management approaches, which would include addressing hazards not regulated by a standard.

For cyanotoxins, this current focus is a dilemma: it means that in many countries, unless they are regulated by standards in national directives, toxic cyanobacteria are not likely to get much attention in monitoring and surveillance. On the other hand, for the reasons discussed above, monitoring compliance to a standard for microcystin-LR (and perhaps a small number of further cyanotoxins), or to a standard for cyanobacterial biovolume, is not a satisfactory approach to protecting public health from this hazard.

Furthermore, as mentioned above, to allow stringent enforcement, regulations are usually prescriptive about the time interval in which a parameter needs to be monitored. However, many cyanobacterial species form surface scums or metalimnetic layers, and these accumulations of cells may change their position within the water-body rapidly, sometimes within hours. While these shifts can be understood and to some extent even predicted on the basis of good knowledge of the specific water-body, they cannot be captured in any universally appropriate time pattern for sampling that could be described in a standard method.

From a limnologist's point of view, the solution to the dilemma seems clear: rather than following a prescriptive monitoring pattern addressing compliance to a debatable selection of standards, the aim would be to understand the hydrology, nutrient loading and budgets, as well as the phytoplankton ecology of the individual water-body (Chapters 6 and 7 in this book). On this basis, optimal monitoring schemes can be tailored for the respective water-body (Chapters 4 and 8), and the control measures can be determined which are most likely to be effective in the specific system, preferably by controlling the causes of the problem. It would be scientifically appropriate to capture this approach in a regulatory system.

Such an approach is not only desirable for assessing and managing the hazard of toxic cyanobacteria, as these have a lot in common with other aquatic health hazards, and the challenges of finding adequate regulatory approaches for these hazards are similar. In many drinking-water supplies and recreational sites, pathogens typically do not occur as a regular phenomenon, but as a consequence of events, such as heavy rainfall flushing high pathogen loads into a reservoir or to a bathing site that normally would not be affected. Such events are easily missed in regular monitoring schemes. Moreover, the large majority of outbreaks of water-borne diseases still being recorded in industrialised countries can be attributed to a poor understanding of the water supply system (or recreational water-body) and of the events that lead to failure of barriers that should keep disease agents out of the water. Consumer safety is a problem of time scales in current regulatory approaches to monitoring: even if microbiological analyses are provided on a daily basis, they typically take 1-2 days to provide results, and by the time a result indicates contamination, the water is usually already at the consumer's tap and consumed.

Thus, we can generalise that, while end-product testing does provide retrospective information about the security with which the system has been functioning and may show trends, drinking-water safety on a day-to-day basis is achieved by good management of the ***processes*** of water provision, from the catchment to the consumer's tap. The same is true for recreational sites which typically are monitored at much larger time intervals – the EU Bathing-water Directive calls for 14-day intervals.

3 A new approach in the WHO Drinking-water Guidelines

In face of the general shortcomings of an approach to water safety driven by end-point monitoring, the revision of the WHO Guidelines for Drinking-water Quality has shifted emphasis towards comprehensive hazard assessment and risk management. The tool introduced for this is '*Water Safety Plans*' (Davison et al., 2004; WHO, 2004). It builds on long established good practice in Drinking-water supply, particularly on the multi-barrier principle, and makes this more effective by integrating elements of an approach successfully used in the food industry (HACCP, i.e. Hazard Assessment and Critical Control Points). Water Safety Plans focus on systematically assessing and controlling the various processes in a drinking-water supply chain, from catchment to consumer.

This new approach begins with taking a step back and revisiting the ***targets for public health***. The risk of contracting disease through environmental exposure is never equal to zero. Society therefore needs to define which level of disease burden through pathogens and substances in water is considered acceptable. This approach is already well established for carcinogens, which are assumed to have no threshold level causing an effect. Rather, a widely used consensus is the target of less than one additional case of cancer caused by the substance to regulate per 100.000 or one million inhabitants. Using dose-effect relationships such targets for public health are translated into maximum acceptable concentrations of a given substance. Quantitative microbial risk analysis has made this approach applicable to pathogenic organisms as well, and by calculating 'DALYs' (Disability-Affected Life Years in a given population; see WHO, 2004) risks for chemicals and infectious organisms may be compared. DALYs further allow comparison of health risks through exposure to drinking-water with those from other exposure routes, and this is important for determining priorities for management interventions.

Setting targets in terms of limiting disease burden thus introduces an element of flexibility for the individual parameter: targets e.g. for cyanotoxin levels would be set in the context of overall reduction of disease burden. For example, in settings with a high morbidity and mortality rate caused by contamination of drinking water with pathogens (e.g. from faecal contamination), as a first step, public health would be most significantly improved by concentrating efforts on mitigating faecal contamination. The health-based target for cyanotoxins in such a setting might therefore not be set as low as in a setting with very low levels of other contamination. Such an approach was taken for setting the Brazilian standard for microcystins: this allows up to 10-fold higher values than the WHO Guideline value up to 3 times per year in order to accommodate for the specific situation in Brazil, where public health problems require other priorities for the time being (Fundação Nacional de Saúde, 2000).

Comparing hazards on the basis of their impact on public health highlights the differences in the rationale behind current standards for different substances. In the European Union, the standard for pesticides in drinking-water in the EU Drinking-water Directive is 0.1 $\mu g\ L^{-1}$ for any individual pesticide, and not more than 0.5 $\mu g\ L^{-1}$ for the sum of all pesticides occurring in the sample (98/83/EU). The rationale for setting this low standard is not health-based in the sense described above, but rather precautionary, based on a consensus that pesticides – i.e. poisons intentionally introduced into the environment – should not occur in drinking-water. The standard was therefore set at the detection limit (achievable at the time). In contrast, WHO derived health-based Guideline Values for individual pesticides from toxicological information, and for many pesticides these typically are much higher, up to 20 (e.g. permethrin) or even 100 $\mu g\ L^{-1}$ (e.g. dichloroprop; see WHO, 1996). The WHO provisional Guideline Value for microcystin-LR of 1 $\mu g\ L^{-1}$ (WHO, 1998) is health-based in the same way, i.e. derived from toxicological considerations. This allows the direct comparison of the public health burden both may cause: envisage microcystin occurring in the concentration range of up to 10 $\mu g\ L^{-1}$ and one of the above-mentioned pesticides found in the same range of up to 10 $\mu g\ L^{-1}$: While we can easily envisage newspaper headlines and public emotions about the pesticide concentration 100-fold higher than allowed, in reality the cyanotoxin concentrations would present the greater public health risk.

Precautionary targets such as the EU standard for pesticides in drinking-water have their justification. It is based on public perception and value judgements such as the guiding principle that drinking-water should be wholesome and free of any hazardous agents. Setting a precautionary standard at a level much lower than the one derived from toxicological knowledge is a rational way of accommodating scientific uncertainty. In face of the crudeness of our current quantitative tools (which are actually at best semi-quantitative) for translating effects seen in animal experiments into Guideline Values or standards, a precautionary approach is prudent, particularly for anthropogenic substances with which mankind has fairly little societal experience, and which have no benefit in drinking-water. Disinfection by-products are different from pesticides in this discussion, because a direct benefit is gained from disinfection, and the assessment of its by-products is the outcome of weighing benefits against risks.

However, for substances in water which have no benefit, the selection of those for which we require a precautionary standard and those for which we feel comfortable with a health-based standard is emotive, i.e. driven by perception of hazards rather than scientific understanding. Thus, the need for regulating the occurrence of cyanotoxins has been questioned in numerous debates, as people are familiar with the phenomenon, have experienced no symptoms after recreational exposure, perceive cyanobacteria as something 'natural' and therefore harmless.

Transparency of the differences in rationale behind target concentrations for different substances is therefore very important. It is essential as a basis both for public communication (see below), and for guiding priorities in planning management responses. For a reservoir where the substantially more relevant health concern is occurrence of toxic cyanobacteria, a priority on action to restrict use of pesticides in the catchment would not be a rational approach to reducing public health risks. Current legislation in Europe would, however, require management interventions at 1 $\mu g\ L^{-1}$ of any pesticide, while a response to several $\mu g\ L^{-1}$ of microcystins would be more difficult to enforce.

4 How does a 'Water Safety Plan' work ?

A Water Safety Plan (WSP) is a systematic quality management tool, in principle specifically designed for each individual water body, to help ensure that the water body meets the health-based targets at all times. In contrast to other quality management systems such as the ISO 9000 series, a Water Safety Plan is based on identification of the hazards specifically relevant in a given water supply system and an assessment of the risks they cause. In contrast to the traditional end-point monitoring approach, depicted in Fig. 1, the Water Safety Plan addresses ***processes***. Figure 2 shows a simplified summary of this way of thinking. The 'review loop' highlights the importance of regular iterations to check whether the relevant hazards are adequately identified and controlled by the management measures specified in the plan.

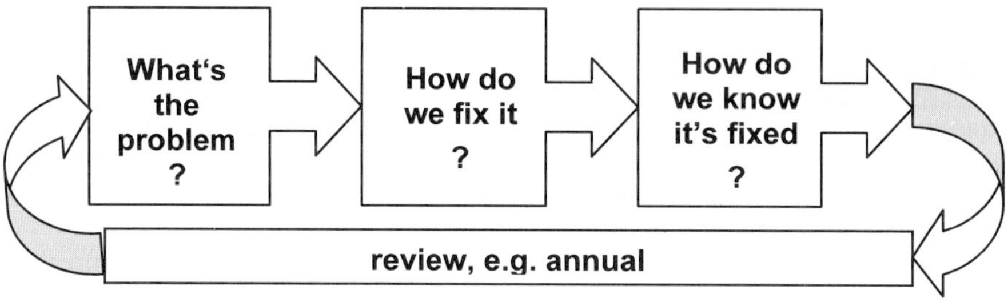

Figure 2. Hazard assessment and risk management approach to water safety (adapted from Davison, 2002).

The Water Safety Plan approach changes the way public authorities fulfil their role of oversight and surveillance: periodic compliance monitoring of the finished product, i.e. drinking-water as it leaves the waterworks or reaches the consumers tap, still remains important to verify that the supply system is working safely, and that the Water Safety Plan is properly designed to control the hazards. However, surveillance also includes assessment of the Water Safety Plan and how it operates in practice. This can be done formally, e.g. through audits, as practiced in some Cantons of Switzerland (Walker, 2003) and cities of Australia (Melbourne; Gold Coast) for similar approaches. The French Drinking-water Ordonance allows a reduction of end-product surveillance for supplies having a HACCP plan to the authority's satisfaction (Metge, 2003). Thus, Water Safety Plans shift the focus of the authorities towards understanding the supply systems for which they are responsible and the processes that are critical for the safety of each. While specifically designed for drinking-water safety, many elements of the Water Safety Plan approach can also be applied to recreational settings.

4.1 'WHAT'S THE PROBLEM ?'

In the context of a Water Safety Plan, hazard assessment is conducted by an interdisciplinary team that understands the system as well as the hazards. Such a ***Water Safety Plan Team*** includes different levels of professionals running the water supply, i.e. managers, engineers, operators and technical staff familiar with details of the system's structure and day-to-day operation. It will also include scientists (microbiologists, chemists, hydrogeologists) to assess the behaviour of hazardous agents in water, such as e.g. cyanobacteria. As many hazards originate from human activities in the catchment, it will often be important to include key players from the catchment in parts of the team's work. Furthermore, the public authorities responsible for the surveillance of the safety of the supply system should be involved. As first step, this team will detail a flow diagram of the water supply, identify the potential hazards and verify both the flow diagram and the hazards through site inspection. Practical experience with site inspection by such teams is unequivocally reported to identify hazards that had previously been overlooked.

The next step for the Water Safety Plan team is to ***identify the water uses***. Though this may at first seem trivial if the obvious use is household supply, the identification of additional uses – such as supply to clinics – may prove important. Also, this step may include identification of particularly vulnerable user groups such as immuno-compromised persons or dialysis patients which may need to be informed about risks specific to them only.

The team then proceeds to estimate the ***human health risk*** associated with each of the hazards identified. This means analysing the probability and frequency of their occurrence and the magnitude of the consequences. Hazard occurrence can be assessed quantitatively from monitoring and historical data, if available, or estimated from an understanding of the catchment's hydrogeology and land-use. Analyses of hazard occurrence result in classification by simple categories such as 'rare', 'unlikely', 'possible', 'likely', 'almost certain'. These also include some measure of the frequency with which the hazard is expected to occur, ranging e.g. from once in five years to daily.

The severity of the impact of a hazardous event is primarily assessed in relation to its effect on human health. This includes estimating the size of the population potentially affected and the severity of illness that the hazard might cause at the concentration ranges expected during an event (potentially using DALYs as quantitative approach; see above). Severity rankings are also used for other impacts, such as the economic impact of hazard occurrence for a water supplier or for a recreational site operator, or the environmental impact (Harding, 1999).

Risk scoring or ranking matrices that plot the severity of impact against the likelihood of the hazards to occur are helpful for identifying priorities for taking action to control the hazard (Davison et al., 2004; WHO, 2004). Table 1 gives an example of how the risks of toxic cyanobacteria occurring in a drinking-water reservoir might be assessed in relation to other hazardous events.

Table 1. Simple risk-assessment matrix. Example: a meso-eutrophic thermally stratifying drinking-water reservoir in a temperate climate with mixed land use in the catchment, including some settlements. Intensive agricultural use has recently been reduced, but some maize fields and cattle breeding remain. TP = total phosphorus.

Hazardous event	***Hazard type***	***Likelihood***	***Severity***	***Health Risk***
Extended period of fine weather with low turbulence of the epilimnion, high irradiance and temperature, TP concentrations of 50-100 μg L^{-1}	Toxic *Microcystis* bloom with heavy surface scum formation	High	Moderate	Moderate
Heavy rainfall causing erosion and surface run-off, overflow of combined sewers as well as short water residence time	High concentration of Cryptosporidia, and human faecal contamination	High	High	Catastrophic
Extended period of fine weather with low turbulence of the epilimnion, high irradiance and temperature, TP concentrations of 30-50 μg L^{-1}	Toxic metalimnetic population of *Planktothrix rubescens*	Moderate	Moderate	Moderate
Spill in chemical plant in the catchment	Chlorinated solvents and other organic pollutants	Very low	Major	Very low
Extended rainfall periods during winter	Nitrate	Moderate	Low	Low
Intensive pesticide application on maize fields shortly before rainfall	Pesticides	High	Low	Low

In this example, the likelihood of a spill in the chemical plant is assessed as very low on the basis of site inspection and surveillance data indicating excellent containment of the facility. The health risk of pesticides is assessed as low, because concentrations were always found to be orders of magnitude below the respective WHO Guideline Values (while the political risk for damaging the drinking-water's public image and the economic risk of fines for non-compliance may nonetheless be assessed as very high). The risk caused by nitrate in the example in Table 1 was assessed as low because only a small fraction of the population would be affected and concentrations were not expected to increase significantly above the standard of 50 mg L^{-1}.

The ranking given in Table 1 is qualitative, or at best semi-quantitative, and as such not a strictly objective exercise. As pointed out by Harding (1999, p. 171):

> *'The quality and comprehensiveness of hazard/risk identification is dependent on the skills, knowledge and imagination of those involved in the task – different teams are unlikely to reach exactly the same conclusions.'*

Furthermore, though gaps and uncertainties in the information and models used to assess both the occurrence of the hazards and their impact on the human population can be reduced, some will always remain. Acknowledging this uncertainty is therefore important.

In such a team, the role of a limnologist or phytoplankton ecologist would be to assess the likelihood of cyanobacteria to occur in the raw water source with respect to dominant taxa, seasonal and spatial patterns, duration and intensity of blooms. As for other hazards, this can be done on very different levels, depending on the options locally available. Potential sources of information are diverse and include observations reported by the local population or public authorities on bloom history and illness, available scientific records, as well as results of programs for sampling and analyses. The latter may range from a single survey (conducted at an appropriately selected time of the year) to regular monitoring. The parameters included will also vary, depending on the resources available for such a program. In some settings, available skills and equipment may lead to a focus on counting cyanobacterial cells and determining biomass. In others, availability of a chemical laboratory might lead to a focus on pigment analysis (as 'marker' for the occurrence of cells), or include cyanotoxin analysis.

All approaches will benefit from including an assessment of the capacity for blooms to develop. Again, this can be done in different ways: a highly useful approach is assessing the catchment with respect to the likelihood of nutrient loading by looking at the hydrogeological conditions and human activities, particularly agricultural practices and disposal of human excreta. A more quantitative approach is to determine the 'carrying capacity' in terms of seasonal patterns of the concentrations of total phosphorus and nitrogen.

This risk assessment does not imply that the hazards will necessarily occur in finished drinking-water. Rather, it is the basis for assessing which hazards risk management needs to address. For example, how much of the biomass of a surface scum of *Microcystis* or a metalimnetic layer of *Planktothrix rubescens* reaches the raw water for the drinking-water supply will strongly depend on off-take depths. The next Water Safety Plan step after hazard assessment is therefore ***system assessment***. This investigates to which extent the system can meet its targets and if not, what investment of human, technical and financial resources would be required to improve the supply. An approach to assessing a supply's potential for eliminating cyanotoxins is given in Bartram et al. (1999). The Alert Levels Framework given by these authors also contains many elements of such an assessment, though in the context of a graduated response to bloom occurrence and development.

For recreational sites, a 'What's the problem' approach is currently being discussed for the revision of the EU's Bathing-water Directive (76/160/EU), which dates back to 1976. The draft includes the obligation to establish a 'bathing water profile' for each site. This investigates all potential sources and circumstances likely to lead to contamination and aims at understanding the risks as a basis for targeted management measures. It includes the proliferation of toxic cyanobacteria as one issue to be addressed in the profile.

4.2 'HOW DO WE FIX IT ?'

'Fixing hazards' may begin with changing the system's structure in order to control the hazard. For example, if system analysis has found a sewage outfall to be the source of high nutrient levels that support cyanobacterial blooms, upgrading the treatment plant to eliminate the nutrient or diverting the outfall might be such a structural change. For a drinking-water supply, such a change might also involve changing the site of the off-take or even the raw water source, e.g. from surface water to bank filtrate or groundwater.

Moreover, the process of supplying drinking water must be managed in such a way that the hazards are controlled. For this purpose, a Water Safety Plan defines ***control measures*** for the hazards identified to be of priority. In the context of Water Safety Plans, these are measures whose operation can be controlled sufficiently tightly to allow for a response before the water reaches the consumer. This requires a ***monitoring system*** showing whether or not the control measures are sufficient to safeguard the ***operational limits*** within which the process is believed to be safely operating, and ***corrective actions*** that can be taken immediately when the monitoring system indicates that the operational limits are not being met.

In this context, the meaning of 'operational monitoring' is entirely different from its meaning in the context of product or end-point monitoring: it is not the analysis of the target water quality parameter such as a cyanotoxin concentration, but rather analysis of some parameter that indicates whether or not the control measure is working. A straightforward example is filtration in a water treatment plant as control measure to remove pathogens, cyanobacteria and particles to which hazardous substances might be adsorbed. Although the health-based target for the entire supply system would be to keep cell density of pathogens and cyanobacteria beneath a previously defined level, these would not be regularly monitored at the filter outlet. Rather, a suitable parameter for operational monitoring of the filter's performance would be particle counting or turbidity reading. For these, operational limits can be set which indicate the boundaries within which the filter is working reliably. Furthermore, these parameters provide timely results that allow corrective action (such as filter backwashing) to be taken immediately once monitoring indicates that the control measure is no longer operating reliably.

A basic principle to drinking-water safety, strongly endorsed by the WHO Guidelines for Drinking-water Quality, is the multi-barrier principle. This aims at ensuring safety through more than one barrier, particularly for hazards with high priority for human health. Figure 3 shows that barriers against hazards may be established through catchment management, managing the raw water source, choice and management of the raw water off-take, drinking-water treatment, and in distribution. While in general, re-contamination in the reticulation network is a major concern for drinking-water safety and control measures to ensure barriers are working in this step are particularly important, they are scarcely relevant for cyanotoxins. The multi-barrier principle increases safety, as the risk of break-through of a hazardous agent is small if one of several barriers proves insufficient. It also provides an option for additive removal effects where one measure alone is not sufficient to control a contaminant. For bathing sites, the multi-barrier approach is restricted to two at best – managing the catchment and managing the water-body. This makes information and warning of recreational site users an important option which often becomes necessary when barriers in the catchment and water-body are insufficient to prevent cyanobacterial proliferation.

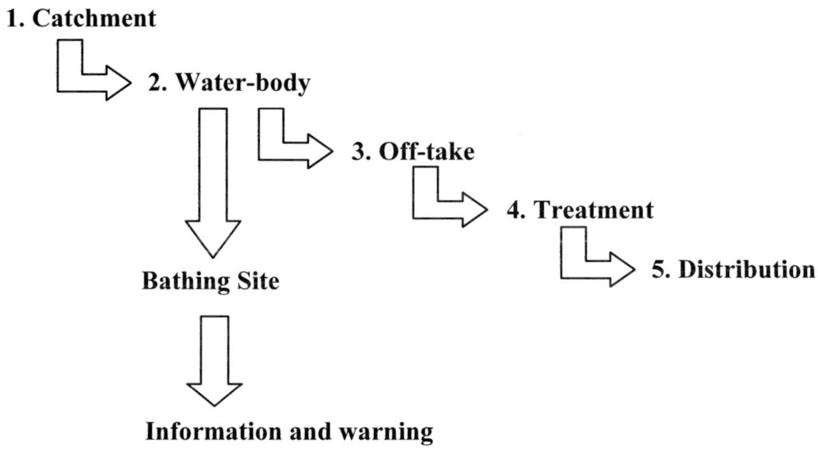

Figure 3. Schematic of a multi-barrier system for recreation and drinking-water.

4.3 'HOW DO WE KNOW IT'S FIXED?'

Knowing that the 'problems are fixed', i.e. the hazards are under control, means having confidence that the system of measures is indeed adequately designed to effectively eliminate the hazardous agents (***validation***), as well as confidence that it is always working as it should (***verification***). Such confidence draws on proper ***documentation*** of assessments, decision criteria and process operation. Public confidence that 'the problems are fixed' requires appropriate ***communication*** of the hazards and risks, criteria for their assessment, operational performance of the supply system and water quality monitoring results. In the following, these aspects of a Water Safety Plan will briefly be discussed in relation to toxic cyanobacteria.

4.3.1 Validation

In the context of a Water Safety Plan, validation means to periodically assess the extent to which the control measures actually eliminate the hazard they are intended to control. Cyanobacterial toxins present a particular challenge to drinking-water supplies as well as to surveillance of recreational sites because of the often highly variable nature of their occurrence: temporal and spatial heterogeneity of cyanobacterial populations, dominant species and their specific toxin content, with variable fractions being cell-bound or dissolved. Validation of control measures for cyanobacteria and cyanotoxins in a Water Safety Plan requires not only an assessment of their reliability in removing cells or toxins, but also whether or not a measure may exacerbate the hazard by causing cell lysis and toxin release (e.g. copper sulfate treatment of blooms, or cell lysis on filters in treatment plants).

Validation includes field experiments such as measuring cell densities and/or toxin concentrations after each treatment step during an extreme bloom event, as well as laboratory-scale or semi-technical scale experiments to test elimination under defined conditions. While the whole range of cyanotoxins potentially occurring in the water source would not be regularly monitored (see '*verification*' below), the validation of control measures would involve analyses of selected 'reference cyanotoxins', i.e. a choice which is representative for the type of problem expected. For example, if hazard assessment identified both microcystins and cylindrospermopsin potentially reaching concentrations causing a health risk, validation would investigate the system's capacity to remove both. This would be necessary because both toxins may respond quite differently to some control measures, as microcystins usually occur cell-bound whereas for cylindrospermopsin a high share of dissolved toxin is frequently observed. In this context, microcystin-LR, which is readily quantifiable in chemical analyses because standard material is commercially available, could serve as 'reference microcystin'. In the context of validation, it is not specific toxicity which is important, but rather the behaviour of the agent in the system with respect to its susceptibility to control measures.

Validation of the efficacy of control measures in mitigating cyanotoxin hazards has a generic research component as well as a strong site-specific component. The generic component involves periodic review of the scientific progress in the understanding of cyanotoxin occurrence and response to control measures, e.g. new results on further toxins identified, on conditions triggering elevated toxin release from cells into the water, on the behaviour of populations under specific ecological conditions, and on the performance of specific treatment steps. However, conditions in each water supply system or recreational site are unique: Limnologists are well-familiar with the individuality of catchments and water-bodies as well as with the uncertainties of predicting how they will react to measures designed to reduce cyanobacterial proliferation (Chapters 6 and 7 in this book). Likewise, though drinking-water treatment plants may use widespread technical components (such as a filter or an ozonation unit), they also are individually designed and combined for the given setting, and will respond differently from site to site. Thus, validation is necessarily also a site-specific exercise with the purpose of developing a sound understanding of the system and its responses to measures under the range of conditions that may be encountered.

The latter aspect – i.e. the range of conditions that occur – is of basic importance for water safety. Most of the cases of water-borne disease recognised in developed countries are not due to regular failure of the barriers in the supply system, but rather to specific events, such as heavy precipitation carrying loading from surface runoff to drinking-water off-takes which normally would not be reached. Such scenarios are easily overlooked when designing a system. Validation places special emphasis on assessing the extreme scenarios as well. One outcome of validation may be to improve control measures and/or their monitoring. Another may be to recognise limits for 'normal' safe day-to-day operation of an existing system and a need for emergency response plans for extreme situations in which the system can no longer operate safely. Populations of toxic cyanobacteria typically wax and wane, and in some systems this is a regular phenomenon, readily recognised in the context of hazard assessment, planning and validating control measures. Specific attention to the cyanotoxin hazard may be warranted particularly for reservoirs and recreational sites in which bloom occurrence is a rare event, e.g. because they are usually dominated by other phytoplankton such as diatoms, and cyanobacteria proliferate only under unusual conditions. The event nature of blooms also challenges validation of removal measures in drinking-water treatment: filters and oxidation steps may fail to cope with the extreme organic load during blooms if they are not laid out for these amounts.

Validation is repeated periodically and particularly when conditions change or the system is changed, or when new scientific results indicate a need to revise the hazard assessment. It is not intended as routine surveillance of the system's operations.

4.3.2 Verification

Verification is the periodic overall surveillance and evaluation of the system's performance, by the water supplier as well as by an independent body such as the public health authority. It operates on a different level than validation by focusing on the efficacy of the system as a whole rather than evaluation of the design of individual control measures.

One important component of verification is inspection of whether the Water Safety Plan is being actively used in practice. This would include assessing records of control measure monitoring, documented corrective action, and whether the management plans included in the Water Safety Plan (such as standard operational procedures) are being adhered to. External auditing may require the water supplier to demonstrate this. Public health authorities in Switzerland, where such plans are in place for drinking-water, have reported improved cost-efficiency through auditing such plans, as they enable the authority to differentiate between well-operated and problematic supplies and their attention can therefore focus on the problematic supplies (Walker et al., 2003).

A further component of verification is monitoring quality parameters in finished drinking water as it leaves the waterworks and/or at the customer's tap. This operates on an overall level in contrast to routine operational monitoring of individual control measure performance (e.g. turbidity reading at filter outlets). Verification in the context of a Water Safety Plan is similar to traditional end-point monitoring for compliance to standards, and it remains critically important as an overall check that the system is working properly. Analyses for verification use a set of pre-determined parameters indicating water quality, such as faecal indicator microorganisms and selected chemicals. Analyses for verification are typically performed regularly by the water supplier and at larger intervals by the public authority responsible for surveillance.

4.3.3 Documentation

Knowing that the 'problems are fixed', i.e. the hazards are under control requires documentation of the initial hazard inventory and risk analysis, the analysis of the system and assessment as to how well it can control the risks, the control measures determined for the system, the management plans (such as standard operating procedures and emergency response plans), and the data from operational monitoring of the control measures as well as corrective action, if taken. Incomplete and fragmented documentation of assessments, decision criteria and processes is a frequent weakness in many water supplies, and those having adopted some quality management system unequivocally report having experienced substantial benefits from improved documentation. These include holding onto experience and knowledge (which otherwise leaves the company or agency as staff leaves) and availability of internal material for training of (new) staff. Documentation of assessments, decision criteria and processes is also important as a basis for internal and external auditing of a Water Safety Plan. Furthermore, it is a useful basis for public communication.

4.3.4 Communication

The importance of communicating cyanotoxin risks to the public using recreational sites is evident and now well recognised in a number of countries, where information and warning of site users has become a major element of regulatory approaches to recreational safety.

However, toxic cyanobacteria are also a prime example for the improvements still needed in public communication of water-borne hazards. Public perception of this hazard in relation to other substances in water, e.g. pesticides, tends to be inverse to the scientific understanding of these risks. Using the example of pesticides outlined earlier, the evidence-based assessment of their relative hazard to health at the concentrations typically found in water indicates substantially less or no risk from exposure to pesticides as compared to that from exposure to cyanotoxins. However, qualitatively pesticides are typically perceived as highly dangerous, often man-made poisons, while cyanotoxins are seen as 'natural'. Quantitatively, the different rationale behind different standards (i.e. precautionary as compared to health-based) has typically not been well communicated and is hardly understood. This easily leads to disproportionate concern when e.g. the precautionary pesticide standard is exceeded as compared to concern about cyanobacterial blooms.

However, when communicating risks and the reasons for the controls chosen to manage these risks, it is important to understand and acknowledge a society's value judgements which underlie the perception of hazards. This includes communicating the uncertainty of scientific assessments, i.e. that we do not comprehensively understand all long-term effects of substances and their combinations, and that our scientific tools for risk assessment are necessarily crude. For cyanotoxins, communicating this uncertainty can be linked to providing guidance on how recreational site users can recognise these organisms and avoid exposure through making their own choices about how and when they use sites potentially affected.

Transparency and communication of the rationales behind standards furthermore facilitates temporary exemptions from compliance. The need for this is currently discussed in relation to permits for temporary exceedance of standards in order to focus the investments on removing the source of a contamination rather than removing the contaminant through treatment – the former being the preferred option where health risks from a contaminant are sufficiently low. This scenario is important for cyanobacteria. In face of an uncertainty factor of 1000 behind the provisional WHO Drinking-water Guideline Value of 1 μg L^{-1} for lifetime exposure to microcystin-LR, a health authority could decide to tolerate occasional occurrence of 2-3 μg L^{-1} for an interim period of a few years while measures in the catchment are being implemented to reduce the reservoir's nutrient load.

5 Control measures for cyanotoxins in a Water Safety Plan

The following text investigates each of the multiple barriers potentially important for mitigating cyanotoxin exposure with respect to the control measures that are conceivable to manage cyanotoxin risks. All of the examples used would only be part of a Water Safety Plan, which is developed to manage the totality of the risks identified for a given supply. Furthermore, it is important to remember that these are just examples, adequate for the one or other specific setting. Water Safety Plans are tailored for specific supply systems, and the combination of control measures deemed most effective in Setting A might be quite different from the 'package' that is best for Setting B. Whether or not a given measure is used as control measure in the context of a Water Safety Plan of a specific supply will depend on whether or not it is assessed to be central to the safety of that specific supply.

5.1 CONTROL MEASURES IN CATCHMENT MANAGEMENT

Measures in the catchment to control cyanobacterial occurrence relate to reducing nutrient loads. In most settings, phosphorus is the critical nutrient to control, although particularly in arid regions, nitrogen is increasingly being identified as the limiting nutrient. However, controlling cyanobacterial growth through limiting nitrogen may fail by leading to the dominance of nitrogen-fixing species, so that even in these settings, control measures are usually more effectively directed towards managing phosphorus loading. The threshold level for the tolerable nutrient load to the water-body is derived from a relationship between the load to the water-body and the resulting nutrient concentration in it. For phosphorus, the Vollenweider (1982) regression provides a generic approach which can be refined for the specific water-body through modelling the load and measuring the corresponding concentration in the water-body. As this model provides only a very rough estimate, such a validation step for the specific setting is important.

Once the target value for maximal allowable loading is defined, determining control measures requires an assessment of the nutrient input from different sources. Loading from point sources such as sewage outfalls is easiest to quantify, technologies to remove phosphorus and nitrogen from them are available, and this renders sewage outfalls a likely control measure for which performance limits can be defined and monitored. However, intuitive approaches often underestimate the share of phosphorus run-off from agricultural areas, and modelling the loads from such diffuse sources in a given catchment may be important for identifying potential control measures. Where run-off from agricultural land is identified as a relevant source of a reservoir's total phosphorus load, more detailed modelling of the loads from smaller areas can identify the most effective control measures. This would show which load reduction can be achieved by specific measures, e.g. (i) by converting particularly erosion-sensitive areas into extensively grazed pasture or even forest, (ii) by maintaining riparian buffer strips covered with shrubs, and (iii) by restricting tillage to specific techniques which cause little erosion on the remaining areas. Typically, both assessing nutrient input from the range of potential sources and identifying control measures is a multi-disciplinary and inter-institutional exercise that requires participation of a range of stakeholders in order to be successful.

Would such measures work as ***control measures*** in the context of a Water Safety Plan? As discussed above, criteria are the availability of an appropriate ***monitoring system***, for which ***operational limits*** can be set that indicate whether or not the measure is operating within bounds, and the availability of timely ***corrective action*** when monitoring shows this not to be the case. For the three examples of catchment management measures mentioned above, such monitoring systems are conceivable: land use can be monitored through periodic site inspection or remote sensing (e.g. by aerial photographs or satellite imaging). Integrity of riparian buffer strips and methods of tillage are amenable to monitoring through similar inspection techniques. Stock density can be monitored by head counts, though this may require more effort. Operational limits can be defined as the percentage of an area to be covered by forest or grassland, the number of stock per hectare, the width and vegetation cover of a riparian buffer strip, and as frequency and type of tillage. Options for corrective action in principle are clear: to enforce the type of vegetation cover, stock density, riparian buffer strips and tillage practices as defined in the management plan. If monitoring and enforcement are sufficiently stringent, corrective action in the case of detected non-compliance could be sufficiently timely, i.e. it would set in before the deficits affect large parts of the area and cause significant increase of erosion and phosphorus loading. Where such tight control is possible, such land use measures can be designated as control measures within the Water Safety Plan.

Is this just an academic exercise, or a useful approach? The advantage of designating specific protection measures in the catchment as elements of a given water supply's Water Safety Plan is that it increases their political profile, i.e. pressure to enforce them. As part of a Water Safety Plan, measures in the catchment are defined as crucial for the safety of the water supply. Thus they would be subject to validation, to documentation of the monitoring results, to regular review, and to inspection or even audit by the authority responsible for the safety of the water supply. Players in the catchment could more easily be legally obliged to respond with corrective action when monitoring shows that a control measure is not keeping the process within the boundaries defined in the plan (e.g. degrading vegetation cover of areas designated to be forested or covered by grassland, degrading riparian buffer strips, stock density exceeding the limit, or deficits in compliance to tillage plans). In practice, such models already exist, though they do not use the Water Safety Plan systematics and terminology. For example, some collaboration models currently practiced in Germany between water suppliers and farmers in their catchment are structured very similarly and could readily be integrated into Water Safety Plans.

In many settings, however, legal tools for enforcing such measures are weak or non-existent, and their successful implementation depends on other approaches such as stakeholder participation, communication, or financial incentives for farmers complying with the management plan. In such settings, control through land use measures would not be reliable. Instead of being part of the Water Safety Plan's set of control measures, these would have the status of 'supporting programmes' which facilitate achieving the water quality targets, but are not relied upon for safety.

Catchment measures targeting nutrient loading obviously need to be validated not only as to whether the water-body responds as predicted and actually reaches the nutrient targets. Furthermore, substantial reductions in nutrient concentrations tend to induce substantial changes in phytoplankton species composition. Validation therefore needs to address whether this leads to new problems. Reducing trophic state may lead to a dominance shift from toxic cyanobacteria to chrysophytes which cause taste-and-odour problems, or to a shift from epilimnetic cyanobacteria to metalimnetic maxima of the cyanobacterium *Planktothrix rubescens*, which typically has a very high cellular microcystin content (although its biomass is usually lower than that of *Microcystis* surface scums).

Control measures targeting phosphorus loading from agriculture are likely to be also effective against pathogens from agriculture, as they will address issues such as stock density, spreading of manure, and reducing erosion. However, their efficacy for each of these hazards needs to be validated separately.

5.2 CONTROL MEASURES IN WATER-BODY MANAGEMENT

Where catchment measures to reduce nutrient loading are not sufficiently effective to achieve the target levels required in the water-body to avoid cyanobacterial proliferation, downstream barriers become particularly important. In some settings, hydrological management of the reservoir is an option to create conditions which are not favourable for cyanobacterial growth, such as augmenting or alternating flow regimes and increasing or alternating mixing depths through artificial mixing (usually by aeration). The design of such measures depends very much on the hydrology of the given setting, and success therefore depends on the quality of system analysis and site-specific planning of the measures. Many aerator operations have proven unsuccessful because they were not sufficiently well tailored to the system (see e.g. Koschel and Jäger, 1995).

In the context of a Water Safety Plan, hydrological measures for controlling cyanobacterial growth would be subject to its stringent systematic approach, i.e. assessment of the extent to which the measure can reduce cyanobacterial growth, which performance targets it must meet at all times and how this can be monitored. An example that would correspond to the Water Safety Plan approach is artificial mixing of Lake Nieuwe Meer in Amsterdam. Here, heavy *Microcystis* blooms were a major nuisance and hazard for recreational use and for inhabitants of houseboats on the lake. Based on a scientific study of the buoyancy characteristics of the *Microcystis* population, an artificial mixing regime was designed to create conditions in which vertical migration through buoyancy regulation would no longer be a competitive advantage of this species over other phytoplankton species, and this proved successful (Visser et al., 1996; Huisman et al., 2004; see Chapters 6 and 7 in this book). In the Water Safety Plan, controlling mixing depth and mixing intensity would be the ***control measures***. The continuous reliable performance of artificial mixing would be monitored by some operational indicators, such as air pressure and operation time of the aerators (***operational monitoring***), for which minimal limits (***operational limits***) would be set. When monitoring indicates that an aerator is not operating properly, this result would trigger ***corrective action*** such as repair within a specified time or switching to a back-up device. The time frame required for this corrective action could be set in the range of days, as growth of *Microcystis* in response to stratification of the water-body would require 1-2 weeks before nuisance or hazardous levels are reached. These time scales are consistent with recent monitoring and simulation studies of the turbulence structure and phytoplankton dynamics in Lake Nieuwe Meer (J. Huisman and K. Jöhnk, personal communication). Thus, this system could be run as a Water Safety Plan control measure, and where the protection of public health depends on its reliable functioning, the stringency of this system would be adequate.

Just as discussed earlier for nutrient load control measures in the catchment, the validation of control measures addressing hydrological conditions would assess whether the hydrological targets can be reliably met, e.g. even during particularly warm summers. Validation would also assess whether the hydrological conditions created are continuously unfavourable for cyanobacterial dominance and actually shift species dominance effectively. As for nutrient load measures, this includes assessment of the new phytoplankton populations with respect to their specific impact on water quality.

In contrast to control measures in the catchment, those addressing cyanobacterial proliferation in the water-body are unlikely to encompass any further hazards, but rather are very specifically tailored to create conditions unfavourable for the growth of these organisms.

5.3 CONTROL MEASURES IN OFF-TAKE MANAGEMENT

Many toxic cyanobacterial species accumulate either in surface scums (e.g. *Microcystis*, *Aphanizomenon*, *Anabaena*) or in metalimnetic layers (e.g. *Plankothrix rubescens*). One example is Deesbach Reservoir in Thuringia, Germany. Preliminary screening showed concentrations of total microcystins at the surface of up to several 100 μg L^{-1}, but less than 1 μg L^{-1} at the off-take depths at about 17 m deep (Table 2; Fastner et al., 2001). In this system, the off-take depth could be designated as *control measure*, and vertical profiles of pigment fluorescence to monitor cell location in relation to off-take location would be effective as *monitoring system*. A threshold value for the fluorescence reading at a defined depth above the off-take would serve as *operational limit*. The *corrective action* would very much depend on the options available and could include increasing the off-take depth (which may not be possible due to other quality problems in deeper, potentially anaerobic layers), a transient switch to an alternative supply, or additional treatment steps.

A further control measure relating to off-take is riverbank filtration. In lowland regions with suitable geology this can be an alternative to impoundments, which often are much more susceptible to cyanobacterial proliferation than the river itself. In many settings, riverbank filtration has proven to be highly effective in removing particles as well as dissolved organic substances. Preliminary results indicate high efficacy in the elimination of cyanobacteria as well as dissolved microcystins through riverbank filtration (Chorus et al., 2001; Grützmacher et al., 2002). Figure 4 shows that bank filtration reduced microcystins in Radeburg Reservoir from 2-18 μg L^{-1} cell-bound and up to 0.5 μg L^{-1} dissolved microcystins in the reservoir to always well below 0.1 μg L^{-1} in the wells surrounding the reservoir for the bloom season of 1997. Current research is addressing whether specific conditions can be identified in which the risk of cyanotoxin breakthrough is elevated (such as low temperatures, lack of biofilm development, slow biodegradation due to anaerobic conditions).

Table 2. Microcystin concentrations at the surface and at the off-take of Deesbach Reservoir (n.d. = not detectable (detection limit 0.01 μg L^{-1}; summarized from Fastner et al., 2001).

Date	***Microcystin fraction***	***Reservoir surface [μg L^{-1}]***	***Off-take at 17 m [μg L^{-1}]***
14.07.98	Cell-bound	75	0.30
	Dissolved	0.01	n.d.
21.07.98	Cell-bound	566	0.10
	Dissolved	0.13	0.02
28.07.98	Cell-bound	0.05	0.23
	Dissolved	3.63	0.07
05.08.98	Cell-bound	0.28	n.d.
	Dissolved	0.11	0.02

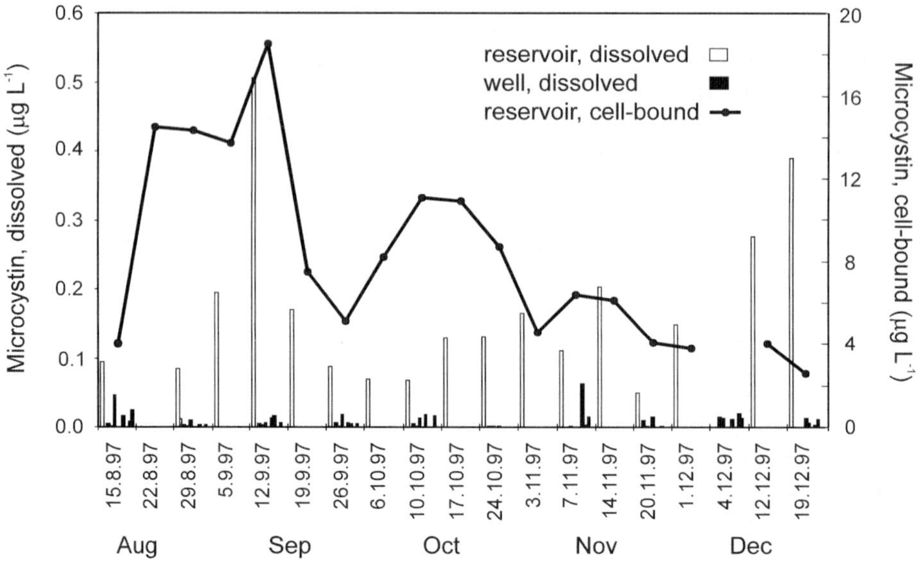

Figure 4. Microcystin concentrations in the Radeburg Reservoir and its bank filtrate in 1997. Curve: Cell-bound microcystins determined by HPLC; bars: dissolved microcystins determined by ELISA – open bars in the reservoir and solid bars in the wells. Note: for 4.12.1997 no results are available for dissolved microcystins in the reservoir. Redrawn from Chorus et al. (2001) with friendly permission from Springer–Verlag.

Using off-take depth as control measure for cyanobacteria and their toxins is unlikely to encompass further hazards, as their distribution in a reservoir is likely to be different. Parasites tend to be entrained in currents of inflowing water, or settle into deeper layers and eventually to the sediments, and quality-impairing chemicals tend to accumulate in hypolimnetic, anaerobic water. Thus, controlling cyanobacterial intake by controlling off-take depth may even compete against optimal depths selected to avoid other hazards. Comprehensive hazard analysis is therefore essential for optimising this measure.

5.4 CONTROL MEASURES IN TREATMENT

Treatment steps in the preparation of drinking-water are quite diverse. Usually, 'conventional' treatment for central drinking-water supplies using surface water typically includes flocculation, filtration, and disinfection. Often, a pre-oxidation step (with ozone or potassium permanganate) is used before flocculation to enhance flocculation of algal and cyanobacterial cells. Further 'advanced' treatment includes ozonation and filtration through granulated activated carbon (GAC), addition of powdered activated carbon and increasingly also membrane filtration. These have been introduced in many supply systems in response to raw water quality problems such as occurrence of pesticides and other pollutants, or high loads of organic substances leading to problems of bacterial regrowth in the distribution system.

All of these steps can be assessed for their impact on reducing cyanobacterial cells and toxins, and an increasing body of knowledge is emerging on their performance. For microcystins, filtration steps tend to be quite effective as long as they remove intact cells still containing the toxin within the cell (Drikas et al., 2001; Bláha and Marsálek, 2001; Gupta et al., 2001; Kruschwitz et al., 2001; Chorus et al., 2001). A concern with pre-oxidation is that while it improves flocculation of cyanobacteria, it may lead to cell damage and release of cell-bound microcystins which are not retained, i.e. break through the filter. This has indeed been demonstrated, particularly for pre-oxidation (Schmidt et al., 2002). A concern with filtration is the accumulation of cells on filters, where they may lyse and lead to toxin release, potentially pulse-wise as also demonstrated in a pilot treatment plant by Schmidt et al. (2002). A further concern is breakthrough of other cyanotoxins of which a large fraction occurs dissolved in water, such as cylindrospermopsin and in some settings neurotoxins. Treatment steps after filtration, such as ozonation and GAC can be very effective in eliminating dissolved cyanotoxins, but operational control is critical to ensure their efficacy.

Conceptually, these treatment steps can readily be managed as ***control measures*** in a Water Safety Plan. This would imply determining ***operational monitoring*** systems with ***operational limits***, within which cyanotoxin removal would be effective. Operational monitoring of treatment steps is already widely implemented as part of 'Good Practice', i.e. continuous monitoring of filter effluents for turbidity or particle concentration, or of disinfection steps for contact times and disinfectant concentration. A Water Safety Plan including cyanotoxins would initially assess the treatment steps for their likely performance in removing cyanotoxins to the level targeted for the finished water, and it would then proceed to validate these assumptions with experiments designed specifically to reflect the conditions on site, and/or with targeted cyanotoxin programmes during bloom events.

Validation of control measures in treatment is particularly important with respect to unintended effects such as cell lysis and microcystin release as a result of pre-oxidation steps or accumulation of cells on filters. For oxidation steps which target degradation of dissolved cyanotoxins, validation would investigate extreme conditions under which efficacy is particularly challenged, e.g. through a high load of other organic material typically occurring in consequence of a bloom. Validation would also address the resulting degradation products. A further important element in validation of cyanotoxin removal is that during bloom events, the treatment system is challenged by an exceptionally high load of organic substances, with 99% of the organic carbon potentially being from cyanobacterial cells, but not being cyanotoxins. This load may quickly exhaust removal steps, while the high load of cells may quickly clogg filters. It may also lead to high concentrations of disinfection by-products.

Though their efficacy needs to be validated specifically for each hazard, control measures in treatments effective for cyanobacteria and dissolved cyanotoxins are also likely to effectively reduce other particle-bound and dissolved contaminants.

6 Discussion – Water Safety Plans as adequate regulatory tools

The presentation above shows that the Water Safety Plan approach is indeed very well applicable to assessing and managing cyanotoxins hazards, suitable to overcome the limitations of the traditional approach of monitoring finished water for compliance to standards. A further aspect when approaching the Water Safety Plan concept from the viewpoint of the cyanotoxin expert is the assessment of cyanotoxins in the context of other hazards. Many of the control measures potentially effective against cyanotoxins are likely to be also suitable for controlling other hazards as well: e.g. catchment measures to control livestock operations will not only reduce nutrient loads, but also loads of pathogens shed by livestock, such as protozoa (Cryptosporidia and Giardia). Likewise, water treatment steps will remove particles of various types, i.e. cyanobacterial cells as well as protozoa and most bacteria, and ozonation or granular activated carbon filtration will address a range of dissolved contaminants, including cyanotoxins. One strength of the Water Safety Plan approach lies in moving away from looking at hazards in isolation towards a comprehensive management package specifically tailored to meet the set of hazards identified to be relevant in the given supply system. This can equally be applied to recreational sites, and the bathing water profile approach to the revision of the EU Bathing-water Directive is such an approach.

Given this positive assessment of the potential of the Water Safety Plan approach to overcome the limitations of the compliance-driven regulatory approach, the logical conclusion would be to base regulations and surveillance on requiring a Water Safety Plan for drinking-water supply systems and on a Bathing-water Profile for recreational sites, rather than on routine surveillance of cyanobacterial cells or toxins in the water. Implementation would be through some form of 'checking' the Water Safety Plan and Bathing-water Profile in relation to the hazards relevant in the system. This could be achieved through an audit, e.g. by the health authority, and penalties for non-compliance could be linked to not having an adequate Water Safety Plan or Bathing-water Profile, or to failing the audit. While compliance to standards in finished drinking-water or at a bathing site would still have a role in the regulatory system in the context of verification, an emphasis on requiring system analysis and management with tools like Water Safety Plans and the Bathing-water Profile will be far more comprehensive and include risks not explicitly regulated by a standard.

The important question is how to achieve such progress in the context of a regulatory culture which currently strongly focuses on monitoring compliance to standards in drinking- or bathing water, and with experience in audit-based approaches only emerging in few countries (e.g. Switzerland and Australia; see Schmoll and Chorus, 2003). Substantial development of an understanding of this approach and capacity building will be necessary before it can take the shape of a legal or regulatory tool.

Currently, in the European Union penalties paid to Brussels for non-compliance to standards currently still strongly drive the priorities for surveillance and action. Understandably, concern is therefore being expressed by many health authority officers that without the pressure of compliance monitoring and penalties paid to the EU, 'nothing will happen' to improve the safety of systems from cyanotoxins, even though they are now recognised as significant health hazard in drinking-water reservoirs and at recreational sites in an increasing number of countries. In the context of this prevalent regulatory culture, two approaches are needed in parallel: (i) setting standards for one or two selected cyanotoxins, and (ii) developing regulatory approaches based on risk assessment and management, such as the WHO Water Safety Plan approach or the EU Bathing-water Profile.

Setting cyanotoxins standards. As discussed in Section 2 of this chapter, setting standards for cyanotoxins is scientifically unsatisfactory because one or two selected substances would insufficiently represent toxicity and occurrence of the wide range of cyanotoxins. Nonetheless, requiring compliance to a cyanotoxin standard would be likely to trigger action towards improving the barriers against their occurrence in water supplies and at recreational sites. In principle, the limitations of such an approach do not appear more severe than those of a number of other standards, e.g. for faecal indicators. Though the presence of faecal indicators also does not satisfactorily indicate all hazards they are intended to represent, the history of their use has nonetheless shown that they very successfully triggered action, resulting in substantial improvement in reducing exposure risks to faecal pathogens. Also, a standard for one or two selected cyanotoxins would be important in the context of verification of a risk-based approach such as a Water Safety Plan.

In Europe, a standard for microcystins would serve this purpose well. Our current scientific knowledge indicates microcystins to be the most relevant group of cyanotoxins, whereas neurotoxins and cylindrospermopsin occur less frequently (Chorus, 2001). A standard for microcystins in drinking-water would therefore be likely to capture a large share of the potentially hazardous situations. Microcystin-LR occurs quite frequently in some countries, whereas other microcystins are dominant in other countries. Therefore, the standard should not be limited to microcystin-LR but include the sum of all microcystins, using concentration equivalents to microcystin-LR for quantification. As this is a worst-case approach (because most other microcystins are less toxic; see above), it would include a precautionary element. Similar to the EU pesticide standard, an EU microcystin standard could be set at 1 $\mu g\ L^{-1}$ for microcystin-LR and X $\mu g\ L^{-1}$ for the sum of all microcystins (where X would yet need to be established). If the data base for chronic toxicity of other microcystins were improved, this worst-case approach could be refined towards using toxicity equivalents (similar to the approach currently used for dioxins; see Wolf and Frank, 2002). For other geographical regions with a more widespread occurrence of cylindrospermopsin, a standard or guideline value for this cyanotoxin would be needed as well, and has been proposed (Humpage and Falconer, 2002).

'Phasing in' the Water Safety Plan approach. The discussion towards setting cyanotoxin standards needs to make their role and limitations transparent, and this can be coupled with discussions promoting the need for risk-based approaches as a more comprehensive way of addressing this (and other) hazards. In parallel, where the WHO Water Safety Plan or similar risk-based approaches are adopted for water safety, an audit-driven approach to surveillance could be gradually 'phased in'. This

is likely to require a transition phase of several years to perhaps a decade. Following the model of the new French Drinking-water Ordonance, compliance monitoring could be reduced for supplies which demonstrate that they have adequate hazard analysis and risk assessment in place, e.g. through an audited Water Safety Plan in which system performance for removing the hazards has been validated.

In many settings the risk of exposure to cyanotoxins is probably rather high when compared to the occurrence of other chemicals in drinking-water and bathing water. Therefore a dual approach is adequate, i.e. to implement a standard either for cyanobacterial biomass or for one to two cyanotoxins while at the same time developing and implementing risk-based approaches such as the Water Safety Plan to the point where we can be confident that hazard assessment is being comprehensively undertaken and process monitoring is controlling the risks.

7 Acknowledgements

The constructive comments of Guy Howard and two anonymous referees, as well as the help of Klaus Jöhnk in redrawing Figure 4 are gratefully acknowledged. This work was supported by the PEPCY program funded by the European Union.

8 References

Bartram, J., Burch, M., Falconer, I.R., Jones, G. and Kuiper-Goodman, T. (1999) Situation assessment, planning and management, in I. Chorus and J. Bartram (eds.), *Toxic Cyanobacteria in Water: a Guide to their Public Health Consequences, Monitoring and Management*, Spon, London, United Kingdom, pp. 179-209.

Bláha, L. and Marsálek, B. (2001) Dissolved microcystins in raw and treated drinking water in the Czech Republic, in I. Chorus (ed.), *Cyanotoxins – Occurrence, Causes, Consequences*, Springer-Verlag, Berlin, Germany, pp. 212-216.

Chorus, I., Schlag, G., Heinze, R., Pütz, K. and Kruspe, U. (2001) Elimination of microcystins through bank filtration at the Radeburg Reservoir, in I. Chorus (ed.), *Cyanotoxins – Occurrence, Causes, Consequences*, Springer-Verlag, Berlin, Germany, pp. 226-228.

Davison, A., et al. (2004) Water Safety Plans. WHO, Geneva, Switzerland.

Drikas, M., Chow, C.W.K., House, J. and Burch, M.D. (2001) Using coagulation, flocculation and settling to remove toxic cyanobacteria, *Journal of the American Waterworks Association* **93**, 100-111.

Falconer, I., Bartram, J., Chorus, I., Kuiper-Goodman, T., Utkilen, H., Burch, M. and Codd, G.A. (1999) Safe levels and practices, in I. Chorus and J. Bartram (eds.), *Toxic Cyanobacteria in Water: a Guide to their Public Health Consequences, Monitoring and Management*, Spon, London, United Kingdom, pp. 155-178.

Fastner, J., Chorus, I., Willmitzer, H. and Rabe, C. (2001) Reducing intake of microcystins at the Deesbach Reservoir drinking-water abstraction system, in I. Chorus (ed.) *Cyanotoxins – Occurrence, Causes, Consequences*, Springer-Verlag, Berlin, Germany, pp. 229-231.

Fundação Nacional de Saúde, N° 1.469/2000 (2000) Controle e Vigilância da Qualidade da Água para consumo Humano e seu Padrão de Potabilidade, Brazil.

Grützmacher, G., Böttcher, G., Chorus, I. and Bartel, H. (2002) Removal of microcystins by slow sand filtration, *Environmental Toxicology* **17**, 386-394.

Gupta, S., Giddings, M. and Sheffer, M. (2001) Cyanobacterial toxins in drinking water: a Canadian perspective, in I. Chorus (ed.), *Cyanotoxins – Occurrence, Causes, Consequences*, Springer-Verlag, Berlin, Germany, pp. 208-212.

Harding, R. (1999) *Environmental Decision-Making – the Roles of Scientists, Engineers and the Public*, The Federation Press, Sydney, Australia.

Huisman, J., Sharples, J., Stroom, J., Visser, P.M., Kardinaal, W.E.A., Verspagen, J.M.H. and Sommeijer, B. (2004) Changes in turbulent mixing shift competition for light between phytoplankton species, *Ecology* **85**, 2960-2970.

Humpage, A. and Falconer, I. (2002) Oral toxicity to Swiss albino mice of the cyanobacterial toxin cylindrospermopsin administered daily over 11 weeks, Canadian Research Centre for Water Quality and Treatment, Research Report No. 13.

Koschel, R. and Jäger, D. (1995) *Verfahren zur Sanierung und Restaurierung stehender Gewässer*, Limnologie Aktuell Band 8, Fischer, Stuttgart, Germany.

Kruschwitz, C., Chorus, I., Heinze, R., Schlag, G. and Gröbe, K. (2001) Elimination of microcystins in the Rostock drinking water treatment plant, in I. Chorus (ed.), *Cyanotoxins – Occurrence, Causes, Consequences*, Springer-Verlag, Berlin, Germany, pp. 217-221.

Kuiper-Goodman, T., Falconer, I. and Fitzgerald, J. (1999) Human health aspects, in I. Chorus and J. Bartram (eds.), *Toxic Cyanobacteria in Water: a Guide to their Public Health Consequences, Monitoring and Management*, Spon, London, United Kingdom, pp. 113-153.

Metge, S., Brodard., E. and Conan, M. (2003) Lyaonnaise des eaux: application of HACCP principles in drinking water, in O. Schmoll and I. Chorus (eds.) *Water Safety*, Federal Environmental Agency, Berlin, Germany, pp. 41-44.

National Rivers Authority (1990) *Toxic Blue-green Algae*, Water Quality Series No. 2, 125 pp.

Oberemm, A., Fastner, J. and Steinberg, C. (1997) Effects of microcystin-LR and cyanobacterial crude extracts on embryo-larval development of zebrafish, *Water Research* **31**, 2918-2921.

Oberemm, A., Heinze, R., Papendorf, O. and Fastner, J. (2001) Significance of unidentified toxic compounds and approaches to their identification, in I. Chorus (ed.), *Cyanotoxins – Occurrence, Causes, Consequences*, Springer-Verlag, Berlin, Germany, pp. 286-295.

Pilotto, L.S., Douglas, R.M., Burch, M.D., Cameron, S., Beers, M., Rouch, G.R., Robinson, P., Kirk, M., Cowie, C.T., Hardiman, S., Moore, C. and Attwell, R.G. (1997) Health effects of exposure to cyanobacteria (blue-green algae) due to recreational water-related activities, *Australian Journal of Public Health* **21**, 562-566.

Ressom, R., Soong, F.S., Fitzgerald, J., Turczynowicz, L., El Saadi, O., Roger, D., Maynard, T. and Falconer, I. (1994) Health effects of toxic cyanobacteria (blue-green algae), National Health and Medical Research Council, Canberra, Australia.

Schmidt, W., Willmitzer, H., Bornmann, K. and Pietsch, J. (2002) Production of drinking water from raw water containing cyanobacteria – pilot plant studies for assessing the risk of microcystin breakthrough, *Environmental Toxicology* **17**, 375-385.

Schmoll, O. and Chorus, I., eds. (2003) *Water Safety*, Federal Environmental Agency, Berlin, Germany.

Visser, P.M., Ibelings, B.W., van der Veer, B., Koedood, J. and Mur, L.R. (1996) Artificial mixing prevents nuisance blooms of the cyanobacterium *Microcystis* in Lake Nieuwe Meer, The Netherlands, *Freshwater Biology* **36**, 435-450.

Vollenweider, R. (1982) *Eutrophication of Waters: Monitoring, Assessment and Control*, OECD, Paris, France.

Walker, H.S. (2003) Safety of drinking water supplies in Switzerland: evaluation by the health authorities, in O. Schmoll and I. Chorus (eds.), *Water Safety*, Federal Environmental Agency, Berlin, Germany.

WHO (1996) *Guidelines for Drinking-water Quality*, Second edition, Volume 2, Health Criteria and Other Supporting Information, World Health Organisation, Geneva, Switzerland.

WHO (1998) *Guidelines for Drinking-water Quality*, Second edition, Addendum to Volume 2, Health Criteria and Other Supporting Information, World Health Organisation, Geneva, Switzerland.

WHO (2003) *Guidelines for Safe Recreational Environments*, Volume 1, Coastal and Fresh Waters. World Health Organisation, Geneva, Switzerland.

Wolf, H.U. and Frank, C. (2002) Toxicity assessment of cyanobacterial toxin mixtures, *Environmental Toxicology* **17**, 395-399.

INDEX

A

B

C

D

E

F

G

H

N

O

P

Q

R

T

U

V

W

X

Z

AQUATIC ECOLOGY BOOK SERIES

http://www.springeronline.com/sgw/cda/frontpage/0,11855,4-40109-69-33114010-0,00.html

Series Editor:
Prof. Robert G. Wetzel, Department of Environmental Sciences and Engineering, The University of North Carolina, Chapel Hill, North Carolina 27599-7431, USA.
E-mail: rwetzel@unc.edu

Current volumes in this series

Volume 1 **Ecology of a Glacial Flood Plain**
J.V. Ward, U. Uehlinger
Hardbound, ISBN 1-4020-1792-8, December 2003, 320 pp.

Volume 2 **The Influence of Primary Producers on Estuarine Nutrient Cycling**
S.L. Nielsen, G.T. Banta, M.F. Pedersen
Hardbound, ISBN 1-4020-2638-2, 2004

Volume 3 **Harmful Cyanobacteria**
J. Huisman, H.C.P. Matthijs, P.M. Visser
Hardbound, ISBN 1-4020-3009-6, 2005

Forthcoming volumes:

Ecological History of the Lowland River Basins of Rhine and Meuse
P.H. Nienhuis

Ecology of Russian Rivers
A.V. Zhulidov, V.V. Khlobystov, R.D. Robarts, A.A. Ischenko
Dr. J. Headley

Biology of Polluted Rivers
T.E.L. Langford

A Continuation Order Plan is available for this series. A continuation order will bring delivery of each new volume immediately upon publication. Volumes are billed only upon actual shipment. For further information please contact the publisher

For new book proposals please contact the Series Editor or the Publisher